Green Energy and Technology

More information about this series at http://www.springer.com/series/8059

Maurizio Tiepolo · Alessandro Pezzoli
Vieri Tarchiani
Editors

Renewing Local Planning to Face Climate Change in the Tropics

Springer Open

Editors
Maurizio Tiepolo
DIST, Politecnico and University of Turin
Turin
Italy

Alessandro Pezzoli
DIST, Politecnico and University of Turin
Turin
Italy

Vieri Tarchiani
IBIMET-CNR
Florence
Italy

ISSN 1865-3529 ISSN 1865-3537 (electronic)
Green Energy and Technology
ISBN 978-3-319-86547-8 ISBN 978-3-319-59096-7 (eBook)
DOI 10.1007/978-3-319-59096-7

Printed on acid-free paper

This Springer imprint is published by Springer Nature
The registered company is Springer International Publishing AG
The registered company address is: Gewerbestrasse 11, 6330 Cham, Switzerland

To Silvia Macchi,
our esteemed colleague at Sapienza University of Rome,
involved in climate planning and development aid,
who passed away during the final preparation of the book.
We would like to dedicate this volume to her memory.

Foreword

The adverse impacts on human-induced climate change are a global problem with different impacts in different localities around the world.

Hence, adapting to those adverse impacts of climate change is both a local and a global problem. At the global level, we have agreed to include tackling climate change as one of the Sustainable Development Goals (SDG13) and also agreed the Paris agreement on climate change.

However, the success of meeting those global goals will depend on every village, town, and city finding its own adaptation solutions and putting them in place. Unfortunately, we are still lagging behind in being able to do this successfully although many efforts are being tried around the world.

This publication tries to learn and share lessons from many of the local-level efforts around the world. These lessons will be very useful for practitioners in towns and cities around the world who are also struggling to find adaptation solutions in their own localities.

Saleemul Huk
IIED-International Institute for Environment
and Development, London, UK
ICCCAD-International Centre for Climate Change
and Development, Independent University, Dhaka, Bangladesh

Preface

This book presents the most recent results of a research started at the end of 2011 in Africa South of the Sahara and then extended to the whole tropics. The focus of our work is climate planning. With this term, we mean all those plans to limit climate change (mitigation and sustainable action plans), to protect human settlements against its impacts (emergency, risk reduction, adaptation plans, and resilience strategies) and to pursue both the medium-term (municipal development plans) and long-term (comprehensive, general, and master plans) aims.

During these years, we have organized three UICCA—Urban Impact of Climate Change in Africa—conferences (Turin 2011, 2013, 2016) to share and discuss the first results of our work with other similar experiences, involving over one hundred researchers, various officials and local administrators, as well as numerous students.

A selection of the reports presented at the 3rd UICCA conference is provided here. The book tackles a topic which is going to be critical in the years to come: How to implement the 11th Sustainable Development Goal—SDG (2015). We are referring to the target which states "By 2020, substantially increase the number of cities and human settlements adopting and implementing integrated policies and plans toward inclusion, resource efficiency, mitigation and adaptation to climate change, resilience to disasters, and develop and implement, in line with the Sendai Framework for disaster risk reduction 2015–2030, holistic disaster risk management at all levels" (United Nations' Sustainable Development Goal 11).

Today, the debate and literature on SDGs is focused on monitoring the achievement of the goals. The aim of our work, on the other hand, is to draw attention on how to reach the above-mentioned target of SDGs at 2030. And the reason lies in the fact the tools used so far to center the target lack efficiency and the cities which will most need to use them have little or no knowledge of them. Our work raises the matter of the quality of climate planning. And this looks at the analyses prior to planning, decision making for planning, and the innovation of climate measures. However, it also looks at transversal topics, such as IT systems and planning methods. For example, we focus on the transition from participated planning built upon traditional knowledge only, which still prevails in the Least Developed Countries (LCDs), to that which integrates this with technical-scientific knowledge, which is better suited to identifying the

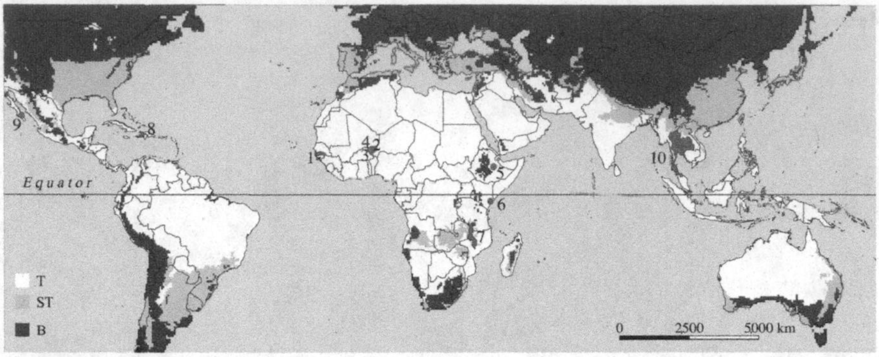

Case studies investigated in the book: Casamance, Senegal (*1*), Tillaberi region, Niger (*2*), Gotheye (*3*) and Ouro Gueladjio (*4*), Arsi region, Ethiopia (*5*), Dar es Salaam, Tanzania (*6*), Malawi (*7*), Haiti (*8*), La Paz, Mexico (*9*), Thailand (*10*), Tropical (*T*), Subtropical (*ST*), Boreal (*B*) zones

nature of the climate change, and the expected impacts of adaptation and risk reduction measures.

The attention of the book is aimed at the tropical LDCs in that they contain the cities less able to limit the emissions responsible for climate change and to cope with the impacts of the latter but which, at the same time, will be the context in which the biggest transformations in human settlements will take place within the next 15 years.

The book has eighteen chapters which examine 10 case studies (see Figure). Chapter 1 (Tiepolo, Pezzoli, and Tarchiani) assesses the state of application of the 11th Sustainable Development Goal in the tropics and the prospects, the lines of research, and the challenges for renewing local planning to face climate change. There are two parts. Part I, centered on climatic monitoring and the assessment of the various components, is made up of eight chapters.

Chapter 2 (Sabatini) discusses some of the main issues to improve climate observation network planning, especially in remote and inhospitable regions with a focus on Niger and Nepal representing the two climatic extremes.

Chapter 3 (Bacci and Mouhaimouni) proposes a comparative analysis of the hazards between present and future, concentrating particularly on the extreme rainfall events and drought on the Western Niger.

Chapter 4 (Bacci) presents an agrometeorological analysis as a tool for characterizing the climatic risks to suit the rice-growing system in southern Senegal (Casamance).

Chapter 5 (Belcore, Calvo, Canessa, and Pezzoli) estimates vulnerability to climate change in 3 woredas of the Oromia region (Ethiopia), whereas data on vulnerability to drought are lacking.

Chapter 6 (Tiepolo and Bacci) presents a method for tracking climate change vulnerability in the 125 rural municipalities of Haiti using open data.

Chapter 7 (Demarchi, Cristofori, and Facello) presents an early warning system for urban Malawi integrating satellite-derived precipitation data and geospatial reference datasets.

Chapter 8 (Vignaroli) proposes a Web-based approach for early drought risk identification using freely available rainfall estimations and forecasts to strengthen the mechanism for the prevention and management of the food crisis in Sahel.

Chapter 9 (Franzetti, Bagliani, and Pezzoli) tackles the climatic characterization of Thailand.

Part II of this book also collects eight chapters which look mainly at decision-making tools for local climate planning and innovation in climatic measures.

Chapter 10 (Tiepolo) presents the state of climate planning in 338 large- and medium-sized cities in the tropics using the QCPI—Quality of Climate Planning Index.

Chapter 11 (Tiepolo and Braccio) presents a case of multirisk analysis and evaluation in rural Niger integrating local and scientific knowledge.

Chapter 12 (Fiorillo and Tarchiani) presents a simplified method for assessing flood hazard and related risks using open-access tools and data in a rural municipality in south Western Niger.

Chapter 13 (Faldi and Macchi) presents an application for forecasting and participatory backcasting methods for assessing urban people's vulnerability to water access in Dar es Salaam, Tanzania.

Chapter 14 (Emperador, Orozco Noriega, Ponte, and Vargas Moreno) presents a method for climate risk reduction mainstreaming at La Paz, Mexico.

Chapter 15 (Bechis) presents an innovative stove for limiting the use of wood as a fuel for cooking in Niger, using farming and forestry residues, estimating the potential impacts on renewable natural resources.

Chapter 16 (Di Marcantonio and Kayitakire) presents a review of the index-based insurance in Africa as a tool to reduce climatic risk at rural level.

Chapter 17 (Schultz and Adler) presents the climate vulnerability reduction credit system applied to rural Niger to assess the outputs of climate adaptation measures to reduce vulnerability to climate change.

Chapter 18 (Tiepolo, Pezzoli, and Tarchiani) gathers conclusions, indicates areas for future research, and supplies numerous recommendations for renewing local planning in the tropics to the main stakeholders.

Turin, Italy — Maurizio Tiepolo
Turin, Italy — Alessandro Pezzoli
Florence, Italy — Vieri Tarchiani

Acknowledgements

Numerous people contributed to the realization of the 3rd UICCA conference and the subsequent book. We would especially like to record Mauro Pedalino, Stefania Cametti, and Bruno Gentile of the Italian development cooperation that allowed the ANADIA Niger project to extend its activities to the conference. We are in debt with Elisabetta Franzé and Marcella Guy for financial management; Cinzia Pagano and Luisa Montobbio for the Internet management; Eleonora d'Elia and Noemi Giraudo for helping us during the conference organization; and Timotheos Vissiliou for photographs and press release. If the works have left a mark on all of us, it is thanks to Prof. Raffaele Paloscia (University of Florence), who led the final roundtable and drew the conclusions of the event. Our sincere thanks go to all those who attended the conference and who asked question, enriched it with contributions or simply listened. All these people gave us the impression that we were part of something important.

We would like to thank Katiellou Gatpia Lawan and Aissa Sitta as respectively the ANADIA national project manager and national technical coordinator. A special thanks to Moussa Labo, director of National Directorate of Meteorology of Niger, for his support in ANADIA activities. We acknowledge the World Climate Research Programme's working group on coupled modeling, which is responsible for CMIP, and we thank the climate modeling groups for producing their model output and making it available. For CMIP, the US Department of Energy's Program for climate model diagnosis and intercomparison that provided coordinating support and led development of software infrastructure in partnership with the Global Organization for Earth System Science portals (Chap. 3), all the collegues of the PAPSEN project especially Andrea Di Vecchia (project coordinator), Institute of Biometeorology-Italian National Research Council. A special thanks goes to the regional directorate for Rural Development of the Sédhiou Region (Senegal) and Marco Manzelli (IBIMET) for information on critical stages for rice crops in Mid-Upper Casamance (Chap. 4) and to University of Turin for International Cooperation project, which supported three young researchers in the Siraro, Shalla, and Shashame Woredas for three months to exchange data and experience with LVIA NGO. Many thanks are also due to the LVIA staff in Italy and in Ethiopia for their precious leading work in field and their active preparation desk work (Chap. 5). We remain indebted to Ottavio Novelli (AESA Group) for having enabled the activity which inspired the open data index for vulnerability tracking in rural Haiti (Chap. 6); Tiziana de Filippis, Leandro Rocchi, Maurizio Bacci, Elena Rapisardi (IBIMET-CNR) for CRZ model implementation and 4 crop system Web application development (Chap. 8); Chayanis Krittasudthacheewa of the Stockholm Environment Institute, Asia Centre for hosting a student in internship and for being a mentor to this student in the research (Chap. 9); Andrea Melillo and Simone Ghibaudo for map design; and Idrissa Mamoudou (CNEDD Niger) for providing valuable information on municipal development plans in Niger (Chap. 10), and Thierry Negre for helping the collection of information (Chap. 16).

The editors would like to thank the head of the DIST, Prof. Patrizia Lombardi, and the head of IBIMET-CNR, Antonio Raschi, for their constant support, encouragement, and unwavering trust.

Contents

Chapter 1
Renewing Climate Planning Locally to Attend the 11th Sustainable Development Goal in the Tropics

Maurizio Tiepolo, Alessandro Pezzoli and Vieri Tarchiani

Abstract In the last seven years, tropical cities with a climate plan have tripled compared to the previous seven years. According to the 11th United Nations' Sustainable Development Goal, climate planning should significantly increase by 2030. The Sendai framework for disaster risk reduction (2015) and the New urban agenda signed in Quito (2016) indicate how to achieve this goal through analysis, categories of plans and specific measures. This chapter identifies the main obstacles to the significant increase in tropical human settlements with a climate plan and the possible solutions. First of all, the distribution and trend at 2030 of tropical human settlements are ascertained. Then local access to information on damage, hazard, exposure, vulnerability and risk, and the consideration of these aspects in the national guides to local climate planning are verified. Lastly, the categories of plans and climate measures recommended by the United Nations are compared with those that are most common today, using a database of 401 climate plans for 338 tropical cities relating to 41 countries. The chapter highlights the fact that the prescription for treating tropical cities affected by climate change has been prepared without an accurate diagnosis. Significantly increasing climate planning must consider that small-medium human settlements in the Tropics will prevail at least until 2030. And most effort will be required from Developing and Least Developed Countries. The recommendations of the United Nations concerning the preliminary analyses ignore the fact that local authorities usually do not have access to the necessary information.

M. Tiepolo is the author of 1.3.3, 1.3.4 and 1.4 sections. A. Pezzoli is the author of 1.1 and 1.2 sections. V. Tarchiani is the author of 1.3.1, 1.3.2 and 1.5 sections.

M. Tiepolo (✉) · A. Pezzoli
DIST, Politecnico and University of Turin, Viale Mattioli 39, 10125 Turin, Italy
e-mail: maurizio.tiepolo@polito.it

A. Pezzoli
e-mail: alessandro.pezzoli@polito.it

V. Tarchiani
National Research Council—Institute of BioMeteorology (IBIMET), Via Giovanni Caproni 8, 50145 Florence, Italy
e-mail: v.tarchiani@ibimet.cnr.it

M. Tiepolo et al. (eds.), *Renewing Local Planning to Face Climate Change in the Tropics*, Green Energy and Technology,
DOI 10.1007/978-3-319-59096-7_1

Climate plans and recommended measures are not those currently in use. We propose three areas of action to facilitate the mainstreaming of the recommendations in the tropical context. They require a renewal of the local planning process if we intend to reach the 11th SDG by 2030.

Keywords Climate change · Climate planning · SDG · Mainstreaming · Open source maps · Open data · Plan guidelines · Vulnerability · Tropics

1.1 Introduction

Today, in the Tropics, one large or medium-size city out of four has a plan to reduce greenhouse gas emissions (mitigation and sustainable plans), the impacts of global warming (emergency, adaptation, resilience plans) or both (general, comprehensive plans). We don't know how frequent these plans are in towns (under 0.1 million pop.) and in rural areas. The resolutions of the United Nations, expressed in the 2030 Agenda for sustainable development (2015), in the Sendai Framework (2015) and in the New urban agenda (2016) supply numerous recommendations to "substantially increase the number of cities and human settlements adopting and implementing integrated policies and plans towards inclusion, resource efficiency, mitigation and adaptation to climate change, resilience to disasters, and develop and implement… holistic disaster risk management at all levels" (UNESC 2016: 28; UN General Assembly 2015).

Getting started with 11th SDG requires more than a simple review of the long-term planning practices, developing need assessment, costing and monitoring (SDSN 2015), this last aspect being the current focus of much reflection on SDGs.

While knowledge of local climatic planning in the Tropics has increased in recent years (Seto et al. 2014), only recently have systematic analyses been developed (Macchi and Tiepolo 2014; Tiepolo et al. 2016; Tiepolo 2017) and vulnerability assessment and risk evaluation have been tested in urban areas (Sakai et al. 2016) and rural locations (Pezzoli and Ponte 2016; Tarchiani and Tiepolo 2016).

The objective of this chapter is to identify the obstacles to the mainstreaming of the United Nations recommendations in the Tropics and how to overcome them.

In the Tropics, the urban population is expected to rise by 43% by 2030: double the increase in the rest of the world (UNDESA 2014). This rapid urbanisation takes place in a context characterised by widespread underdevelopment (96% of Least Developed Countries-LDCs are in the Tropics) and the presence of advanced economies (Brazil, China, France, Mexico, Singapore, South Africa, Taiwan, USA). In advanced economies, plans and policies to support adaptation and the mitigation of climate change are consolidated, while in Developing Countries and especially in LDCs, there is a prevalence of community-based adaptation undertaken by Non-Governmental Organizations-NGOs and partners in development. This said, it is in the Tropics that certain measures to face climate change practiced in advanced economies can be adapted and reproposed to LDCs, because they are conceived to cope with the same hazards.

The following sections analyse the 11th SDG, the recommendations relating to climate planning in the Sendai Framework and in the New urban agenda. Then identify today's distribution of tropical human settlements by demographic class and that expected by 2030. The next step is the consideration of access to the information necessary to develop the analyses of exposure, vulnerability, and risk recommended by the United Nations. Then the analyses recommended by the national guidelines to local climate planning are identified. Lastly, the most popular plan categories and climate measures in the Tropics are analysed. This highlights what is missing from the mainstreaming of the recommendations of the United Nations.

1.2 Materials and Methods

The international agenda on city climate planning is based on the 11th Sustainable development goal of the United Nations (UN General Assembly 2015), on the recommendations formulated in the New urban agenda (2016) and in the Sendai framework for disaster risk reduction (2015).

The tropical climate zone derives from Koppen Geiger according to the Trewartha ranking (Belda et al. 2014) (Fig. 1.1). Settlements with a population of less than 0.3 million are classed as towns, those with a population of between 0.3 and 1 million are defined as medium-size cities and those with over a million people are considered large cities. The 2014 revision of the World urbanization prospects (UNDESA 2014) supplies the urban populations at 2015 and those estimated at 2030 by size class of human settlement and by country. The figures are not supplied by climate zone. The tropical area comprises 113 countries, 22 of which also belong

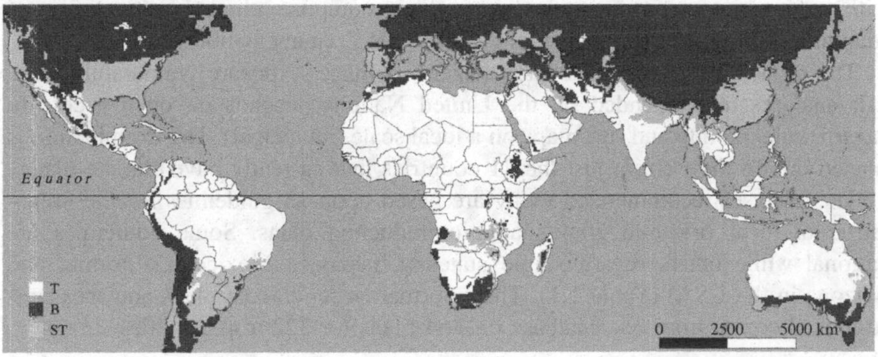

Fig. 1.1 The tropical zone (T)

to other areas (subtropical and, sometimes, boreal) like the Andine countries, and various countries in Southern Asia and Southern Africa. The UNDESA information allows us to identify the inhabitants of the tropics only for settlements 0.1–0.3 million inhabitants. We don't know how many people live in settlements with less than 0.3 million inhabitants in the tropical area. We have assumed that these settlements have the same share of cities (>0.3 million pop.) falling within the tropical zone of the country in question. This gave us the population of the towns in the tropics. This information on the population by category of human settlements allows us to highlight the characteristics of tropical human settlements and their trend at 2030.

This chapter, however, aims to appreciate the current relevance of climate planning, so we need to compare the number of human settlements with the number of climate plans. We have to pass from the population of the urban agglomerations to the number of administrative jurisdictions, conserving the articulation by class of demographic size. The UNDESA database cannot help us in this analysis.

Consequently, we have prepared a specific database starting from the national censuses of the tropical countries. It has enabled us not only to identify the number of tropical human settlements with more than 0.1 million inhabitants, but also to identify the transition that human settlements will make from one class to another in the years to come, due to the urbanisation process, an aspect which is not considered by the UNDESA statistics. The new entries have been identified by applying the average growth rate 2010–15 of the cities of every tropical country (UNDESA 2014) to towns with less than 0.1 million inhabitants and considering all the towns expected to pass the population threshold of 0.1 million by 2030. The rates vary from 0.2% (Australia, Venezuela) to 3.7% per annum (Rwanda) which, in this last case, would take all towns that have more than 60 thousand inhabitants nowadays above 100 thousand by 2030.

The relevance of local climate planning and of the relative climate measures is obtained from two previous surveys (Tiepolo and Cristofori 2016; Tiepolo 2017) on climate plans in the large and medium size cities of the Tropics. The database used contains 364 plans related to 322 tropical cities in 41 countries, updated to December 2016. With the term local we intend the minimum administrative level with safety and environmental tasks, which comprise climate mitigation and adaptation. In most cases, this level corresponds to municipalities.

The capacity of cities to develop hazard, damage, exposure, vulnerability and risk analyses recommended by the United Nations depends on open access to hazard, vulnerability and risk maps on a local scale (municipal). The examination is carried out on the web, in just the 41 countries that currently have climate plans.

The analyses recommended today are based upon 13 guidelines for the preparation of local risk mitigation/adaptation/reduction plans. Some guidelines are national while others regard single states (Chiapas in Mexico, California, and Arizona in the USA) (Table 1.1). The information on climate plans and measures contained comes from our database on 364 plans for 322 tropical cities.

Table 1.1 Climate planning mainstreaming snapshot method

United Nations	Existing conditions			
Recommendations	Size class of settlement	D, E, H, R, V maps	Guidelines for local climate planning	Municipal climate plans
Information	–	Open source	Analysis required	–
Climate plan	Most frequent	–	–	Most frequent
Climate measures	–	–	–	Most frequent

D Damage, *E* Exposure, *H* Hazard, *R* Risk, *V* Vulnerability

1.3 Results

1.3.1 Future Climate Planning According to the UN Vision

The resolutions adopted by the United Nations General Assembly on the 2030 agenda for sustainable development (2015), the Sendai framework for disaster risk reduction and the New urban agenda (NUA) trace the path for improving and extending climate planning (UN General Assembly 2015; UNISDR 2015, UN Conference 2016).

The 11th SDG envisages a substantial increase in climate plans for cities and human settlements. The Framework and the NUA contain 25 recommendations related to climate planning: type of analysis, plan categories, type of climate planning. Emergency and risk reduction plans, zoning maps and development codes are recommended as well as seven climate measures: four for adaptation and three for mitigation. Next steps are: first, to check whether planning tools recommended by the United Nations are suited to the main type of settlement of the Tropics; second, to assess to what extent it is possible to increase the dissemination of climate planning given the planning capacity demonstrated over the past seven years; third, to learn if the analyses recommended can be implemented with the information available today; fourth, to learn if the types of plans and measures recommended have already been implemented or are brand new.

1.3.2 Importance of the Tropical Human Settlements

A third of the world's urban population lives in the tropics and this share is expected to rise by 2030. This will change the breakdown of human settlements by category of demographic size. Large cities (over one million inhabitants) are expected to pass from 38 to 44% of the urban population.

Table 1.2 Expected trend 2015–30 of large and medium size cities of the Tropics compared with the rest of the World (Tiepolo on UNDESA 2014)

Category of settlements	2015			2030		
	World	Tropics	World-T	World	Tropics	World-T
Large, over 1 million pop., %	41	38	43	45	44	46
Medium, 0.3–1 million pop., %	16	15	17	16	10	20
Small, under 0.3 million pop., %	43	48	40	38	46	33
Urban, %	100	100	100	100	100	100
Urban, million pop.	3926	1317	2609	5058	1887	3170

In 2030, six of the ten most populated cities in the world will be in the Tropics: Delhi (36 million population), Mumbai (28), Dhaka (27), Karachi and Cairo (25), and Lagos (24). Medium-size cities and towns (less than one million inhabitants) will drop but will continue to be the main settlement in the Tropics (56% of urban inhabitants by 2030) (Table 1.2).

These figures, obtained from UNDESA statistics, are produced by urban agglomeration, which, for large cities are made up of several administrative jurisdictions. Moreover, they do not consider that, as time goes by, the urbanisation process will add new settlements to the urban class. These aspects are essential to ascertain the relevance of the plans.

We have directly drawn data from the censuses of the tropical countries and considered the dynamics within the single categories of demographic dimension generated by the urbanisation process, which will bring 65 new cities among the large cities and 213 towns among the medium-size cities by 2030 (Fig. 1.2). These shifts will take large and medium-size cities to 1591 by 2030. In short, in the Tropics, medium-size cities and towns will continue to prevail until. This differentiates this climate zone from the others, in which large cities will take a clear lead in the next decade.

1.3.3 *Relevance of the Effort Required*

If climate planning succeeds in maintaining the pace of the last seven years until 2030 (31.6 plans/year) cities with climate plans will increase from today's 25 to 47%. We could consider this a "substantial increase" in the number of cities implementing climate plans. To reach complete coverage would require 90 plans/year: this value has never been reached before, not even in 2012, when a total of 70 plans were implemented (Fig. 1.3).

However, we need to consider that today's climate planning is the result of an outstanding commitment by the member countries of the OECD and the BRICS over the last seven years, which has made it possible to bring the respective

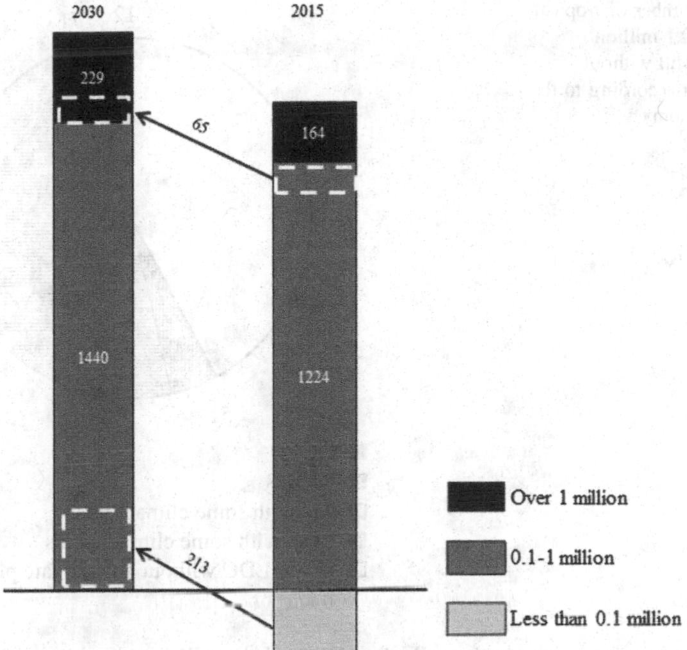

Fig. 1.2 Large and medium-size cities of the Tropics trends between 2015 and 2030

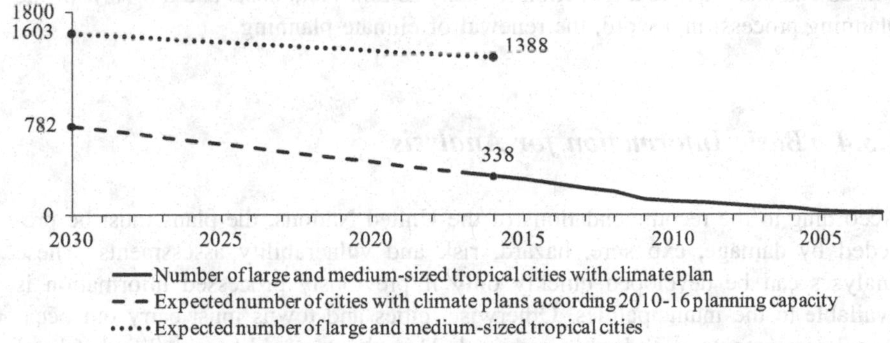

Fig. 1.3 Number of climate plans for tropical large and medium-size cities, 2003–30

contexts to 79% and to 62% the cities with a climate plan. In DCs and LDCs, cities with climate plans reach 22 and 35% respectively.

The 1062 tropical cities that still have no climate plan belong mainly to DCs and LDCs countries which, with just a few exceptions, have no urban climate planning experience. This will prevent maintenance of the pace that has characterised the last seven years in the coming fourteen (Fig. 1.4).

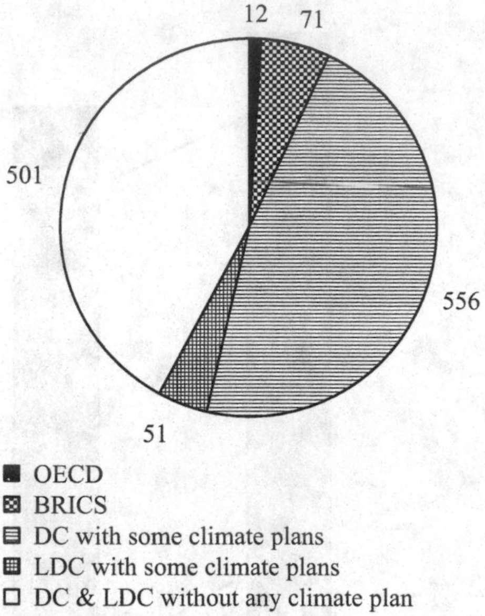

Fig. 1.4 Number of tropical cities over 0.1 million population still without climate plan according to the type of economy

Lastly, we must consider that all of today's plans will be obsolete in 14 years' time, needing a complete makeover. Consequently, the number of cities without a plan will be many more than today's 1062. Maintaining the pace of local climate planning in the future and ensuring that it includes the recommendations of the United Nations requires a reduction in analysis times and costs and a review of the planning process: in a word, the renewal of climate planning.

1.3.4 Basic Information for Analysis

According to the recommendations of the United Nations, the plans must be preceded by damage, exposure, hazard, risk and vulnerability assessments. These analyses can be developed quickly only if previously processed information is available to the municipalities. Otherwise, cities and towns must carry out occasional assessments as and when required. This phase would be simplified if local administrations had open access to a national system of vulnerability or risk tracking. This paragraph verifies the existence of this condition.

Today, of the 41 tropical countries that have cities with climate plans, 12 have open access information (hazard prone areas, vulnerability and risk) that help with the analytical phase.

There are two types of open access information. The first type is vulnerability and risk maps at administrative jurisdiction level, like risk map at municipal level in

Mexico. The second type localises hazard, vulnerability and risk on a topographic map (or on Google Earth) on a large scale (1:2000 in the case of Brazil, 1:5000 in the case of Réunion) or on a medium scale (1:50000 for the Northern territory, Australia). In other cases, the scale is too small or the resolution of the pictures is too low (Honduras, Peru). In some cases, in which web GIS is used, the user can choose the level of detail (Ecuador, Guyana, USA/Arizona, California, Texas) and the thematics (Malawi, Philippines), and export the information generated in numeric format (Table 1.3).

A special mention should go to climatic and hydrological data. While projects to provide open access to data observed and estimated have multiplied at global level (remote sensing, reanalysis, models, etc.), at local level, climate and hydrological data is still jealously guarded (apart from research applications) by the national hydrometeorological services due to the interest by governments in selling intellectual property for commercial applications (Overpeck et al. 2011).

This said, the big data-sets currently available offer a dual challenge: on the quality of the data used and on the use of such data locally for diachronic analyses (Goodge 2003; Faghmous and Kumar 2014).

All this information is essential to local planning because it allows the identification of flood, landslide or tsunami prone locations within the area of interest, sometimes in considerable detail. In particular, it is vital in risk reduction or adaptation plans to localise and quantify measures, including the risk transfer (insurance). It was with this in mind that maps were drawn up by FEEMA/USA. Producing these maps at local level would have been a long and costly job, especially for medium-size cities with few resources, and would generate different information in terms of scale, presentation and assessment from one city to another.

An effort is being made to encourage the tropical countries that still don't have an open source system to set one up (GFDRR, RCMRD, etc.) and to overcome the concept of confidentiality. United Nations' recommendations on essential knowledge for planning (Table 1.4) have little consistence with the knowledge required by guidelines for local climate planning.

1.3.5 Planning

UN-Habitat and UNISDR recommend emergency, risk reduction, land use and master plans, and building codes. Climate measures are expected to be mainstreamed in comprehensive, general or master plans and in zoning maps and development code quite common in large tropical cities but rare in medium-size cities and towns. These human settlements currently house 2/3 of the urban population of the Tropics and mainly use municipal development and emergency plans (Table 1.5). These plans still have a low quality if we consider the climatic characterisation, number, quantification, importance, potential impact, cost of measures, identification of sources of financing, the clear identification of a responsible of execution, the existence of a monitoring and evaluation device and the organisation

Table 1.3 Open access information on local risk and vulnerability

Country	Zone	Information	Hazard	Scale	Provider	www	Year
AUS	Northern Territory	Prone area	F	1:20,000	NT Gov.	https://denr.nt.gov.au	2006–…
BRA	Country	RV	FL	1:2,000	MIN		2013–14
ECU	Country	E	FLT	1:20,000	SNRiesgo	http://gestionriesgosec.maps.arcgis.com	2016
FRA	Guadelupe	R	CFL	M	Prefecture	http://pprn971guadeloupe.fr	2012
	Martinique	E	F		Prefecture	www.martinique.developpement-durable.gouv.fr/carto-graphie-des-risques-sur-les-territoires-a-a572.html	…
	Réunion	H	F	1:5,000	Prefecture	www.reunion.gouv.fr/politiques-publiques-r2.html	…
	Guyane	E	F	1:100,000	MEDD	http://cartelie.application.developpement-durable.gouv.fr	2009
HND	Country	RV	F	M	CIDBIMENA	http://cidbimena.desastres.hn	2016
MEX	Country	V	CC	M	INECC	www.sicc.amarellodev.com	2013
MLW	National	HV		Di, TA	RCMRD	http://tools.rcmrd/vulnerabilitytool/	2015
NIC	Country				Marena	www.sinia.net.ni	various
PER	Country and some regions	R	F	1:220,000	MinAm SINIA	geoservidor.minam.gob.pe	2015
PHL	National	H	F	1:50,000	Geoportal	www.geoportal.gov.ph	unk
SAL	Country	R	D	M	MMARN	www.snet.gob.sv	2003–07
USA	CA	R	FT	Lo	CalOES	http://myhazards.caloes.ca.gov	unk
	AZ, CA, TX	E	F		FEEMA		unk

C Cyclon, *D* Damage, *E* Exposure, *Di* District, *F* Flood, *L* Landslide, *Lo* Locality, *M* Municipal, *R* Risk, *T* Tsunami, *V* Vulnerability

Table 1.4 Knowledge desired according to international recommendations (left) and real according to climate plan national guidelines (right)

Desired knowledge	Rank	Freq	Required knowledge
Potential impacts of CC	1	3	CC impact
R information and maps dissemination	2	2	R and H maps
	2	2	Population in high R zones
	2	2	Receptors id.
	3	1	Climate characterization
D, E, H, R, V assessment	3	1	H index
	3	1	Meteo stations linked with EWS
	3	1	R scenario
	3	1	R index
	3	1	V index
	3	1	V priority
Disaster losses record and share	0	0	
H-E, V, R, D information open access	0	0	
DRM capacity assessment	0	0	

Sources ADNE (2009), CEMA (2012), CMC (1992), Estado (2016), Gobierno (2015), Local (2014), RC (2012), RN –, RG (2014), RM (2014), SFRPC (2015)

Table 1.5 Plans categories according to international agenda (left) and reality (right)

Vision	Real	
Category	Category	Frequence %
	Municipal/Integrated development	45
Land use and Building codes Spatial development planning	Comprehensive, General, Master	14
Emergency	Emergency	12
Disaster risk reduction	Risk reduction	7
	Mitigation	6
	Sustainable	5
	Adaptation	3
	Smart city	4
	Other	3
	Resilience	2

of measures according to a time table (Tiepolo 2017). Large cities on the other hand, have sufficient resources to be able to implement a stand-alone plan (mitigation, adaptation, sustainable plan) with a zoning map and admitted land uses for every area.

The recommendations of the United Nations indicate seven measures: tree planting, pedestrian streets, resettlement, low-carbon urban form, risk transfer, disaster resistant critical facilities and increased resilience of vulnerable/informal areas (Table 1.6). In the 366 plans in force in the medium-large cities consulted,

Table 1.6 Measures according the international agenda (left) and foreseen by 271 climate plans (right)

Vision	Real		
Measure	Rank	Freq	Measure
Trees planting	1	109	Tree planting
Design streets for pedestrians	4	67	Pedestrian and cycle mobility
Resettlement	6	44	Resettlement
Low-carbon urban form	8	32	Compact city
Risk transfer and insurance	38	9	Insurance
Disaster resilient critical facilities/infrastructure	20–99	20–2	Civil protection, fire department, evacuation centers/sites/paths, water pumps, emergency tools, risk management office, water reservoir
Increasing resilience vulnerable/informal areas	55	6	Vulnerable communities information

only four of these recommended measures figure among the eight most frequently implemented: tree planting, pedestrian streets, resettlement and low carbon urban form. Insurance is in 38th place. The resilience of critical facilities (agriculture and food, water, public health, emergency services, industries, telecommunications, energy, transportation, banking, chemicals and hazardous materials, postal and shipping) (ADPC 2015: 15) in civil protection strengthening, fire department, evacuation sites, cereal banks, pumping water, emergency evacuation centre, emergency tools and rescue services, evacuation paths, risk management office, water tanks, occupy between twentieth and ninety-ninth place, depending on the measure. Lastly, the resilience of vulnerable communities is in 55th place.

The expected impact of these measures is little known. Some of them are not successful in every context. This is the case of the compact (low carbon) urban form, which is not advisable where densities are already high (Tiepolo and Braccio 2017).

1.4 Discussion

This chapter compares the resolutions and recommendations of the United Nations relating to climate planning with the state and trend of human settlements and with climatic planning in the Tropics, to understand how international orientations are compatible with reality.

It has emerged that the 11th Sustainable development goal, not being quantified, will require at least the same progress as in the last seven years to reach 2030 with

47% of cities equipped with a climate plan. This goal won't be easy to reach because countries with more experience and resources (OECD, BRICS) have almost completed (62–79%) their local planning, while 67% of cities without a plan are in DCs and LDCs.

The first obstacle lies in the lack of urban climate planning experience in half of the cities that lack a climate plan today. The second obstacle lies in the lack of access to essential information to assess damage, hazard, exposure, vulnerability and risk, since open access systems that allow knowledge of hazards, risks and vulnerability are operational in few tropical countries only. The third obstacle is dual. Firstly, tropical medium-size cities and towns do not use planning tools with zoning maps and building codes to include the climatic measures according to the recommendations of the United Nations. Secondly, the seven climate measures recommended by the United Nations are, with just one exception, quite uncommon.

This general framework shadows the considerable effort made by certain countries: Colombia, with its risk management municipal plans, Mexico with its municipal climate action plans, the USA with its climate action plans, India with the smart city program, South Africa with risk reduction plans and integrated development plans, to mention just a few examples. But best practices in a few dozen cities are not yet able to generate a propulsive effect on the remaining 1388 tropical large and medium size cities.

We have found that the recommendations regarding the implementation of the SDGs do not raise the matter of climate planning quality. We find this to be an important aspect. Significantly increasing the number of plans isn't enough. Quality of these plans matter. This aspect should be considered in 11th SDG monitoring.

We can see four possible solutions. First, hazard assessment. Hazard assessment is not in particular demand from the national guides to local planning. Planning is rarely consequent to climate characterisation. The definition of climate measures does not consider that drought-prone locations are increasingly being flooded and will be so more and more in the future (e.g. Niger) and that wet areas also suffer droughts which have catastrophic effects on farming and human needs (e.g. Paraguay, Haiti).

Second, inclusion of the concept of potential impact on the planning process. The process of prioritisation of the measures is based on potential impact only in mitigation plans. Adaptation plans do not base the prioritisation of measures on potential impacts. A device that estimates potential impacts, consequently selects climate measures and priorities, and monitors the results achieved, communicating them to stakeholders to adjust the aim and make them accessible open source, would seem to be helpful. The analytical part, that aimed at ascertaining the risk level for example, is made up of exclusively participative methods which originate in the participated rural appraisal introduced by Chambers (1992), the best qualities of which were the involvement/mobilisation of local communities and the identification of measures compatible with weak local capacities. If, twentyfive years ago, remote rural areas offered few alternatives to indigenous knowledge skills, this have long since changed: villages have become towns, administrative decentralisation has now been consolidated almost everywhere, basic infrastructures (water, sewers) are often present, high resolution daily precipitation gridded datasets and Google earth images are now available open source throughout the Tropics and for a series of years sufficiently long as to allow the appreciation of broad scale climate trends and their impacts. What is limiting, particularly in LDCs, is the capacity to access and use this information outside the capital city.

Third, access to open source information for analysis. We should switch from occasional assessments to the open source tracking systems of vulnerability and risk, local needs and action, to address national resources and those of official development aid towards the communities most vulnerable to CC and coordinate operations among the myriad of development partners, avoiding repetitions. This information cannot be produced locally with acceptable costs and standards. But it can be produced by central or intermediate bodies and made open access.

Fourth, integrate scientific-technical and local knowledge. The effort required is dual. On one hand, simplifying and contextualising the former in order to make it easier to be used locally. On the other, fostering learning for adaptation and mitigation among rural people (Nyong et al. 2007). Reference is made to so-called climate services: meaning climate information targeted to assist decision-making. This type of information has to be based on scientifically credible climatic information and expertise but requires local knowledge of user needs and mutual engagement between users and providers in order to be effective (WMO 2011).

Among the limitations and weaknesses of this chapter, the first thing that should be mentioned is the state of planning in towns (less than 0.1 million inhabitants) which, on a global scale, contain 43% of the urban population.

As regards the implications of our findings, we show that mainstreaming, as recommended by the United Nations, cannot take place in medium size cities and towns because this category of human settlements rarely uses physical planning tools. A revolution in the local diagnosis is necessary, switching to the use of open access information, collected using a tracking process, integrated with local knowledge and standardized. This will increase the relevance of the plans as well as their quality.

The recommendations for further research are to extend the diagnosis to towns (rural world) and make it permanent. Another important area of research for the future will be the better understanding of the urban-rural relationship, its evolution and management under climate change. The interconnections between rural and urban areas are complex on a series of interlinked plans: migration, pollution, food supply. Many strategic decisions which significantly affect rural communities could not be subject to the jurisdiction of local governments, and vice versa, the planning of neighbouring or peri-urban rural areas could influence both the risk and the quality of life of urban populations.

1.5 Conclusions

This chapter highlights a paradoxical situation related to climate planning in the Tropics: we are in the presence of prescriptions (11th United Nations SDG and recommendations) without diagnosis (state and trend of human settlements, of local climate change and climate planning).

We have investigated the relevance of the challenge: much greater than it might seem, due to the tropical urbanisation, which will add 213 new medium-size cities in the next 14 years. And then the fact that mostly of the cities that will have to have climate plans are in DCs and in LDCs, which often have no climate planning at all.

A gap still exists between the vision of the United Nations and the real state of climate planning today: cities and towns need to adopt climate plans at a pace of the past seven years, but not the plans produced so far, and not using the measures envisaged up to now, leaving the job to LDCs because the member countries of the OECD and BRICS have already done the job.

Climate planning recommendations regard the analysis and the planning process (categories of plans, measures). The mainstreaming of these recommendations encounters three obstacles today.

First, the local knowledge of hazards, exposures, damage, vulnerability, and risk is absent or occasional, and is not freely accessible in most (70%) of the 41 tropical countries observed. The tracking of these last elements and their free accessibility is essential to extend climate planning, to reduce costs, considering the range of intervenants (multi-bilateral development aid, coalitions of local governments, central and intermediate governments, NGOs, etc.) and to produce information which is really helpful to the local government. In some cases (Malawi), the albeit commendable initiative is practiced without taking into account these two aspects: too many indicators to produce the vulnerability index, and an excessively rough geographic definition to be useful at local level. Second, the physical plans on which the mainstreaming recommended by the United Nations focuses are rare in tropical small-medium settlements, DCs and, particularly, in LDCs. Third, the measures recommended aren't successful in all tropical contexts.

Mainstreaming UN recommendations on local scale requires to fill different cognitive and procedural gaps, despite the fact that, in recent years, knowledge of climate planning has been strengthened.

Analysis. Knowledge of relevant climate trends for urban and rural sectors (agriculture) in contexts which don't have a network of weather stations on the ground or a sufficiently long series of records, vulnerability to climate change assessment in contests lacking information on a local scale and, transition to open access vulnerability tracking, early warning systems.

Planning process. Assessment of the quality of climate plans, estimate of the potential of the risk reduction measures, risk reduction methods, backcasting.

Measures. Risk transfer, fuel for cooking, adaptation and adaptation credits.

The 18 chapters that follow tackle these matters.

References

ADME-Agence de l'Environnement et de la Maîtrise de l'Energie. 2009. *Construire et mettre en oeuvre un Plan climat-energie territorial. Guide méthodologique.*

ADPC-Asian Disaster Preparedness Centre. 2015. *Disaster recovery toolkit*. ADPC: Guidance on critical facilities. Bangkok.

Belda, M., E. Haltanová, T. Holenke, and J. Kavlová. 2014. Climate classification revisited: From Koëppen to Trewartha. *Climate Research* 59: 1–13. doi:10.3354/cr01204.

CEMA-California Emergency Management Agency. 2012. *California adaptation planning guide.*

Chambers, R. 1992. Rural appraisal: Rapid, relaxed and participatory. *IDS Discussion Paper* 311.

Estado plurinacional de Bolivia. Ministerio de planificación del desarrollo. 2016. *Lineamentos metodólogicos para la formulación de planes territoriales de desarrollo integral para vivir bien PTDI.*

Faghmous, J.H., and V. Kumar. 2014. Spatio-temporal data mining for climate data: Advances, challenges and opportunities. In *Data mining and knowledge discovery for big data—Methodologies, challenge and opportunity, studies in big data 1*, ed. W.W. Chu, 83–116. Berlin: Springer. doi:10.1007/978-3-642-40837-3_3.

Gobierno del estado de Chiapas. Secretaría de gestión pública y programa de gobierno. 2015. *Guía metodológica para la elaboración del plan municipal de desarrollo*. Tuxtla Gutiérrez.

Goodge, G.W. 2003. Data validity in the National archive. In *Handbook of weather, climate and water*, ed. T.D. Potter, and B.R. Colman. Hoboken, New Jersey: Wiley.

Local government academy. Department of interior and local government. 2014. *LGU guidebook on the formulation of local climate change action plan (LCCAP) Book 2*. Manila: LGA.

Macchi, S. and M. Tiepolo (eds.). 2014. *Climate change vulnerability in southern African cities. Building knowledge for adaptation*. Cham, Heidelberg, New York, Dorderecht, London: Springer. doi:10.1007/978-3-319-00672-7.

Nyong, A., F. Adesina, and B.O. Elasha. 2007. The value of indigenous knowledge in climate change mitigation and adaptation strategies in the African Sahel. *Mitigation and Adaptation Strategies for Global Change* 12 (5): 787–797. doi:10.1007/s11027-007-9099-0.

Overpeck, J.T., G.A. Meehl, S. Bony, and D.R. Easterling. 2011. Climate data challenges in the 21st Century. *Science* 331 (6018): 700–702. doi:10.1126/science.1197869.

Pezzoli, A., and E. Ponte. 2016. Vulnerability and resilience to drought in the Chaco, Paraguay. In *Planning to cope with tropical and subtropical climate change*, ed. Tiepolo, M., E. Ponte, and E. Cristofori, 63–88. Berlin: De Gruyter Open. doi:10.1515/9783110480795-005.

RC-República de Colombia. Unidad Nacional para la Gestion del Riesgo de Desastres. 2012. *Guía para la formulación del plan municipal de gestión del riesgo de desastre*. Bogotá.

RN-República de Nicaragua, Ministerio del ambiente y los recursos naturales, Dirección general de cambio climático. - *Guía fácil para la elaboración de los planes municipales de adaptación ante el cambio climático.*

RG-République de Guinée, Ministère de l'administration du territoire et de la decentralisation, Direction nationale diu Développement local. 2014. *Guide méthodologique d'élaboration du plan de développement local.*

RM-Royaume du Maroc, Ministère déléfué auprès du Ministre de l'énergie, des mines, de l'eau et de l'environnement charge de l'environnement, Ministère de l'interieur. Direction générale des collectivités locales. 2014. *Guide pour l'intégration de l'environnement dans la planification communale*.

Sakai, de Oliveira R., D.L. Cartacho, E. Arasaki, P. Alfredini, M. Rosso, A. Pezzoli, and W.C. de Souza. 2016. Extreme events assessment methodology as a tool for engineering adaptation measures—Case study of North Coast of Sao Paulo State (SP), Brazil. In *Planning to cope with tropical and subtropical climate change*, ed. M. Tiepolo, E. Ponte, and E. Cristofori, 42–62. Berlin: De Gruyter Open. doi:10.1515/9783110480795-004.

Seto, K.C., S. Dhakal, A. Bigio, H. Blanco, G.C. Delgado, D. Dewar, L. Huang, A. Inaba, A. Kansal, S. Lwasa, J.E. McMahon, D.B. Müller, J. Murakami, H. Nagendra, and A. Ramaswami. 2014. Human settlements, infrastructure and spatial planning. In *Climate change 2014: Mitigation of climate change. Contribution of working group III to the fifth assessment report of the Intergovernmental Panel on Climate Change*, ed. O. Edenhofer, R. Pichs-Madruga, Y. Sokona, E. Farahani, S. Kadner, K. Seyboth, A. Adler, I. Baum, S. Brunner, P. Eickemeier, B. Kriemann, J. Savolainen, S. Schlömer, C. von Stechow, T. Zwickel, and J.C. Minx. Cambridge, UK and New York, USA: Cambridge University Press.

SFRPC-South Florida Regional planning council. 2015. *Adaptation action areas. A planning guidebook for Florida's local governments*.

SDSN-Sustainable Development Solutions Network. 2015. *Getting started with the sustainable development goals. A guide for stakeholdes*.

Tarchiani, V., and M. Tiepolo (eds.). 2016. *Risque et adaptation climatique dans la region Tillabéri, Niger. Pour renforcer les capacités d'analyse et d'évaluation*. Paris: L'Harmattan.

Tiepolo, M., and S. Braccio. In press. Urban form dynamics in three Sahelian cities: Consequences for climate change vulnerability and sustainable measures. In *Adaptation planning in a mutable environment: From observed changes to desired futures*, ed. L. Ricci, S. Macchi. Berlin: Springer.

Tiepolo, M. 2017. Relevance and quantity of climate planning in large and medium size cities of the Tropics. In *Renewing climate planning to cope with climate change*, ed. M. Tiepolo, A. Pezzoli, and V. Tarchiani. Cham, Heidelberg, New York, Dorderecht, London: Springer.

Tiepolo, M., E. Ponte, and E. Cristofori. (eds.). 2016. *Planning to cope with tropical and subtropical climate change*. Berlin: De Gruyter Open Ltd. doi:10.1515/9783110480795.

Tiepolo, M., and E. Cristofori. 2016. Climate change characterization and planning in large tropical and subtropical cities. In *Planning to cope with tropical and subtropical climate change*, 42–62. Berlin: De Gruyter Open. doi:10.1515/9783110480795-003.

UN Conference on Housing and Sustainable Urban Development. 2016. Habitat III. New urban agenda. Draft outcome document for adoption in Quito, October 2016.

UNESC-United Nations Economic and Social Council. 2016. Report of the inter-agency and expert group on sustainable development goal indicators at forty seventh session of the Statistical commission, 8–11-March 2016.

UN General Assembly. 2015. *Resolution adopted by the general assembly on 25 September 2015 transforming our world: The 2030 Agenda for sustainable development*.

UNDESA-United Nations Department of Economic and Social Affairs, Population Division. 2014. *World urbanization prospects. The 2014 revision*.

UNISDR. 2015. *Sendai framework for disaster risk reduction 2015–2030*. Geneva: UNISDR.

WMO-World Meteorological Organization. 2011. *Climate knowledge for action: a global framework for climate services*. Geneva: WMO.

Part I
Analysis for Planning

Chapter 2
Setting up and Managing Automatic Weather Stations for Remote Sites Monitoring: From Niger to Nepal

Francesco Sabatini

Abstract Surface weather observations are widely expanding for multiple reasons: availability of new technologies, enhanced data transmission features, transition from manual to automatic equipment, early warning for critical climate risks. One of the main objective is to rehabilitate/increase the density of existing network, by providing data from new sites and from sites that are difficult to access and inhospitable. Despite the increasing number of AWS's deployed, many remote sites are still not covered by surface observations. The goal is to improve AWS network planning, especially in regions where the scarcity of local trained personnel and funding availability to manage the instrumentation are relevant issues. Some consultancies performed in the past aimed to support, remotely and/or locally, National weather services, Public agencies, Local authorities and International organizations in defining and evaluating AWS's siting and selection. The efficacy of the results mainly depends on the accurate choice of the sites of installation (network plan), on the correct selection and description of instrumentation type to prepare the international tenders, on the training process to improve the AWS's management efficiency. The present chapter would discuss some of the main issues arisen from the experience gained during the institutional activities and consultancies in international projects.

Keywords Remote · Automatic weather station · Sensor

2.1 Introduction

An automatic weather station (AWS) is defined as a "meteorological station at which observations are made and transmitted automatically" (WMO 1992a). Despite the increasing number of AWS's deployed, many remote areas are not

F. Sabatini (✉)
National Research Council—Institute of BioMeteorology (IBIMET),
Via Giovanni Caproni 8, 50145 Florence, Italy
e-mail: f.sabatini@ibimet.cnr.it

M. Tiepolo et al. (eds.), *Renewing Local Planning to Face Climate Change in the Tropics*, Green Energy and Technology,
DOI 10.1007/978-3-319-59096-7_2

covered yet by surface observations. The lack of local trained personnel and funding availability to manage the instrumentation, together with the risks associated with the safety of the equipment in remote and possibly insecure areas, represent the most relevant constraints. All the National weather services, public agencies, domestic organizations share these problems resulting more acute in developing countries (DC).

Furthermore to introduce and understand the role that AWS's can play in DC, we cannot limit ourselves to evaluate the performances of a standard AWS, but we should put them in the local specific context. As such the AWS "per se" do not change the efficiency of a Meteorological/Hydrological/AgroMeteorological Service if:

- They are not combined with others technologies such as computer science, remote sensing, GIS;
- the preparation of personnel is not very different from the past in terms of ability to deal with computers, electronic instrumentation, data transmission, data analysis, data combining on a geographical basis, choice and adaptation of models;
- there is not an international effort to prepare the tools such as manuals, software and learning procedures, to train personnel in an efficient and modern way (Maracchi et al. 2000).

FAO, GIZ, ICIMOD, AghryMet, World Bank, are international organizations that promote projects and cooperation, to sustain local Authorities in developing affordable weather observing networks. The efficacy of the results mainly depends on the sensors/systems selection (technical specifications), the accurate choice of the sites of installation and the evaluation of the observation network sustainability (economic and technical).

2.2 Why RAWS?

In the international scenario of AWS's, most of the developed countries have gone in for automation in their surface observational system since early 1970s. Countries like USA, Australia, Canada, China, India and many European countries are in the forefront. At the same time, it should be noted that no country has completely withdrawn their existing set-up of conventional surface observatories (WMO ET-AWS Final report 2008). AWS are being installed to enhance their meteorological observational network and many countries have their AWS installed in Antarctica or other reference sites (i.e. Himalaya, Nepal—EVK2CNR, Pyramid laboratory–Italy) for research purposes. Though AWS is unmanned by design, to check its security, exposure conditions and to do preventive maintenance, inspections to the sites are required for the upkeep of the AWS.

Moreover the availability of though, low power electronic equipment, encouraged the installation of RAWS's. In the past years, consultancies that I provided for

FAO to draw technical specifications for AWS and AHyS to be installed in Northern Iraq, Somalia and Afghanistan confirmed this trend.

Example of RAWS networks are HKH-HYCOS (Regional Flood Information System Hindu Kush Himalayan Region (http://www.icimod.org/hycos), SHARE (Station at High Altitude for Research on the Environment), SNOTEL (Snow Telemetry, http://www.wcc.nrcs.usda.gov/snow/), THAMO (Trans-African HydroMeteorological Observatory (www.thamo.org).

WMO Integrated Global Observing System (WIGOS) aims at an improved collaboration and coordination between NMHSs and relevant national and regional organizations. It embraces the Global Observing System (GOS), the observing component of the Global Atmosphere Watch (GAW), the WMO Hydrological Observing Systems (including the World Hydrological Cycle Observing System (WHYCOS)) and the observing component of the Global Cryosphere Watch (GCW)—(https://www.wmo.int/pages/prog/www/wigos/wir/index_en.html).

Snowfall/snow depth measurements, undisturbed air quality observations, climate change studies, lead to select remote sites for ideal displacement of sensors (Niţu and Wong 2010). Even agrometeorology and hydrology applications are often dealing with the remoteness of rivers, valleys, mountains.

2.3 Transition from Manual to AWS

The WMO ET-AWS-5 Geneva 2008 (WMO report 2008), agreed that human observers were able to integrate and classify a wide range of information needed for full description of weather events. With the advent of AWS, particular areas such as visual observations, cloud classification and weather type identification are handled poorly or not at all, by an AWS unattended by observer. On the other hand, AWS provide benefits in frequent, regular, objective and consistent measurements and can be located in any environment, including extremely remote and harsh conditions.

The transition to AWS is often instigated by a perception that these systems are cheaper to operate and easier to manage than human observers. This was not the experience of a number of member countries. Therefore, it identified a number of responsibilities and costs that may not be immediately apparent to those that adopt automatic systems. (Shanko 2015).

The change for automation is a reality in all networks (India, Nepal, Lebanon, Niger, Ethiopia, etc.) and the careful management of the transition process is needed to protect data user needs. Canada created a Change Management Board (CMB), to provide a forum for network planners, operational managers and data stakeholders to discuss strategic issues regarding the transition process.

Before embarking on automation of their network, Members and/or local organizations should consider (i) the resource requirements, (ii) potential gains and losses, (iii) how to manage this transition, (iv) Training/professional improvement, (v) homogenization of different datasets.

M.J. Molineaux (Molineaux 2010) suggested a preliminary list of guidelines which would assist in the process of converting to automatic systems:

- Management of network change
- Defining and assigning responsibilities
- System costing
- Parallel testing (Traditional vs. automatic)
- Metadata
- Data quality and reliability
- User requirements
- Access to data and metadata

In the last years, many AWS manufacturers have improved their products, allowing researchers, field assistants and operators to set up unattended data collection with several options depending on the field of application. Despite the availability of relatively cost-effective automatic stations, traditional instruments can still play an important role, as an alternative or back-up for the automatic network, especially in areas where a lack of personnel trained in electronic equipment is a limiting factor (Table 2.1).

The WMO ET-AWS 6—2010 included the following considerations in the Final report. The meeting considered the report prepared by the CCl ET on Climate Database Management Systems (ET-CDBMS), in particular the observing requirements and standards for climate. The importance of retaining some manual stations for optimizing the complementary aspects of AWS and manual stations was

Table 2.1 Comparison between traditional and automatic system

Type	Advantages	Disadvantages
Manual recording (Direct reading)	Simple use Rapid installation Immediate utilisation Reliability No power required	It needs an operator for data observations Slow data transfer from paper reports. Errors due to incorrect transcription. Low number of readings available
Mechanical (Chart recording)	Simple use Simple maintenance Reliability Rapid installation No power required	It needs an operator for chart replacement Errors during chart replacement and transcription Slow data transcription from strip diagrams High sensitivity to mechanical vibration
Automatic recording	Fast data acquisition Data real time visualization and/or transmission Large amounts of observations Data pre-processing Control output (alarms, water supply systems, etc.) In some cases low cost systems can be exploited	Power supply required High professional maintenance Expensive sensor management Specialised laboratories for sensor calibration Electromagnetic interference on electronics

emphasized. For a GIZ consultancies in Nepal I suggested to install RAWS together with a new set of manual sensors in order to facilitate the transfer process from traditional observing procedures to new technologies (Figs. 2.1 and 2.2). The

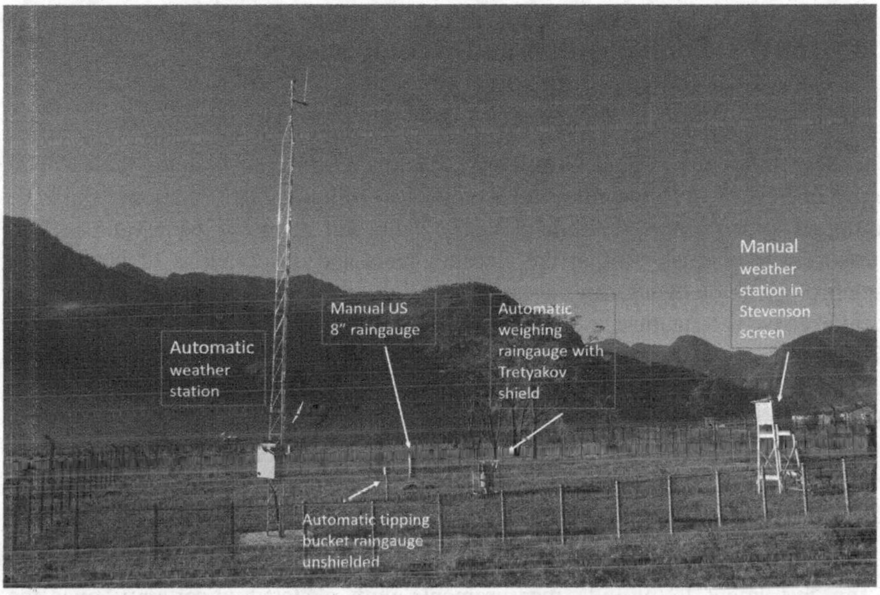

Fig. 2.1 Chainpur Nepal—Manual and AWS

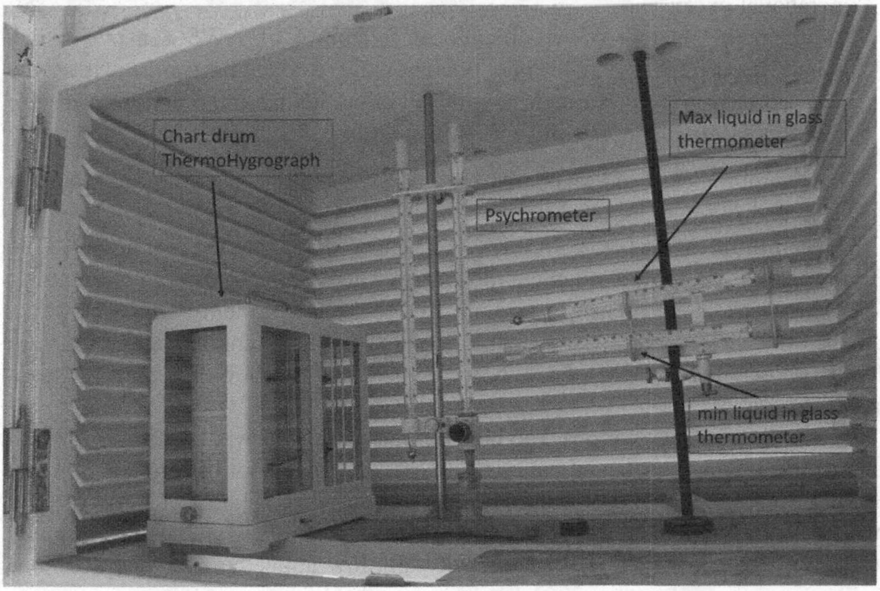

Fig. 2.2 Rara Nepal—Manual station in Stevenson Screen

observer should be kept in charge until the met service is not ready to rely on automatic equipment and to build up a dataset from manual and automatic stations in order to evaluate how this process may affects the observations.

2.4 RAWS Configuration and Requirements (Technical Specifications)

The Manual on the Global Observing System (WMO 2013), states the General requirements of a meteorological station as follows: "All stations shall be equipped with properly calibrated instruments and adequate observational and measuring techniques, so that the measurements and observations of the various meteorological elements, are accurate enough to meet the needs of synoptic meteorology, aeronautical meteorology, climatology and of other meteorological disciplines."

The requirements for AWS instruments have to take into consideration all aspects related to the ability to provide relevant and representative measurements over their entire life cycle.

The requirements for sensors installed at AWS can be ranked into three basic categories, all contributing to the long term sustainability of AWS data:

- Requirements related to the measuring performance of instruments: the ability of an instrument to provide measurements with a stated uncertainty over the specified operating range and condition;
- Requirements related to maintaining the traceability of measurements over the operational cycle; and
- Requirements related to the operational reliability of AWS sensors, which include features that enable their operation for extended periods, within the expected measuring performance, with minimum human intervention over their entire operating range. (WMO ET-AWS-5 2008)

These requirements and standards for AWSs operating in remote regions fall under several different broad aspects.

Telecommunications. GSM/GPRS is cheaper than Satellite communication. These networks are now available almost everywhere in many emerging countries. GPRS hardware is much cheaper as compared to satellite systems. GPRS networks may be used wherever they are available whereas satellite communications (such as IRIDIUM or ORBCOMM) have been used where GSM/GPRS networks are unavailable.

Power. Solar panels with backup batteries are a useful power source for AWS at remote areas. Solar panels and batteries are easily available and affordable in many regions. Sometime the use of wind turbines has been investigated at some sites, to integrate solar power (i.e. in CMA, China network). For specific sites, especially during wintertime, with low solar intensity and low temperatures, solar solutions alone do not supply reliable off-grid power, especially in critical weather and the dark seasons, so the use of fuel cell deserve mention. (i.e. New-Zealand remote snowfall stations)

Sensors. Robust and high quality sensors are suggested for remote areas. Sensors and the platform employed should be suitable for remote diagnostics and troubleshooting in order to avoid frequent inspection visits..

AWS equipment. The cost of the data loggers and communication devices has come down drastically with recent advances in electronics technologies, BUT equipment, which can work over an extended temperature range, are still expensive. Robust data loggers and communication devices are required which can work in extreme environments e.g., temperature and prolonged time with high humidity values.

AWS enclosure. In extreme weather areas waterproof box should be rust proof and salt resistant. Normally we go for standard IP65-66 or NEMA 4 protection grade. These enclosures should be made of suitable material or properly shielded so that inside temperature does not increase considerably, causing malfunction to the electronic equipment or batteries. High rainfall rates can cause water infiltration, so the connector and core hitches exposure to the environment mast be minimal.

Radiation shields. Radiation shields should be rugged and allow free airflow over the sensors, so wooden or plastic Stevenson screens should be employed. Normally fan-aspirated shield are not suggested unless strictly required.

Earthing. AWS equipment is often damaged by lightning strikes. Appropriate conventional or maintenance-free earthing, could be very effective in the long run. State-of-the-art lightning arrestors together with maintenance-free earthing at the site could reduce AWS faults due to lightning.

Calibration of sensors and maintenance of stations. Tropical; mountain and dry regions have peculiar problems in maintenance of AWS sensors due to dust deposition or frost/snow damages. Though AWS are generally unmanned by design, regular visits to a site are required to check its security, exposure conditions and for performing preventive maintenance. The costs of maintenance, calibration and running expenses for an operating AWS network far outweigh the initial purchase expense, so these expenses should be kept in mind before planning the siting and installation of an AWS network. Calibration of sensors should be performed at least once or twice per year. Availability of manpower and funds are major constraints in accomplishing these tasks.

Safety. In remote regions security of AWS equipment has become a major concern. It is quite common that there are many thefts of solar panels and batteries from the AWS sites. I had experienced 2 solar panels stolen and 1 cup anemometer broken for pure vandalism, The general public now understand that solar panels and batteries can be used in domestic or mobile applications (i.e. Van, Campers). Public participation of local people and awareness programs can help to reduce these events. The participation of Non-Governmental Organizations (NGO) can be also considered. The installation near military camps or police stations is often adopted as a suitable compromise between safety vs RAWS representativeness. Adoption of WMO siting classification guidelines, is important for exhaustive description of sensors exposure. M. Leroy of Météo-France proposed a standardised method to describe siting classification for the main variables measured at AWS, that are now incorporated in the WMO-CIMO reference guide (WMO Technical Guide n. 8 2012)

When deciding to buy an AWS the first step is to design the exact configuration (i.e. the type of sensors, amount of memory of the data-logger, power required, etc.). The main characteristics should be: Reliability; Robustness (toughness); Simple operation; Reduced level of maintenance; Installation of few sensors and estimation of other parameters.

Some examples:

- Instead of using a standard "A" Pan evaporimeter tank, which requires frequent interventions, it is better to estimate the evaporation from the other meteorological variables such as air temperature and humidity, wind speed, solar radiation, etc.
- For wind speed and direction, bi-dimensional sonic anemometers can be used which can replace traditional cup anemometers and vane sensors. Sonic anemometers are now comparable in terms of costs and signal output features; as they have no moving parts the maintenance operations are greatly reduced. Virtually no recalibration procedures are required.
- The net radiation can be estimated from the global radiation.
- If soil moisture sensors have to be installed, Time Domain Reflectometry sensors (i.e. Delta_T Thetaprobe, Campbell CS610)[1] offers low maintenance operations and long term stability. Alternatively, this parameter could be estimated from soil water balance methods.

Furthermore, if the AWS is not part of a reference climate network, the adoption of a passive (natural ventilation) solar radiation shield for Air temperature and Relative humidity sensors is suggested (no power required for the fan and no fan failures). Possible errors on air temperature observation during unfavorable conditions (wind speed ≤ 1 m/s and global solar radiation >800 W/m^2) are expected. The accuracy of thermometer can be of ±0.3 °C instead of ±0.1 °C.

The data logger (Fig. 2.3) has to meet the following generic requirements: portability, prolonged field deployment, flexibility, rapid sampling over multiple channels, small size and weight, low power drain, non-volatile data storage, and battery-backed real time clock. ADC (Analog-to-Digital converter) should have at least 14–16 bit resolution and a voltage range of 0–2.5 Vdc, 0–5Vdc, −5/+5 Vdc or autorange. SDI-12 port, RS-485 and RS-232 input, Digital inputs for tipping rain gauge bucket, pulse anemometer and digital leaf wetness sensors. Normally up-to-date systems have these features. One of the following arrangements can be considered:

[1]The names are necessary to report factually on available data; however, the author neither guarantees nor warrants the standard of the product of specific manufacturer, and the use of the name by the author implies no approval of the product and manufacturer to the exclusion of others that may also be suitable.

Fig. 2.3 Rara Nepal. OTT data logger and data transmission modem in IP65 enclosure

1. A complete AWS as a turnkey system (data-logger plus sensors, telemetry option, wires, software, power device, mast, sensors brackets, etc.), assembled and configured upon request by the manufacturer, or
2. A composite arrangement of one or more items (i.e. the data-logger, plus specific sensors chosen from other company's catalogues + telemetry system), configured and customized by the user.

2.5 Data Transmission

Remote data transmission is an important feature which adds more complexity to the system (Fig. 2.4). In particular it requires a careful evaluation in terms of:

- Power. It may be one of the option of the system that drains the most current, especially during data transmission).
- Cost of the equipment (modem, antenna, improved power supply…).
- Cost of data transmission (i.e. future sustainability).
- Routine checks towards provider for GSM/GPRS or satellite services, i.e. available credit on the SIM for each contract (each station + base station); GSM/GPRS tariffs, comparison of the expected cost to the invoices received.

Fig. 2.4 Available data transmission systems

Remote data transmission
• Modem, landline phone
• Modem, cell phone
• Radio
• Meteor burst
• Geostationary satellites (Meteosat, Goes, Immarsat, Thuraya)
• Low orbit satellites (Orbcomm, Iridium…)

Once the site of installation has been accurately identified, a specific selection of the telemetry option can be based upon:

- Presence of an affordable phone landline at the site of installation.
- GSM-GPRS/CDMA cell phone coverage at both the site of observation and at the base station.
- Radio frequency availability and license restrictions. Distance of the observation sites from the base station. (Signal repeaters required?).

Normally we choose the satellite option to overcome the disadvantages of the other systems. Coverage of the service as stated by the providers is normally affordable. It can be more complex and/or expensive with respect to the other telemetry options (unless it is possible to access a Meteosat platform as a recognized National Meteorological Service) (Fox et al. 2007).

Meteor Burst wide range data transmission systems deserve mention. This technology is free from transmission fees, but the hardware is *very expensive* (especially the base station). It relies on the phenomenon of reflecting radio waves off the ionized trails left by micrometeors as they enter the atmosphere and disintegrate. These trails exist in the 80–120 km region of the earth's atmosphere, and reflect the RF energy between two stations. The height of the trails allows over-the-horizon communication at distances up to 2000 km. This option is particularly well suited for long-range, low data rate applications for both messaging and data acquisition, like SNOTEL network. (Schaefer and Paetzold 2000).

Common communication methods can be via dial-up modems, leased-lines, radio (licensed or unlicensed frequencies) and satellite. Combinations of these can be used as well. For example, RTS Kathmandu (Nepal) developed a device equipped with CDMA, GPRS and Satellite modems: if the CDMA and/or GPRS line fails the systems automatically switch to satellite modem to ensure data transmission. The negative aspect is that the customer has to keep two contracts active instead one, but when a reliable early warning system is required, this option may result of great interest.

Each data transmission method has its advantages and disadvantages (Table 2.2). Frequency of polling and speed should be balanced with cost and power supply availability at the remote station. Site requirements may dictate that communications be close to real-time or not, depending on the applications. With radio systems, it is especially important that enough time and money be allocated to design and implement the best system possible. Generally, communication speed

Table 2.2 Comparison between different types of data transmission systems

Type	Advantages	Disadvantages
Modem, landline phone (PSTN)	PSTN = Public Switched Telephone Network Utilises the normal landline phone service Data transmission speed starts from 1200 bps Possibility to manage a network from a main site Low modem cost	Possibility of line interruption service not available in remote locations Expensive installation in remote locations
Modem, GSM Cell phone	GSM = Global System for Mobile Communication) Low cost of modem installation also in remote locations Speed of transmission up to 9600 bps Normally several types of contracts are available Cost of the modems is reducing SMS service for data transmission	Coverage not complete in several zones Possibility of interruptions due to electromagnetic interference Phone calls more expensive Voice call priority higher with respect to data call
Modem, GPRS Cell phone	GPRS = General Packet Radio Service Low cost of modem installation also in remote locations Higher speed of transmission with respect GSM Data sent to a remote server (also a third party server, i.e. of AWS manufacturer) Sometime no software required for data receiving (just an ftp connection)	Coverage not complete in several zones SIM card maintenance (expiration date, credit available, etc.)
Modem, CDMA Cell phone	CDMA = Code Division Multiple Access Low cost of modem installation also in remote locations Comparable speed of transmission with respect GSM Data link to internet. They can be sent to a remote server (also a third party server, i.e. of AWS manufacturer or foreign server) no specific software required for data receiving (just an internet connection)	Service developed only in few countries Some remote areas relays only on 1 antenna. If it fails no service is provided for long time
Radio	No charges for communications Possibility of real-time connection Alternative in absence of landline or GSM service phone	License is required in many countries for long distances High cost of radio relays

(continued)

Table 2.2 (continued)

Type	Advantages	Disadvantages
Geost. Satellites (Meteosat, GOES, Inmarsat, Thuraya)	No charges for communications for some organisations (Meteosat) Continuous coverage in European and African countries (Meteosat), Arabian countries (Thuraya) Modem device works like a common phone (Thuraya) Alternative in remote locations in absence of landline or GSM phone service	Time window assigned for data transmission not aligned with data logger clock (Meteosat) High cost for channel loan High cost of phone communications (Thuraya)
Low orbit satellites Orbcomm, Iridium	Bi-directional communication via email Good coverage for many countries Every terminal (modem) has its own email address Simple contract for terminal activation with the provider SMS service for data transmission SBD service for data transmission (bill in bytes) Modems are now cost effective	Quite high cost of data transmission The modem cost is double with respect to a GSM modem
Meteor bursts	No on-going communication service fees Both point-to-point services and multiple station networks for wide ranges (up to 1600 km) Upgrading of the network by adding other remote stations (modems). Supports large networks Operates on a single frequency in the low VHF band (40–50 MHz) Modem (i.e. MCC-545B) is dynamically configurable as base station, mobile transponder, or repeater Independent of satellite or cellular infrastructure; self-contained	Very high cost for initial system setup May require end-users to acquire their own approved frequency (rules differ in each country) Trained technical personnel required for the installation of the network

and availability relate directly to cost. Some modems support a communication scheme that can substantially reduce power requirements, resulting in a battery and panel size reduction of over 50%.

A poorly designed system can waste money in terms of troubleshooting and maintenance cost. Mr. Karl Blauvelt in his technical paper on AWS maintenance (Blauvelt 2000) talking about cell phones reported "…When these phones work, they are great. When they do not, they are a headache." I think that things have improved nowadays but accept the fact that is never fault of the phone company or of the network provider. Moreover, if you setup a GSM/GPRS configuration in a country it may be different in another country, so preliminary local field tests are mandatory before going for the final installation.

2.6 RAWS Management

The WMO ET-AWS 5 (2010 established a list of advances and limitations in AWS technology. The main advances concern telecommunications means and ability of internal diagnostic to optimize the maintenance.

Rodica Nitu of Environment Canada reported that the decreasing cost of an AWS make them more affordable and attractive, however, it has to be recognized that the cost of AWS stations remains marginal compared to the initial and running costs of a network. It is mandatory to evaluate this aspect to avoid useless investments by lack of subsequent network management, maintenance, calibration and training.

There is a wide range of configurations of AWSs for surface measurements available; high end systems (e.g. for airport, climate, micrometeorological applications) are most often user specific configured. This foresees trained local personnel to manage the equipment.

The customers may tend to choose a turnkey system managed by a third party (manufacturer, local agent/distributor), if they cannot rely on trained personnel. The RAWS maintenance is the first aspect to be taken into account; type and number of sensors to be choose is fundamental in the planning phase. As far as I know, there is no maintenance-free automatic weather equipment but we can work at reducing the inspections operations selecting appropriate sensors, as discussed in the RAWS configuration paragraph above.

No moving parts is one of the best features of a sonic anemometer because it reduces failures typical of the traditional cup anemometers. Non-catchment rain gauge type represents an alternative to standard rainfall collector for long time deployment in remote sites, even if the scientist are still validating their responses against field reference systems. Intercomparison campaigns organized by WMO show interesting perspectives for their use in multiple precipitation regimes: rainfall, snow, mixed rain-snow, drizzle, etc. The absence of a collector and/or moving parts led to more reliable observations, maybe of less accuracy, but less affected by some error source.

The operational procedures within a NMHS is a set of processes relating to both institutional and technical management aspects. In this context, the leadership shall be aware of the importance of implementation of the remote sites monitoring network in their observation network (Hall 2007). To help improve AWS's management I would recommend the following to those who are responsible of the observation network.

Management:

1. To establish clear working processes (related to the RAWS implementation) that are described and documented. The process descriptions identify responsible persons and staff activities, as well as relevant documentation and information flow.
2. To officially appoint local agent/company for the maintenance of the AWSs of the monitoring network, at least in the start-up phase.
3. To nominate one person who shall actively interact regarding technical aspects with local company/distributor and relevant organizations engaged in the monitoring network: system maintenance, routine checks, data management, data processing. He/She shall have basic experience in computer science and hydrology/weather data management and will participate in the inspection visits to the sites. He/She should be responsible for compiling metadata about the sites/stations.
4. To provide equipment and train the personnel identified in point 3) on the use and practice of AWSs and AHySs and software management.
5. To set a calendar of meetings with local company/distributor and relevant organizations personnel (i.e. every 6 months) to evaluate and discuss the data received, to prepare and schedule both extraordinary and ordinary maintenance inspections to the sites, to periodically check telemetry costs and performances.
6. To fully support capacity building efforts to help develop its own service. Some help would be provided immediately through the distribution of WMO guidance material and promoting activities such as training, publications and technical advice.

Technical issues:

1. To keep and maintain in good condition a complete AWS in their laboratories, to perform tests, routine checks, training and related activities.
2. To keep and maintain in good condition spare parts or new accessories such as batteries, solar panels, data logger, GPRS modem, sensors, and cables for the network of automatic stations. At least one complete system for each type of station should be available in stock as a back-up in case of failures or recalibration.
3. A Notebook equipped with USB and/or USB-RS232 adapter and the communication and elaboration software is fundamental for sensor and system checkup.
4. To setup one portable sensor kit to be used as a "travelling standard" for comparison against the deployed sensors.

Field inspections/maintenance:

1. Give consistency to sensor sampling and data acquisition intervals to the AWS and RAWS configurations.
2. Double-check all the system configurations (i.e. sensor conversion factors, height of installation, sensor exposure).
3. Implement their "Field Form Report" with data comparison against portable and/or traditional sensors, where applicable.
4. Set into the configuration one or more "awakening intervals" for remote modems (i.e. two or three intervals of 1 h each per day), in order to set communications for routine checks or remote configuration modifications.
5. Interact with relevant organizations to strengthen practical training on AWS management and data evaluation. This will be supportive for the person(s) in charge of network maintenance.
6. Develop a document that describes work processes typical for observation generation, making reference to the relevant WMO documents (Sabatini 2010).

2.7 Low-Cost RAWS: A Suitable Choice?

The WMO ET-AWS-6 2010 agreed that there is a need for continuous monitoring of advances of AWS technology for timely and comprehensive advice to WMO Members. Research Institutes, Universities, and international organizations are following the AWS technologies evolution as well, in order to suggest up to date solutions for meteorological and micrometeorological monitoring. Advances in technology and the decrease of costs of sensors and data loggers led to the development and integration by manufacturers of compact/low-cost weather stations. This is significantly driven by the ability to integrate and display data from multiple sources, in real time and near real time, primarily through the internet. (WMO ET_AWS-6-Final report 2010)

Computer science, real-time data and metadata transmission, Smartphones, Internet, electronic devices, led to the foundation of open access monitoring platforms.

Data-sharing, participative approach, citizens science, data interoperability are the keywords of this process. A common base is that these platforms are based on low-cost (or very low-cost) equipment because it should serves as much volunteers/observers as possible.

The compact AWSs are increasingly used for agricultural, urban meteorology. They normally integrate inexpensive sensors of performance of measurement yet to be assessed. Most of these AWS include pre-programmed data processing algorithms that are the property of the integrator.

Given the affordable cost, including installation, the small foot print required, and the increased availability, these platforms have a good chance to capture a large share of the market for producing meteorological data for various applications.

Advantages:

- compact, self-contained;
- low-cost;
- easily deployable;
- integrating up to date forms of communication which allow networking and spatial integration of data.
- ability to establish a denser network;
- real-time and on site storage of data;
- potential to be used as a subset of a fully equipped AWS (e.g. temperature and precipitation);

Disadvantages:

- Reduced interaction (Black box) between the system and the customer
- Third-party sensors or transmission devices cannot be implemented
- Metadata absent or poorly reported
- Lack of standardized observations procedures (WMO ET-AWS-6-2010)

Manufacturers like Davis (USA, CA), Pessl (Austria), Caipos (Austria), Decagon (USA), Onset (USA), Netsens (Italy), mainly propose low cost (or cost-effective) equipment for environmental monitoring and agrometeorological applications. Some of them provide to customer an AWS with some optional configurations, powered by a Solar Panel, plus an internal battery and transmits weather data in quasi real time (i.e. 15 to 60 min) over GSM/GPRS.

The system can also send SMS Alarms upon specific events like frost, rain, temperature, etc.

Data are regularly uploaded to a web platform developed by the manufacturer—(password protected). The user besides checking the weather data, can access additional packages for Weather Forecast, Disease prediction models and Irrigation Management The user pays a flat yearly fee that includes the Internet traffic and the web domain management.

More advanced research institutes or public organizations tend to build-up AWS's based on very low cost acquisition board (i.e. Arduino, Raspberry) that offer a large number of options and devices (multiplexer, Wi-Fi boards, GPRS modem, applications notes developed by the community). These systems are normally limited to specific projects for specific applications for a defined amount of time (i.e. 1 to 3 years), because the overall toughness of the system is not normally comparable to the traditional AWS's even if examples of services based on this approach exist. The power consumption is also an issue since low-cost systems normally drain more power with respect advanced data loggers. This then has an impact on the size of the batteries and of solar panels.

Info4Dourou and Info4Dourou 2.0—sustainable management project (http://cooperation.epfl.ch/Info4Dourou) aims at increasing the availability of natural resource information for local people of Burkina Faso and improve environmental data collection using low-cost local-made automatic stations. The EPFL—CODEV Lausanne, Switzerland developed a low cost hydro-meteorological observation network to reinforce and increase water and natural resource management capacity by engaging new technology for development and to increase the ability of the rural community to adapt to changing environmental pressures.

CNR-IBIMET developed its own infrastructure (www.sensorwebhub.org), to receive data from multiple type of homemade sensors/stations, even mobile systems, in order to provide an open common platform compliant to OGC approach towards the world of IoT (De Filippis et al. 2015).

The adoption of the low-cost approach for remote sites meteorological monitoring deserve attention, since it can integrate conventional networks providing useful information otherwise not available.

2.8 Conclusions

Surface meteorological monitoring at remote sites requires a decision process that result in a cross-cutting activity among a given Organization. The applications, the site accessibility, harsh environments, lack of data transmission platforms and power supply options, may led to a very difficult selection procedure.

As already pointed out by the present chapter the most important constraints are represented by: (i) the limited technical capacity in the local organizations to operate and maintain the systems and (ii) the risks associated with the safety of the equipment in remote and possibly insecure areas.

Upon my experience the best approach would be represented by a step by step, progressive installation, of AWS at the remote sites. Since the skill of personnel grows normally up according to the complexity of the network, the multiple setup of AWS's can pose problem of sustainability at the beginning.

Whatever the sites position an expensive part of the whole plan is also represented by the software/hardware infrastructure and the processing center for data management (normally installed at the organization headquarter), but the aspect of maintenance of several surface observation sites represents a very important aspect.

Unfortunately, this approach rarely matches those of the national/international implementation projects: even well-conceived initiatives are normally focused on a specified amount of years (3–5 years), without an extended follow-up process. Once the project expires, the network is no longer appropriately managed and missing inspections lead to lack of data in the observations series and hardware failures. Sometime the result is a set of several AWS's generations, not interoperable, installed for different projects and from different manufacturer.

WMO-RTC's should accompanying this transition process addressing specific training course on AWS management and IT integration, but again the synchronization among national and international bodies in not that efficient.

Anyway the RAWS's are expanding despite their difficult management. This process will continue in the next years for multiple applications: water resource monitoring, solid precipitations, climate change studies, basic climatology, ground truth for remote sensing analysis.

The selection and management of robust and well maintained RAWS is a real challenge for whatever country is planning their implementation in the present observing system. WMO, FAO, AghryMet, ICIMOD and related international projects could play an important role in this process, in order to adopt appropriate approaches compliant to up-to-date techniques.

References

Blauvelt, K. 2000. Maintenance and calibration manual for the automated weather data network. Calibration facility, High plains climate center. University of Nebraska–Lincoln USA.

De Filippis, T., L. Rocchi, and E. Rapisardi. 2015. SensorWeb Hub infrastructure for open access to scientific research data. *Geophysical Research Abstracts* 17. EGU2015-7847-2.

Fox, J., P. Geissler, and R. Worland. 2007. Versatile Iridium Campbell link,. (1) British Antarctic Survey (j.fox@bas.ac.uk). *Geophysical Research Abstracts* 9. 01810. SRef-ID: 1607–7962/gra/EGU2007-A-01810. © European Geosciences Union.

Hall, B.L., and T.J. Brown. 2007. Comparison of weather data from the remote automated weather station network and the North American Regional Reanalysis. In 87th AMS Annual Meeting.

Maracchi, G., F. Sabatini, and M.V.K. Sivakumar. 2000. The role of automated weather stations in Developing Countries. In *Proceedings of the International workshop on "Automated weather stations for applications in agriculture and water resources management: Current use and future perspectives". 6–10 March 2000*. Lincoln, Nebraska, USA.

Molineaux, M.J. 2010. Automation of surface observations. Development of guidelines and procedures for the transition from manual to automatic weather stations. WMO CBS—Opag on Integrated observing systems. Expert team on requirements and implementation aws platforms (Et-Aws). Sixth session. Final report. Geneva, Switzerland, 20–23 April 2010.

Nitu, R., and K. Wong. 2010. Measurement of solid precipitation at automatic weather stations, challenges and opportunities. Meteorological Service of Canada, 4905 Dufferin St, Toronto, ON, M3H 5T4, Canada.

Sabatini F. 2010. Recommendations for the design of a system for acquisition, transmission and storage of Hydro-Meteorological data on the Kailash sacred landscape in Nepal. Project title: Supporting climate change adaptation and biodiversity conservation in the Kailash landscape. Final report.

Schaefer, G.L., and R.F. Paetzold. 2000. SNOTEL (SNOpack TELemetry) and SCAN (Soil Climate Analysis Network). In *Proceedings of the International workshop on "Automated weather stations for applications in agriculture and water resources management: Current use and future perspectives". 6–10 March 2000*, Lincoln, Nebraska, USA.

Shanko, D. 2015. Use of automatic weather stations in Ethiopia. National Meteorological Agency (NMA) report, Addis Ababa, Ethiopia.

WMO. 2010. *Guide to meteorological instruments and methods of observation, Technical Manual n° 8*. 7th Edition 2008 updated 2010. Geneva, Switzerland: World Meteorological Organization.

WMO. 2008. Final report. World Meteorological Organization—Commission For Basic Systems - Opag On Integrated Observing Systems. Expert Team On Requirements for Data From Automatic Weather Stations—Fifth Session—Geneva, Switzerland, 5–9 May 2008.

WMO. 2010. Final report. World Meteorological Organization—Commission For Basic Systems - Opag On Integrated Observing Systems. Expert team on requirements for data from automatic weather stations—Sixth session - Geneva, Switzerland, 22–25 June 2010.

WMO. 2013. *Manual on the global observing system. Revised latest edition 2010.*

Chapter 3
Hazard Events Characterization in Tillaberi Region, Niger: Present and Future Projections

Maurizio Bacci and Moussa Mouhaïmouni

Abstract Niger is one of the countries most vulnerable to climatic risks. An adaptation to meet these threats is urgent and supported by politicians and decision makers, as stressed in the *Programme d'Action National pour l'Adaptation aux changements climatiques* (PANA) of Niger. The main aim of this paper is to provide an assessment of the current and future scenario of natural hazards in Tillaberi Region (Niger). The mapping of hazard changes in the study area is done comparing the probability of recurrence of severe meteorological conditions for droughts and floods between present and future climate using several projections from the Coupled Model Intercomparison Project Phase 5 (CMIP5). The result is a hazard characterization highlighting the need for urgent interventions. The natural hazard information with exposure and vulnerability assessment indicators can help decision makers in prioritizing interventions in the Tillabéri Region using an objective approach. This methodology has been proposed within the framework of the ANADIA Project that aims to support disaster prevention activities, from national to local scale, helping institutions in the design and implementation of disaster risk management strategies.

Keywords Climate change · Natural hazard · Drought · Extreme rainfall · Niger

3.1 Introduction

Natural disasters, notably floods and drought, are becoming a major concern in West African countries. These are caused mainly by natural hazards such as extremely intense rainfall events or extensive dry periods. With the current rapid

---

M. Bacci (✉)
National Research Council—Institute of BioMeteorology (IBIMET),
Via Giovanni Caproni 8, 50145 Florence, Italy
e-mail: m.bacci@ibimet.cnr.it

M. Mouhaïmouni
National Meteorological Service of Niger, Niamey, Niger
e-mail: mouh_moussa@yahoo.fr

M. Tiepolo et al. (eds.), *Renewing Local Planning to Face Climate Change in the Tropics*, Green Energy and Technology,
DOI 10.1007/978-3-319-59096-7_3

climate change conditions attention should be paid to natural disasters in all development planning at national, regional and local levels to minimize the risks. The *Programme d'Action National pour l'Adaptation aux changements climatiques* (PANA) of Niger sets adaptation policies as urgent and necessary to cope with future climate risk.

Tillaberi Region, the case study area, is located in the westernmost part of Niger. It consists of 45 municipalities with an estimated 2,722,482 inhabitants in 2012 (INS 2012). Its economy is mainly based on subsistence agricultural and livestock farming. 84.9% of its population live in rural area and 71.7% live below the poverty threshold (INS 2011).

Niger is extremely vulnerable to extreme climate events. The country faced heavy losses due to increased flooding and drought episodes in the last decade. In 2013 alone, the number of victims from floods in the Tillaberi Region was estimated at 39,681 and 2902 hectares of fields were flooded (OCHA 2013).

Numerous factors contribute to increasing local exposure and vulnerability to natural disasters: (i) the changes in river flows and discharges caused by human activities (Mahe et al. 2013), (ii) population growth and related environmental degradation (Leblanc et al. 2008) and (iii) changes in hydrography dynamics (Descroix et al. 2009, 2012). It is clear that some actions are needed to face climate risks. We consider a bottom-up approach in the climate change adaptation process as a key of measures' effectiveness. Adaptation capacity is linked with some local elements such as population food security, education and knowledge of climatic risk. A decrease of vulnerability to climate risk can also be achieved knowing what the future could hold.

Tillaberi Region belongs largely to the Sahelian zone characterized by difficult climatic conditions such as low rainfall and high temperatures that can reach 45 °C. It has a single rainy season from June till September averaging an annual precipitation from 250 mm/year in Abalak, in the north, to 800 mm/year in Gaya, in the south.

The interannual variability is quite significant and it is possible to identify three different phases in the last 30 years: (1) a dry period in the 1980s, (2) a transition decade in the 1990s with a high interannual variability and (3) a wet phase in the first decade of 2000 (Fig. 3.1).

Despite this increasing trend in total rainfall amount, changes in extreme events distribution are slightly different and not immediately related to the seasonal sum. In this case we integrate different climate indexes, such as the Standardized Precipitation Index (SPI) by McKee et al. (1993) at monthly scale, to achieve a more complete climatic characterization.

We analyze changes in rainfall distribution with particular emphasis on extreme events. As described in the Special Report from Intergovernmental Panel on Climate Change (IPCC) *Managing the risks of extreme events and disasters to advance climate change adaptation* (IPCC 2012), the near future will likely see an increasing number of extreme events.

The study analyzes historical climatic trends and future scenarios in Tillaberi Region, identifying hotspots and discusses how observed trends could impact

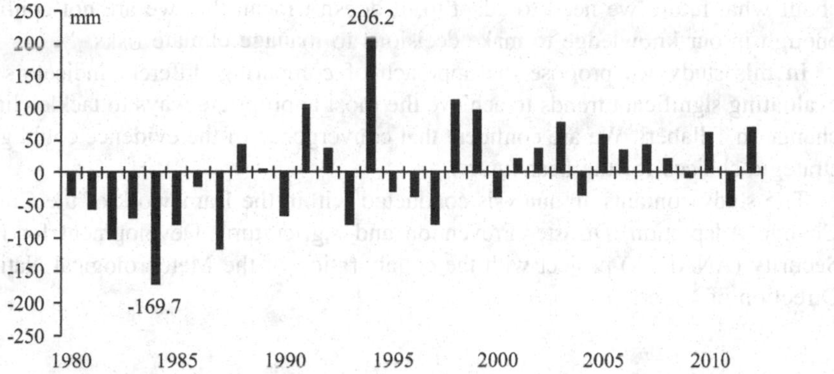

Fig. 3.1 Mean seasonal rainfall in Tillaberi Region (1980–2012). Average over Tillaberi gauge network

human activities. Climate analysis aims to support the quantification of natural risk highlighting the most sensitive zones in the region for drought and floods.

The analysis is split in two steps:

- the evaluation of current trends, using data from the rain gauge station network of the National Meteorological Direction of Niger (DMN), the national body in charge of meteorological monitoring;
- the evaluation of some future scenarios using climate model outputs from the Coupled Model Intercomparison Project Phase 5 (CMIP5) at different Representative Concentration Pathways (RCPs). This information is compared with a rainfall estimation dataset, the Climate Hazards Group InfraRed Precipitation with Station data (CHIRPS), to quantify future changes compared to present climate.

The hazards are characterized by a selection of climate indexes able to intercept drought and extremely wet conditions. In particular, we have used local experience to choose those most representative of dangerous conditions for the local community aiming to determine the magnitude and location of natural hazards in the region. In fact, the aim of the study is to prepare a minimum set of indexes, simple to calculate, which can intercept local criticalities. This collaborative approach with local institutions has been guided by the need to make the process sustainable by giving local authorities the possibility to reproduce this procedure independently.

Uncertainties are still present in climate modeling. How can this source of uncertainties be managed in the decision making process? The awareness of error must guide the interpretation of the study findings. Uncertainty is frequently seen as a not manageable factor, which limits adaptation process and local communities often consider that the best approach would be to carry on as normal, and respond to climate risks as they occur. Nevertheless, decisions may need to be taken and waiting for a reduction in uncertainty is not the right solution because this approach implies reducing the time available for adaptation measures. If we are uncertain about what future we need to adapt to, it doesn't mean that we are not confident enough in our knowledge to make decisions to manage climate risks.

In this study we propose the approach of comparing different indicators and evaluating significant trends to achieve the most appropriate ways to tackle climate change in Tillaberi. We are confident that convergence of the evidence could guide strategic choices for the near future.

The study contains an analysis conducted within the framework of the Climate Change Adaptation, Disaster Prevention and Agricultural Development for Food Security (ANADIA) project with the collaboration of the Meteorological National Direction of Niger.

3.2 Materials and Methods

3.2.1 Current Conditions

Daily rainfall data, from the official DMN database, come from the rain gauge station network in the Tillaberi Region. This information is used to assess current climatic conditions. In the analysis we retain stations with at least 30 years of records, considering this period as the minimum requirement to produce a statistical analysis with a common climate reference (1981–2010). The stations fulfill requirements are 29.

Using daily data for each station, we evaluate the total number of dry days (days without rain or with less than 1 mm) and the maximum consecutive dry days recorded in the June–September period identifying drought conditions and assessing their magnitude and distribution. The number of wet days with rain equal to or more than 1 mm and the number of episodes that exceed the threshold of 10 and 20 mm/day are selected identifying wet conditions. Finally, we choose the highest 1-day precipitation, the yearly number of days with rain exceeding the 95th and 99th percentile to describe heavy rains distribution.

Therefore, each station is characterized by the following yearly index (basing on 1981–2010 period):

- R95p, Very wet days (days), Number of days >95th percentile calculated for wet days;
- R99p, Extremely wet days (days), Number of days >99th percentile;
- RR, Precipitation sum (mm);
- RR1, Wet days (days), Number of days ≥ 1 mm;
- R20 mm, Very heavy precipitation days (days) Number of days ≥ 20 mm;
- R10 mm, Heavy precipitation days (days), Number of days ≥ 10 mm;
- RX1 day, Highest 1-day precipitation (mm) Maximum sum for 1 day.

Here, for concision needs, we present only the regional averaged value computed over these 29 stations for each index. This mean value allows the evaluation of linear regression over time series, intercepting climatic trends and their magnitude at regional scale.

The Pearson product-moment correlation coefficient has been computed as a measure of the linear correlation of the index variation over time and the non-parametric Kendall's tau estimation for monotonic trends. Kendall's tau measures the strength of the relationship between the two variables making no assumptions about the data distribution or linearity of any trends. Like other measures of correlation, Kendall's tau assumes a value between −1 and +1. A positive correlation signifies that the values of both variables are increasing. Instead, a negative correlation signifies that as one variable increases, the other one decreases. Confidence intervals can be calculated and hypotheses tested with the help of Kendall's tau. The main advantage of using this measurement is that the index distribution has better statistical properties, and Kendall's tau can be interpreted directly in terms of the probability of observing concordant and discordant pairs. On the contrary, outliers, unequal variances, non-normality and nonlinearity unduly influence the Pearson correlation.

3.2.2 Standardized Precipitation Index (SPI)

The Standardized Precipitation Index (SPI) describes the magnitude of dry and wet conditions in a desired location. It allows an analyst to determine the rarity of an episode at a given time scale (e.g. Monthly resolution).

The SPI index is recommended by the World Meteorological Organization (WMO) to be used by all National Meteorological and Hydrological Services around the world to characterize meteorological droughts (WMO 2012). The SPI is simple to calculate because precipitation is the only required input parameter and it is just as effective in intercepting wet periods as dry periods.

The SPI formula is based on the long-term precipitation record. This record is fitted to a probability distribution, which is then transformed into a normal distribution. Positive SPI values indicate greater than median precipitation (wet conditions) and negative values indicate less than median precipitation (drought conditions). Because the SPI is normalized, wetter and drier climates can be represented in the same way.

The classification system shows SPI values to define drought and wet intensities resulting from the SPI as proposed by WMO (Table 3.1).

The SPI was designed to quantify the precipitation deficit for multiple timescales. In fact these timescales reflect the impact of drought on the availability of the different water resources. For instance, soil moisture conditions respond to precipitation anomalies on a relatively short timescale, groundwater, streamflow and reservoir storage reflect the longer-term precipitation anomalies. In this case we are looking for a meteorological characterization and use the 1-month SPI to characterize each single month in the rainy season (June, July, August and

Table 3.1 SPI classification (WMO 2012)

SPI values	Conditions
2.0 and more	Extremely wet
1.5 to 1.99	Very wet
1.0 to 1.49	Moderately wet
−0.99 to 0.99	Near normal
−1.0 to −1.49	Moderately dry
−1.5 to −1.99	Severely dry
−2 and less	Extremely dry

September) and a 4-months SPI to characterize the entire rainy season (from June to September).

3.2.3 Future Climate Scenarios

Data for future climate scenarios come from phase five of the Coupled Model Intercomparison Project (CMIP5), which is a set of coordinated climate model experiments. These products are freely available for scientific purposes and there are projections of future climate change on two timescales, near-term (up to about 2035) and long-term (up to 2100 and beyond). So it is possible to download the chosen variable from a list of available models and RCP for the selected period of investigation.

Coordinated Regional climate Downscaling Experiment (CORDEX) is a World Climate Research Programme-sponsored program to organize an international coordinated framework to produce an improved generation of regional climate change projections worldwide for input into impact and adaptation studies within the AR5 timeline and beyond. CORDEX will produce an ensemble of multiple dynamic and statistical downscaling models considering multiple forcing GCMs from the CMIP5 archive.

In this case we evaluate the near-term scenario, until 2035, using daily data on the 2025–2035 period. A subset of these models is evaluated at two RCPs:

- CNRM-CM5 is the CMIP5 version of the ESM developed jointly by CNRM-GAME (Météo-France/CNRS) and CERFACS;
- EC-Earth is an Earth system model that is being developed by a number of European National Weather Services and elaborated by ICHEC (Irish Centre for High End Computing);
- HadGEM2-ES Hadley Global Environment Model 2—Earth System by Met Office Hadley Centre (MOHC);
- MPI-ESM is the Earth System Model running on medium resolution grid of the Max-Planck-Institut für Meteorologie.

The climate change projections depend on greenhouse gas concentrations and emissions and they normally produce different outputs using different RCPs.

The scientific community defines some global anthropogenic Radiative Forcing for the high RCP8.5, the medium-high RCP6, the medium-low RCP4.5 and low RCP2.6 scenarios, namely based on the respectively radiative forcing (+8.5, +6.0, +4.5 and +2.6 W/m^2).

For this study we select the medium-low RCP 45 and high RCP 85 scenarios outputs to perform the analysis with a sufficient number of different simulations to represent future climate variability.

In order to evaluate how realistic the models are in simulating rainfall distribution in the Tillaberi Region, we perform a comparison analysis with the CHIRPS v. 2.0 identifying bias in models outputs. We use the time series overlapping among CHIRPS and General Circulation Models (GCMs) datasets, the 2006–2014 period, performing a simplified bias correction. We evaluate an average correction coefficient for each model output. The CHIRPS dataset has a resolution of 0.05′, so applying this bias correction to the GCM dataset, namely at 0.5′ resolution, we also perform a downscaling of information for future scenarios. This represents an added value in determining different dynamics in the region at very local scale.

The scientific literature gives a variety of bias correction methods that are focused on determining the best transfer function to convert model data into realistic ones. In this case, we are evaluating extreme values of rainfall distribution, dry days and the wettest condition. So the Delta change approach is not useful in these ranges, because the method applies the additive coefficient independently of the amount of daily rainfall. The more complex transfer functions such as exponential, logarithmic and quantile mapping methods are more adequate in creating transfer functions applicable over the entire range of rainfall. In this case we adopt a multiplicative correction transfer function determined by averaging yearly accumulation of two datasets.

So for each year we compare the yearly amount of rainfall by CHIRPS and the GCM output:

$$K_m = \frac{\sum_{i=1}^{n} CHIRPS\, daily_i}{\sum_{i=1}^{n} GCM\, daily_i}$$

where m is the year.

We then average the yearly coefficient on the overlapping period among the two datasets, in this case the 2006–2014 period.

$$\overline{K} = \frac{1}{n}\sum_{m=1}^{n} K_m$$

Finally we produce the bias corrected series of the model applying the multiplicative coefficient to each single day.

$$Daily\ CGM_{corr} = K\ daily\ GCM$$

This method has the advantage that GCM corrected series has a yearly amount of rain coherent with the rainfall estimation value. Furthermore it is an easy bias correction method giving local institutions the possibility to perform the analysis on their own contributing to process sustainability.

With the corrected data series the synthetic index has been calculated for drought and extreme rainfall events over 2025–2035 period in order to make a comparison with the current climatic conditions defined by CHIRPS in 1981–2010 climate reference period.

3.3 Results

The spatial distribution of rainfall in Tillaberi is characterized by a clear south-north gradient. Rainfall ranges between 600 and 800 mm in the southern part of the Region, and between 250 and 350 mm in the north. Rainfall distribution averaged over the entire Region demonstrates that some trends are intercepted (Table 3.2). Tillaberi climate is characterized by a slight increase in number of rainy days, above 10 and 20 mm, reflecting changes in rainfall regime recorded in the first decade of 2000. Moreover, regarding extremes, we observe a clear increase in the number of episodes greater than the 95th and 99th percentile.

The regression coefficients for these indexes are quite consistent for all indicators, intercepting significant increases in yearly precipitation, in number of wet days, in number of very heavy precipitation days (RR10 and RR20) and in number of very wet days (R95p). A slightly increase is recorded in number of days >99th percentile and in the highest 1-day precipitation.

Table 3.2 Tillaberi region

	RR	RR1	RR10	RR20	R95p	R99p	RX1
Average	400.3	29.1	13.6	6.9	1.6	0.3	55.4
Standard deviation	77.2	3.9	2.4	1.5	0.6	0.2	7.5
Max	574.3	37.2	18.1	9.9	3.1	1.1	76.6
Min	238.9	21.9	7.5	2.7	0.6	0.1	39.9
Regression coefficient	4.2	0.2	0.2	0.1	0	0	0.2
correlation	0.48	0.51	0.57	0.48	0.32	0.13	0.2
Tau	0.322	0.241	0.339	0.313	0.258	0.117	0.144
2side_pvalue	0.01	0.05	0.01	0.01	0.03	0.31	0.24
Probability (%)	99	95	99	99	97	69	76

Observed trends in the 1981–2010 period averaging all gauge stations of the Region. RR = Rainfall rate, RR1 = N. rainy days, RR10 = N. rainy days > 10 mm; RR20 = N. rainy days > 20 mm; R95p = N. Rainy days > 95p; R99p = N. Rainy days > 99p; RX1 = highest amount of 1-day rainfall

3.3.1 Standardized Precipitation Index

Using daily rainfall data from the DMN gauge network we calculated the SPI index for each station. We then averaged the SPI values over the entire region giving an overview of the climate in Tillaberi region in 33 years. SPI has been produced at two time scales: (i) 4-months resolution, in order to intercept dry or wet conditions over the entire rainy season, (ii) monthly resolution, describing monthly climatic conditions of the region at higher temporal resolution. Figure 3.2 presents the regional SPI graph. The comparison of these two time scales intercepts criticalities in different stages of the rainy season evolution (onset, central part, end) and gives an idea of the rainfall regime recorded in each year. The graph shows extremely dry conditions recorded in 1984 with a persistent negative SPI in June, July and August. Extremely wet conditions are recorded in 1994 with the highest value of SPI in August. We could associate June SPI to the interannual variability of the rainy season's onset strength. No extreme episodes are recorded in this month even if conditions are moderately dry in 1987. In July, SPI shows a lack of significant fluctuations, especially after 2000 when a sequence of normal conditions is observed. August records the maximum rainfall amount so it is the month most influencing the SPI value of the rainy season. In August there are extreme values over the entire period of monthly SPI: 1984 with a significant episode of drought and 1994 with the wettest conditions.

The 2000s are almost characterized by normal conditions. September marks the end of the rainy season and in the time series the SPI values for this month show a slight trend toward wet conditions. In summary SPI shows a quasi-normal situation in the last decade that means a globally favorable condition in rainfall distribution

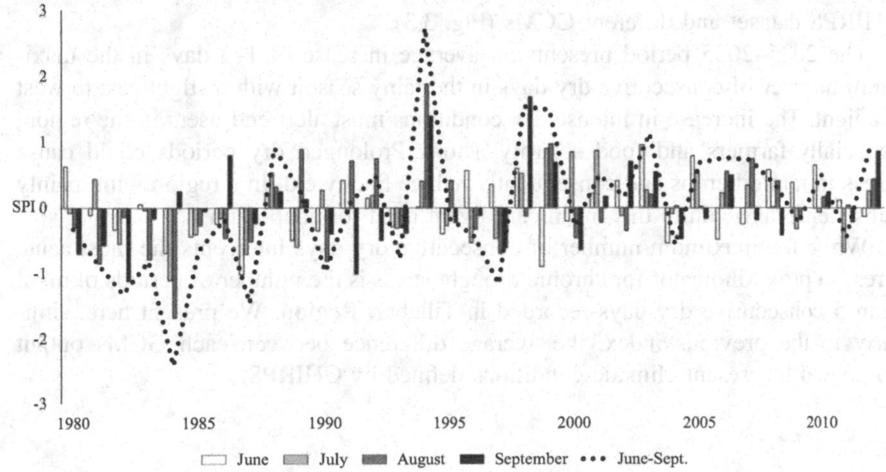

Fig. 3.2 June, July, August and September monthly SPI distribution and 4-months SPI in 1981–2012 period

comparing the driest period of the 1980s and wettest period of the 1990s. SPI is not conceived to intercept changes in daily extremes distributions, in fact in the last decade several floods are recorded in the region, but it could intercept dry and wet periods in rainfall distribution. SPI must be considered as complementary information of the daily extreme events distribution in classifying target period rainfall distribution with respect to time series. The increase of extreme rainfall events in a SPI wet period is normal, but as in the case of Tillaberi, the increase of extreme rainfall events in SPI normal period indicates an increase of extremes in rainfall distribution.

3.3.2 Future Scenarios

The last IPCC AR5 affirms: *The frequency and intensity of heavy precipitation events over land will likely increase on average in the near term.* (Kirtman et al. 2013). This statement is valid at global scale, so this study focuses on the Tillaberi Region analyzing the near-future conditions and extreme events likelihood.

We computed a very large set of indicators and here we present the most representative ones. We found that rainfall distribution is predicted to very likely increase drought and flood risk in Tillaberi Region. In fact several climate trends and climate model simulations from CORDEX-CMIP5 show: likely increase in maximum consecutive dry days; likely increase in number of dry periods of more than 5 days; likely increase in number of very heavy precipitation days; and likely increase in highest 1-day precipitation (mm)—Absolute value in the period.

The projections of future climate, describing drought conditions, have been produced and we show here the result of the comparison between the maximum number of consecutive dry days (RR < 1 mm) in June-September period between CHIRPS dataset and different GCMs (Fig. 3.3).

The 2025–2035 period presents an average increase of 1–3 days in the maximum number of consecutive dry days in the rainy season with a slight east to west gradient. The increase in intense dry conditions must alert end users in the region, especially farmers and food security actors. Prolonged dry periods could cause stress in rainfed crops and consequently reduce final yield. In a region with mainly subsistence agriculture this means an urgent need for adaptation.

While the maximum number of consecutive dry days intercepts the most acute stress, a proxy indicator for chronic drought stress is the number of periods of more than 5 consecutive dry days recorded in Tillaberi Region. We present here, similarly to the previous index, the average difference between each GCMs output compared to present climatic conditions defined by CHIRPS.

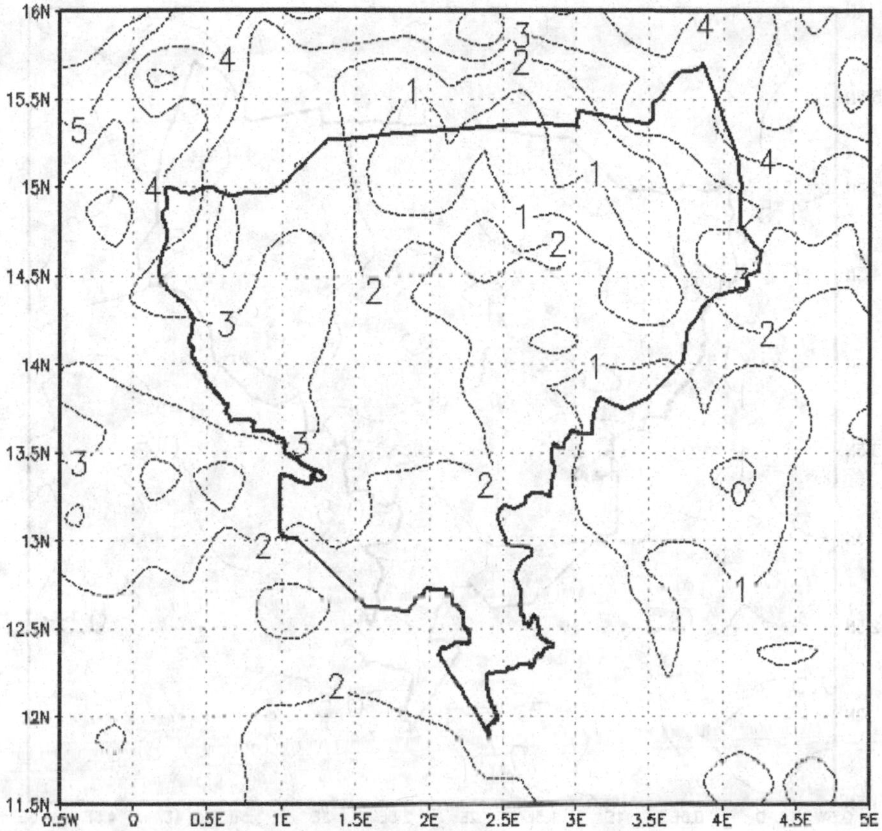

Fig. 3.3 Maximum number of consecutive dry days (RR < 1 mm)—June–September (rainy season)—Difference 2025–2035 (GCMs) versus 1981–2010 (CHIRPS)

The map (Fig. 3.4) shows that in Tillaberi Region we could expect an increase in the number of periods of more than 5 dry days that ranges between 0.5 and 1.5.

Coupling this information with the previous, the increase in the maximum length of dry period, we can affirm that more acute and chronic drought conditions are likely in the near future. We also produced some heavy rains scenarios using GCMs. We focused attention on the very heavy precipitation days, the ones with rainfall of more than 20 mm, making a comparison between GCMs and CHIRPS dataset. The averaged result is presented in Fig. 3.5.

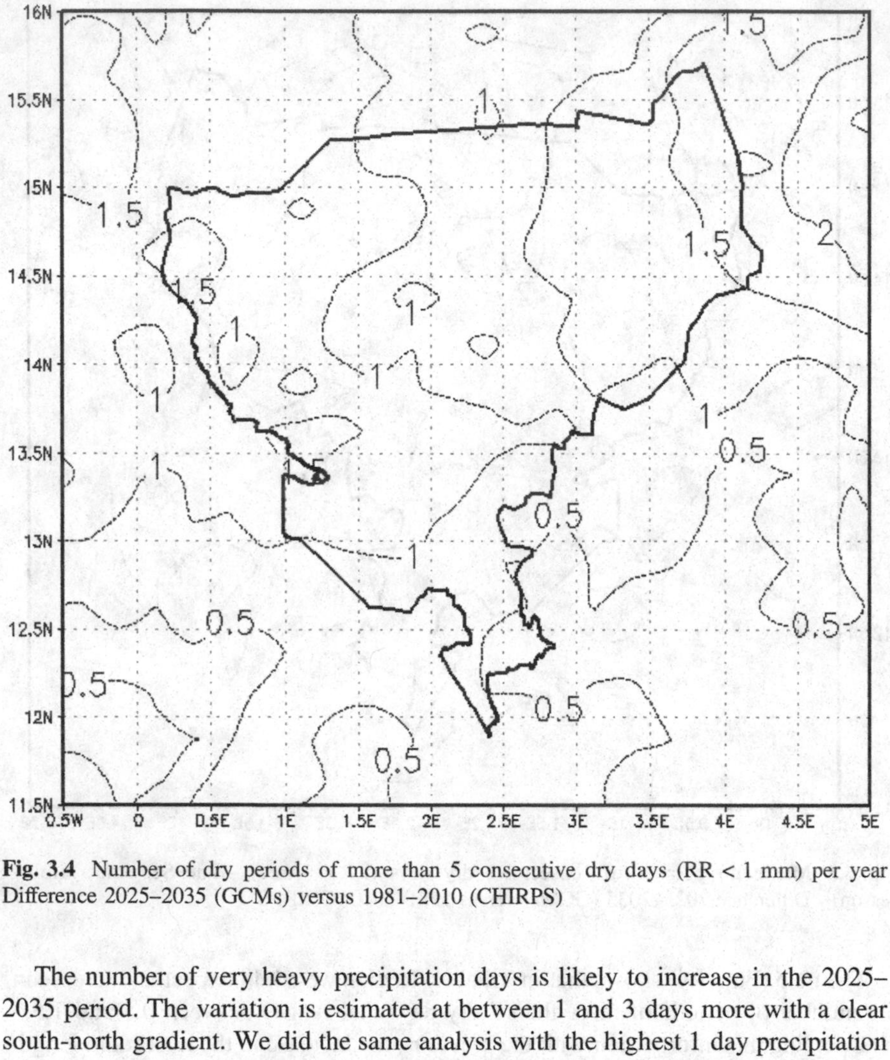

Fig. 3.4 Number of dry periods of more than 5 consecutive dry days (RR < 1 mm) per year Difference 2025–2035 (GCMs) versus 1981–2010 (CHIRPS)

The number of very heavy precipitation days is likely to increase in the 2025–2035 period. The variation is estimated at between 1 and 3 days more with a clear south-north gradient. We did the same analysis with the highest 1 day precipitation amount evaluating changes in the most intense phenomena (Fig. 3.6).

The map confirms a clear increase in the maximum of rainfall in the highest 1-day precipitation with a rise ranging between +30 and +70 mm. We want to underline the fact that an absolute value in changes could be affected by unbiased

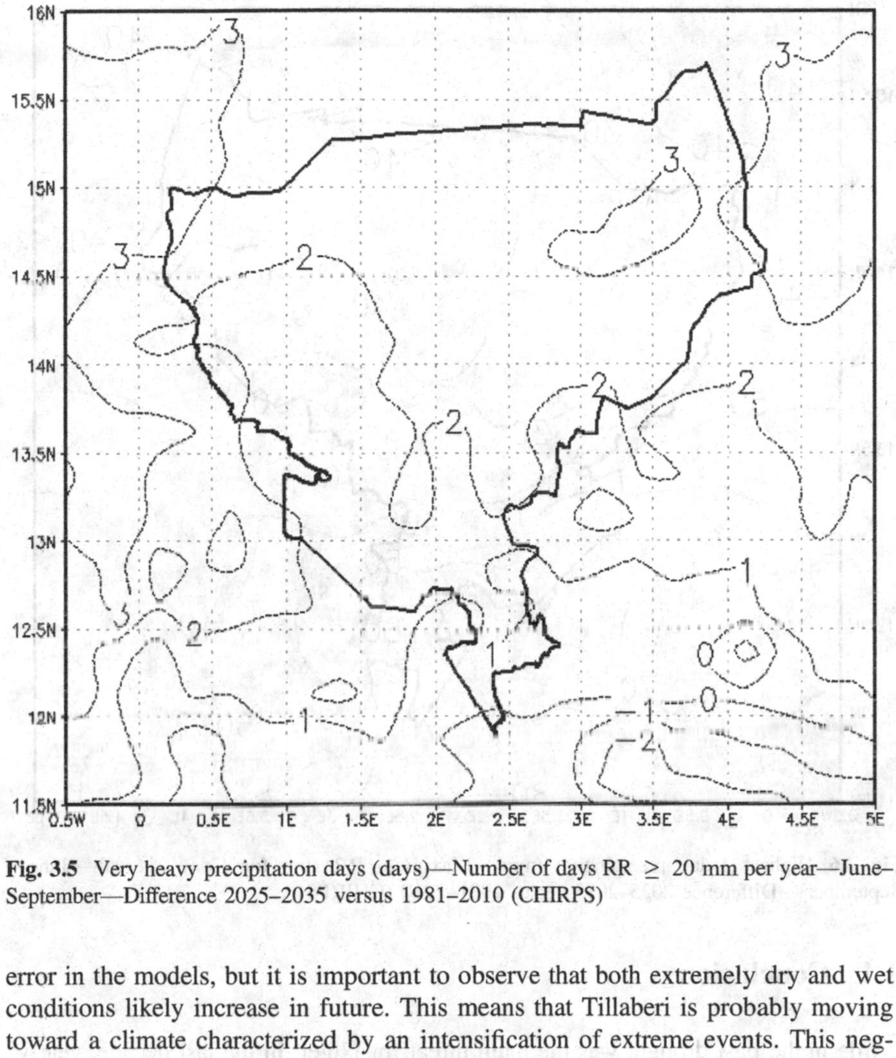

Fig. 3.5 Very heavy precipitation days (days)—Number of days RR $\geq$ 20 mm per year—June–September—Difference 2025–2035 versus 1981–2010 (CHIRPS)

error in the models, but it is important to observe that both extremely dry and wet conditions likely increase in future. This means that Tillaberi is probably moving toward a climate characterized by an intensification of extreme events. This negative scenario means more favorable condition for floods and droughts and the expected exacerbation of intrannual variability could lead to a reduction in crop yields, soil erosion phenomena and an increased population exposure to climate risks.

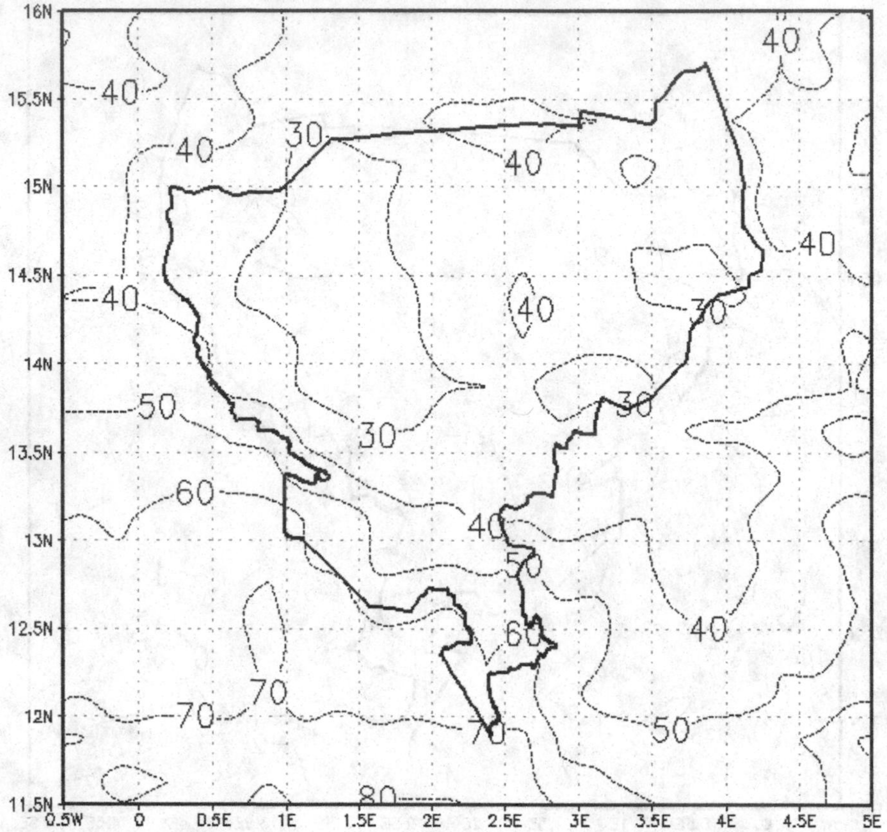

Fig. 3.6 Highest 1-day precipitation (mm)—Maximum RR sum for 1 day interval—(June–September)—Difference 2025–2035 versus 1981–2010 (CHIRPS)

3.4 Conclusions

While in the past drought was the main threat for Niger, in the last decade, yearly records for Tillaberi show the loss of many lives and production due to floods. Hazardous events frequency and intensity seem likely to increase in the future.

In this paper we characterize the distribution of extreme events evaluating current trends and future scenarios. Attention is placed on time and space distribution performing a daily resolution analysis because the pattern of intense rains can be detected only at high temporal resolution.

The findings of this study show that a serious menace is the extreme rainfall events exacerbation as confirmed by future scenarios from several bias corrected GCMs. We applied a simplified bias correction method mostly dedicated to intercept changes in rainfall distribution extremes using the overlapping period (2006–2014) of GCMs and CHIRPS rainfall estimate.

The most significant current trend intercepted in Tillaberi Region is the increase in intense rainfall phenomena. SPI analysis shows an absence of significant dry and wet periods in the recent past. The combination of these two pieces of information implies an increase of extreme events in a normal rainfall regime. Near-future projections for rainfall distribution confirm the intensification of rainfall extremes. GCMs outputs show a consistent increase in very intense rainfall phenomena and an increase in the highest 1-day rainfall amount. At the same time for drought characterization, GCMs outputs show an increase in dry spells intensity. Both maximum number of consecutive dry days and the number of periods of more than 5 consecutive dry days are predicted to increase. The combination of this information gives a not favorable picture of what we could expect for near-future climate conditions in Tillaberi and it must alert decision makers to urgently take some action.

The findings of the study demonstrate the importance of tailored climate analysis when we move toward local scale (from sub-national to municipalities). Criticalities could be intercepted only by the study of most intense phenomena at a daily resolution because these changes in extremes distribution are not detectable by typical seasonal climatic trends analysis, distributed within IPCC reports or other studies at national level

Mapping climate changes highlights criticalities at very local scale and it could contribute to selecting priorities for the Tillaberi climatic risk adaptation process. In addition, maps support the risk assessment process giving the possibility of superposing the quantification of natural hazard likelihood with other vulnerability and exposure indicators.

Considering Tillaberi's current population growth and environmental degradation combined with the higher probability of negative climatic extreme conditions must alert decision makers to apply opportune and urgent measures in order to reduce climate risk.

As a last statement, we suggest that whenever new GCMs and rainfall estimation datasets are distributed, to perform an up-to-date analysis giving decision makers the most accurate information on future climate conditions.

References

Descroix, L., G. Mahé, T. Lebel, G. Favreau, S. Galle, E. Gautier, J.C. Olivry, et al. 2009. Spatio-temporal variability of hydrological regimes around the boundaries between Sahelian and Sudanian areas of West Africa: A synthesis. *Journal of Hydrology* 375 (1–2): 90–102. doi:10.1016/j.jhydrol.2008.12.012.

Descroix, L., P. Genthon, O. Amogu, J.L. Rajot, D. Sighomnou, and M. Vauclin. 2012. Change in Sahelian rivers hydrograph: The case of recent red floods of the Niger river in the Niamey region. *Global and Planetary Change* 98–99: 18–30. doi:10.1016/j.gloplacha.2012.07.009.

INS-Institut National de la Statistique. 2011. Tillaberi en chiffres 2011. http://www.stat-niger.org/statistique/file/Regions/TillaberienChiffres2011.PDF. Accessed 13 June 2016.

INS-Institut National de la Statistique de la Republique du Niger. 2012. Résultats définitifs du quatrième (4ème) Recensement général de la population et de l'habitat (RGP/H) 2012, par région et par department. Niamey, Niger.

IPCC. 2012. Managing the risks of extreme events and disasters to advance climate change adaptation. A special report of working groups I and II of the Intergovernmental Panel on Climate Change, ed Field, C.B., V. Barros, T.F. Stocker, D. Qin, D.J. Dokken, K.L. Ebi, M.D. Mastrandrea, K.J. Mach, G.-K. Plattner, S.K. Allen, M. Tignor, and P.M. Midgley. Cambridge, UK, and New York: Cambridge University Press.

Kirtman, B., S.B. Power, J.A. Adedoyin, G.J. Boer, R. Bojariu, I. Camilloni, F.J. Doblas-Reyes, A.M. Fiore, M. Kimoto, G.A. Meehl, M. Prather, A. Sarr, C. Schär, R. Sutton, G.J. van Oldenborgh, G. Vecchi, and H.J. Wang. 2013. Near-term climate change: projections and predictability. In Climate change 2013: The physical science basis. Contribution of working group I to the Fifth assessment report of the Intergovernmental Panel on Climate Change, ed. T.F. Stocker, D. Qin, G.-K. Plattner, M. Tignor, S.K. Allen, J. Boschung, A. Nauels, Y. Xia, V. Bex and P.M. Midgley. Cambridge, United Kingdom and New York: Cambridge University Press.

Leblanc, M.J., G. Favreau, S. Massuel, S.O. Tweed, M. Loireau, and B. Cappelaere. 2008. Land clearance and hydrological change in the Sahel: SW Niger. *Global and Planetary Change* 61 (3–4): 135–150. doi:10.1016/j.gloplacha.2007.08.011.

Mahe, G., G. Lienou, L. Descroix, F. Bamba, J.E. Paturel, A. Laraque, M. Meddi, H. Habaieb, O. Adeaga, C. Dieulin, F. Chahnez Kotti, and K. Khomsi. 2013. The rivers of Africa: Witness of climate change and human impact on the environment. *Hydrological Processes* 27: 2105–2114. doi:10.1002/hyp.9813.

McKee, T.B., N.J. Doesken, and J. Kleist. 1993. The relationship of drought frequency and duration to time scale. In *Proceedings of the Eighth Conference on Applied Climatology, Anaheim, California*, 17–22 January 1993. Boston: American Meteorological Society, 179–184.

OCHA-United Nations Office for the Coordination of Humanitarian Affairs. 2013. Niger and natural disaster—Flooding, Niger: aperçu humanitaire provisoire sur les inondations (au 25 Septembre 2013) - Cellule de Coordination Humanitaire/Cabinet du Premier Ministre avec l'appui technique de OCHA. http://reliefweb.int/report/niger/niger-aper%C3%A7u-humanitaire-provisoire-sur-les-inondations-au-25-septembre-2013.

WMO-World Meteorological Organization, M. Svoboda, M. Hayes, and D. Wood. 2012. *Standardized precipitation index—User guide*. Geneva: World Meteorological Organization. ISBN 978-92-63-11091-6.

Chapter 4
Characterization of Climate Risks for Rice Crop in Casamance, Senegal

Maurizio Bacci

Abstract This chapter contributes to a global reflection on climate change and its implications for agricultural production. We present a case study aiming to quantify trends of climate risks for rice crop in Casamance (Senegal). We evaluate the recurrence of drought and extreme rainfall conditions in the most sensitive phases of plant life and identify trends in the rainy season distribution. To overcome the low quality of climate records from gauge stations in the Region we use the rainfall estimation Climate Hazards group with InfraRed Precipitation Stations (CHIRPS), a daily gridded dataset with 0.05′ resolution over the period 1981–2013. The analysis is centered on the critical aspects that determine rice final yield such as: availability of water in the (i) plant germination and (ii) flowering phases, and (iii) the dynamics of the rainy season. We use the return period method to identify extreme events probability in rice crop's sensitive phases. Lastly, we identify the dynamics of the three parameters of the growing season: start, end and length by highlighting significant changes recorded in the study period (1981–2010). These outputs aim to support strategic agronomic choices in the Region.

Keywords Climate change · Rural development · Rice crop · Casamance

4.1 Introduction

Casamance Region is an enclave territory in southern Senegal (Fig. 4.1). The historical sociopolitical instability of the Region contributed to a regression of its rural development process (Evans 2003). Nowadays, the changes in the social and economic context of the country and the more secure conditions in Casamance lead the Region to a new rural development phase. According to the National Program

M. Bacci (✉)
National Research Council—Institute of BioMeteorology (IBIMET),
Via Giovanni Caproni 8, 50145 Firenze, Italy
e-mail: m.bacci@ibimet.cnr.it

M. Tiepolo et al. (eds.), *Renewing Local Planning to Face Climate Change in the Tropics*, Green Energy and Technology,
DOI 10.1007/978-3-319-59096-7_4

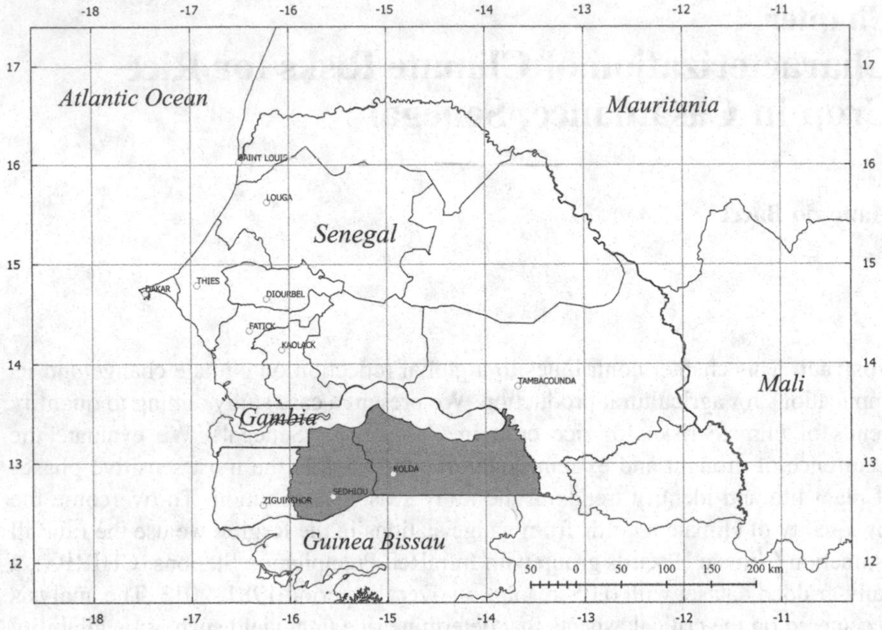

Fig. 4.1 Senegal. The Mid-Upper Casamance, the Sédhiou and Kolda regions (*grey*)

for Self-Sufficiency of Rice (PNAR) of Senegal, Ministerial Act No. 5042 dated June 8th, 2010, the goal of which is to help Senegal to achieve rice self-sufficiency and reduce poverty, the lowland rice cropping system of Casamance Region is designated as the one of the priority zones for an increase in rice productivity.

The climate of the Region is favorable to agriculture. Rains reach an average amount of 1000–1200 mm/year distributed in one rainy season from June till October. This allows the production of several rainfed crops; among these the most important are millet, sorghum, peanuts, maize and rice. Nowadays the agricultural production of the Region is far from reaching its potentials for different reasons.

These include the slow mechanization process in local agriculture that would be difficult to make sustainable and the limited market access for smallholders that discourages economic investments in the production process (Manzelli et al. 2015).

Yield is strictly dependent on the rainfall distribution and climate change impacts directly on farmers' activity. Knowledge of these changes support the adaptation strategies to face climate risk.

Risk is a clear concept for farmers and decision makers, they have to face a risk in every choice they make. But, how much climate risk? The answer depends on the local climate conditions that Climate General Circulation Models or information produced at national level cannot describe. To bridge this gap the study aims to produce a quantification of climate risks for rice crops in the Region.

The results are mainly addressed to the entire agricultural community, from decision makers to smallholders, who must receive clear and pertinent information for their agronomic choices.

In this paper attention is placed on the analysis of the agrometeorological conditions that most influence the rice yield in the Region. In Mid-Upper Casamance there are two different types of rice crop: (i) upland rainfed rice (*plateau*) and (ii) lowland rainfed rice (*vallée*).

The upland rice is a rainfed crop grown on dry soil without any water management. This activity is carried out by men and mechanization is sometimes used to aid their labor.

The lowland rainfed rice is grown in the natural depression near Casamance River. This activity uses traditional practices by which farmers, mostly women, seek to support the family's food balance. So this essentially remains a subsistence crop where women try reduce economic risks rather than intensify productivity. The effects of climate on this farming system are linked to the poor management of water in the field. Heavy rains and dry spells in some specific periods of the plant growing cycle could cause stress to plants and reduce yield potentials.

The other important factor, determining a good or bad harvest, is the dynamics of the West African Monsoon. In other words how an early or delayed start and end of the growing season impacts on yield. Each crop and variety is characterized by an optimal cycle length, so if there is no water available to complete the crop cycle there is a risk of the loss of the entire harvest (Roudier et al. 2011; Sultan et al. 2005). Indeed seasonality influences farmers' decisions about when to sow and harvest, and ultimately the success or failure of their crops (Graef et al. 2001).

The labor organization in the field is dependent on rainfall distributions, in the case of a late start to the season farming could be in competition with other activities and farmers risk not having enough time for field preparation and sowing, leaving some plots uncultivated (Manzelli et al. 2015b).

A longer growing season allows farmers to diversify crops or, as in this case, rice varieties. Instead, a shorter growing season and delayed start to the season reduce available options. Traditional agriculture is characterized by local varieties that women select in the years, mostly ecotypes naturally adapted to the local environment, but in the case of rapid climate change they aren't able to independently adjust seeds selection to new conditions.

Based on the observed climate, we try to identify local climate trends aiming to support the evaluation process of risk in agricultural production and the climate change adaptation process in the region. A set of synthetic indicators useful for the local community has been defined based on local knowledge, trying to highlight the factors most influencing final yield. A collaborative approach with the local agricultural service allowed us to select the following risk indicators: the presence of extreme events, both rainfall and drought, in (i) the germination and (ii) flowering period, and (iii) the dynamics of the growing season.

The probability of negative events is immediately comprehensible by decision makers and the results of the study can support strategic choices in the short- and mid-term.

Raw climatic data from the ground measurement network are poor and unevenly distributed, so we chose for the analysis a daily rainfall estimation images dataset: the Climate Hazards Group InfraRed Precipitation with Station data (CHIRPS). The selected dataset has a spatial resolution of 0.05°, sufficient to describe dynamics of rainfall in the Region with more of 30 years of data available to define a climatic index. This approach has the advantage of describing phenomena over the entire domain and producing hazard maps capable of intercepting criticalities at a very local level (i.e. municipality level).

4.2 Materials and Methods

The rainfall estimation dataset used in this work is the Climate Hazards Group InfraRed Precipitation with Station data (CHIRPS) v1p8 by the U.S. Geological Survey and Earth Resources Observation and Science (Funk et al. 2015). CHIRPS is a quasi-global daily dataset spanning 50°S–50°N, starting in 1981 to near-present. CHIRPS incorporates 0.05° resolution satellite imagery with in situ station data to create gridded rainfall time series for trend analysis and seasonal drought monitoring.

CHIRPS rainfall estimation daily images are distributed in NetCDF (Network Common Data Form) format, one of the most common climatic gridded data formats. We used the Climate Data Operators (CDO) to deal with this dataset. CDO is a large open source tool set for working on climate and model data. Using this tool we selected the domain over Casamance Region from the original dataset and computed the different indexes starting from the CHIRPS daily data.

4.2.1 Climate Extremes

Climate extremes may be time averages or frequencies of events above a given threshold of a single meteorological variable. In this case study we focused on detecting climatic stress in rice yield through the likelihood of recurrence of extreme events in the most vulnerable crop stages. The proposed approach identified criticalities supporting agronomic strategic choices that involve short- to medium-term economic investment. In Casamance, according to rice plant physiology and local knowledge, the most critical periods for rice yield are during emergence and flowering. We therefore chose the months of June-July and October as the most critical stages. In both periods we evaluated a set of indexes describing drought and extreme rainfall phenomena, these are:

- the maximum consecutive dry days, i.e. the greatest number of consecutive days per time period with daily precipitation below 1 mm; and the number of dry periods of more than 5 days to assess drought periods;
- the number of days when rainfall rate is at least 10 and 20 mm plus the highest one day precipitation amount per time period to assess heavy rainfall.

Using the yearly index in the time series it is possible to produce statistics and trends. The approach is useful in describing climate context as it gives a quantification of the probability that an extreme event will occur.

We evaluated the probability at a given threshold applying the return period analysis approach for each selected climatic index.

The return period is defined as the estimated time interval between events of a similar size or intensity, normally it is used to estimate the probability that a hazardous event such as a flood, drought or extreme rains will occur. The method proposed in this study was formulated by Gumbel (1954). The Gumbel equation is given by:

$$F(x) = \exp\left[-\exp\left(-\frac{x-u}{\alpha}\right)\right]$$

where F(x) is the probability of an annual maximum and u and α are parameters. Denoting the mean of the distribution by $\bar{x}$ and the standard deviation by s then the parameters u and α are given by the following:

$$\alpha = \frac{\sqrt{6}s}{\pi}$$

$$u = \bar{x} - 0.5772\alpha$$

In order to estimate the T-year extreme events it is possible to reverse the procedure to estimate the annual maximum for a given return period. So giving:

$$y_T = -\ln\left[\ln\left(\frac{T}{T-1}\right)\right]$$

It is possible to apply to the substitution to the original formula calculating u and α using the sample mean and standard deviation, as estimates of the population values, then the estimates of x_T may be obtained by:

$$\begin{aligned} x_T &= u + \alpha y_T \\ &= \bar{x} - 0.5772\frac{\sqrt{6}}{\pi}s + \frac{\sqrt{6}}{\pi}s\left\{-\ln\left[\ln\left(\frac{T}{T-1}\right)\right]\right\} \\ &= \bar{x} - \frac{\sqrt{6}}{\pi}\left\{0.5772 + \ln\left[\ln\left(\frac{T}{T-1}\right)\right]\right\}s \end{aligned}$$

So we can simplify in the following formula:

$$x_T = \bar{x} + K_T s$$

where:

$$K_T = -\frac{\sqrt{6}}{\pi}\left\{0.5772 + \ln\left[\ln\left(\frac{T}{T-1}\right)\right]\right\}$$

K_T is the frequency factor, which is a function only of the return period T. Thus if an estimate of the annual maximum consecutive dry period for a return period of 20 years is required, if T = 20 years in the formula the result is an estimate of the likelihood of an event with 20 years return period. In synthesis, it is a statistical measurement based on historic data denoting the average recurrence interval over an extended period of time, and it is usually used for risk analysis, especially flood risk analysis. The associated yearly probability is the inverse of the Return Period (RP). For instance the probability (P) of an event with 10 years return period is P = 1/RP = 10%, this means that each year there is a 10% chance of recording a value greater than or equal to the indicated value of extreme event.

Whilst the method generally relies on using gauged rainfall records, in this case we applied the method on the CHIRPS gridded rainfall estimation dataset. For each single gridded point in the domain we quantify the probability of an extreme event removing incertitude by surface interpolation methods. It is clear that rainfall estimation by models implies bias errors, but the representation of dynamics of the phenomenon over space is not warped by systematic errors of the model.

4.2.2 Growing Season

Being part of the sub-humid regions of West Africa, the rainfall distribution in the Casamance area is typically unimodal from May-June till October–November.

The definition of start of the growing season is debated in the literature. Regional and local onset determine a different approach and analysis (Fitzpatrick et al. 2015). Despite an accurate review of the existing method, in this case we focused on determining dynamics and trends over a specific territory for a specific crop. We chose the method proposed by Sivakumar (1988) as analyzed in Ati et al. (2002) where the onset is defined as the first date after May 1st when three consecutive days accumulated rainfall exceeds 20 mm with no seven days dry spell in the subsequent 30 days.

The end of the growing season is the date after September 1st when the soil water content (set to 100 mm) is nil with a daily ETP of 5 mm (Maikano 2006; Traoré et al. 2000). The season length is derived from these two dates.

The limit of this method is its reliance on the threshold approach chosen by the two agronomic definitions of start and end of the season. Rain estimates incur sampling and retrieval errors that affect the accuracy of the satellite-inferred rain information provided to users. In this case the method, based on a rainfall threshold, could introduce an error in defining the date due to the uncertainty linked to the rainfall amount recorded on that day. Studies in the literature on similar gridded datasets (Ficher 2007) demonstrate that error is randomly distributed in rainfall estimation images so it is possible to assume that statistics in the trend analysis are not significantly biased by this kind of error.

The dates of the start and end of the rainy season were determined for each year applying the methods to the CHIRPS daily rainfall estimation dataset. We used the resulting time series from 1981 to 2013 as input in growing season parameters trend analysis.

The linear regression coefficient evaluated from the three values for each grid point in the domain allowed us to plot current dynamics in the modulation of the growing season in the Casamance Region.

4.3 Results

Germination and flowering periods are the most vulnerable to rainfall anomalies. In the plants' installation stage, the roots are weakly developed and they are not able to absorb moisture in the deepest part of the ground to avoid water stress. Flowering time is a critical stage of development in the life cycle of most plants because seed number is determined. Abiotic stresses of temperature and water deficit, and biotic (pest and disease) constraints could reduce the number and quality of seeds. Hence drought or very wet conditions in these stages mainly influence the rice crop final yield.

4.3.1 Rice Crop Germination

Historical data for Casamance region show high inter-annual variability with a heterogeneous distribution of dry and wet periods (Fig. 4.2). The graph has been produced averaging values from CHIRPS gridded dataset over the Mid-Upper Casamance using a mask.

Until the early 1990s a very prolonged dry period was recorded with extreme events beyond the normal statistical distribution, particularly 1981 and 1986 are the years characterized by the greatest length of consecutive dry days and the number of episodes of more than 5 consecutive dry days respectively. The second part of the 1990s until 2010 was characterized by wet conditions with a substantial absence of prolonged drought periods in the early stages of the rainy season. Decision makers are normally very interested in the quantification of the probability of negative events. To meet this need we determined drought risk through the evaluation of several return periods (2, 5, 10 and 20 years) for the two indexes in this study: (i) the maximum consecutive dry days and (ii) the number of dry periods of

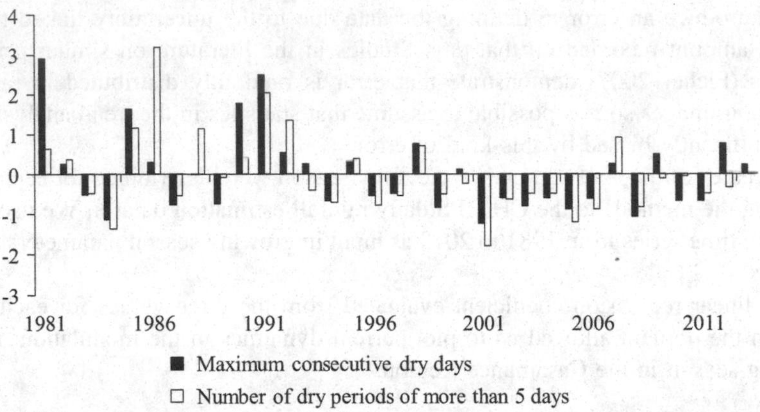

Fig. 4.2 Standardized distribution of maximum consecutive dry days and the number of dry periods of more than 5 days over Mid-Upper Casamance in June and July (CHIRPS data)

more than 5 days recorded in June–July over 1981–2013 period. The values are calculated and presented as a set of comparison maps with the plots of the dry periods' probability distribution over the area (Fig. 4.3).

As shown in the maps, there is a southeast northwest gradient, with a significant likelihood of a long series of consecutive dry days near the Gambia. In the eastern part of the Kolda region it seems that very long episodes of drought conditions never normally occur.

Regarding the distribution of the average number of periods with at least 5 consecutive dry days it can be observed how the distribution is irregular and follows a south north gradient.

The maximum number of consecutive dry days per year in the Region is expected to be from 6 to 10 but there is a 5% probability of recording 10–18 days, which is almost double.

July is when seedlings emerge. This is a critical stage for the plant development. Heavy rains can cause prolonged submersion of small plants and soil erosion phenomena. A graph is presented in Fig. 4.4 with a yearly distribution of rains in July over Mid-Upper Casamance.

July seems to have a downward trend in extreme events of rain. 1984 and 2004 were the years with the most intense rain phenomena and maximum variability. Starting from 2006 the maximum value distribution recorded has been quite regular reducing risks for crop submersion.

A map of the return period of the maximum daily amount of rain in the month of July has also been produced (Fig. 4.5). The map shows that values of over 80 mm/day can be reached in 1 year every 20 in the central part of the Mid-Upper Casamance. This high amount of daily rainfall might also create problems in well-drained valleys. The distributions of rainfall extremes do not follow the normal north south gradient. It is possible to observe a cone of maximum values between the eastern part of the Sédhiou Region and western part of the Kolda Region.

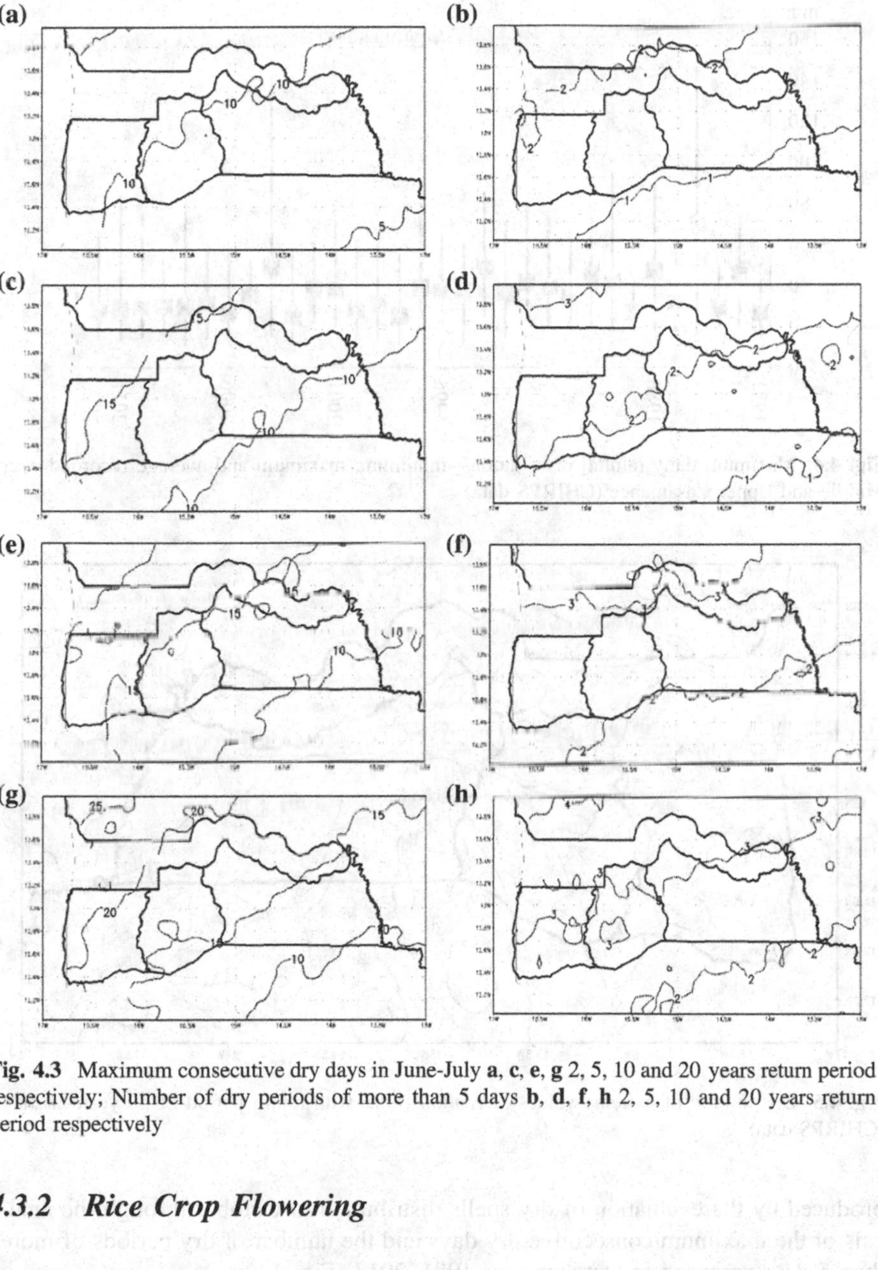

Fig. 4.3 Maximum consecutive dry days in June-July **a**, **c**, **e**, **g** 2, 5, 10 and 20 years return period respectively; Number of dry periods of more than 5 days **b**, **d**, **f**, **h** 2, 5, 10 and 20 years return period respectively

4.3.2 Rice Crop Flowering

Regarding flowering, the other vulnerable stage of rice crop, in Casamance this normally occurs in October. At this stage it is very important that plants receive a sufficient amount of rain. In fact drought stress in this period could generate 80% of yield loss. The characterization of drought risk in the flowering period was

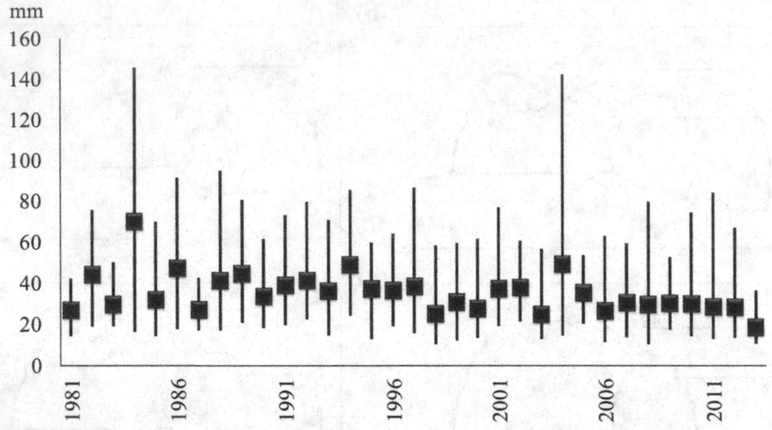

Fig. 4.4 Maximum daily rainfall distribution—minimum, maximum and average recorded over Middle and Upper Casamance (CHIRPS data)

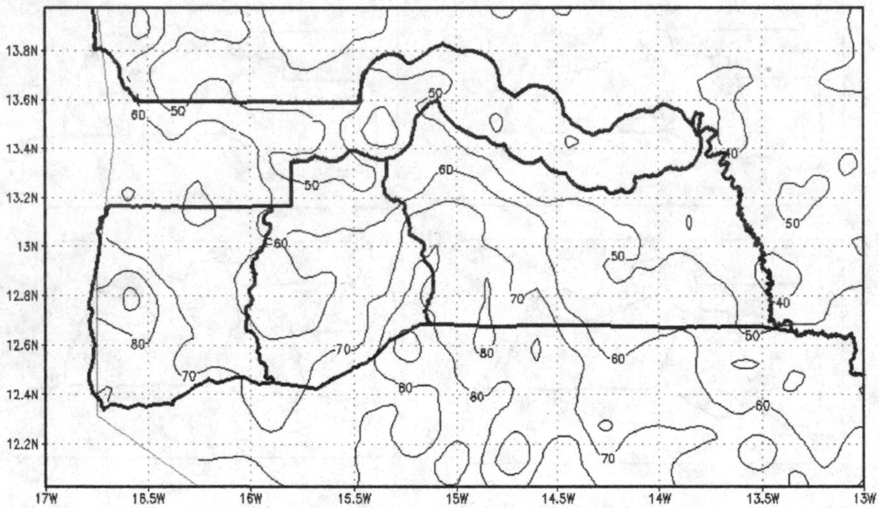

Fig. 4.5 Distribution of extreme values of rainfall (mm) with return period of 20 years in July (CHIRPS data)

produced by the evaluation of dry spells distribution in October through the analysis of the maximum consecutive dry days and the number of dry periods of more than 5 days recorded in October over 1981–2013 (Fig. 4.6).

The probability of having an average of 1.2–1.4 periods of 5 consecutive dry days per year is almost normal, while over 20 years 2.4–2.8 dry periods per year can be expected with an uneven distribution over the Middle and Upper Casamance Region.

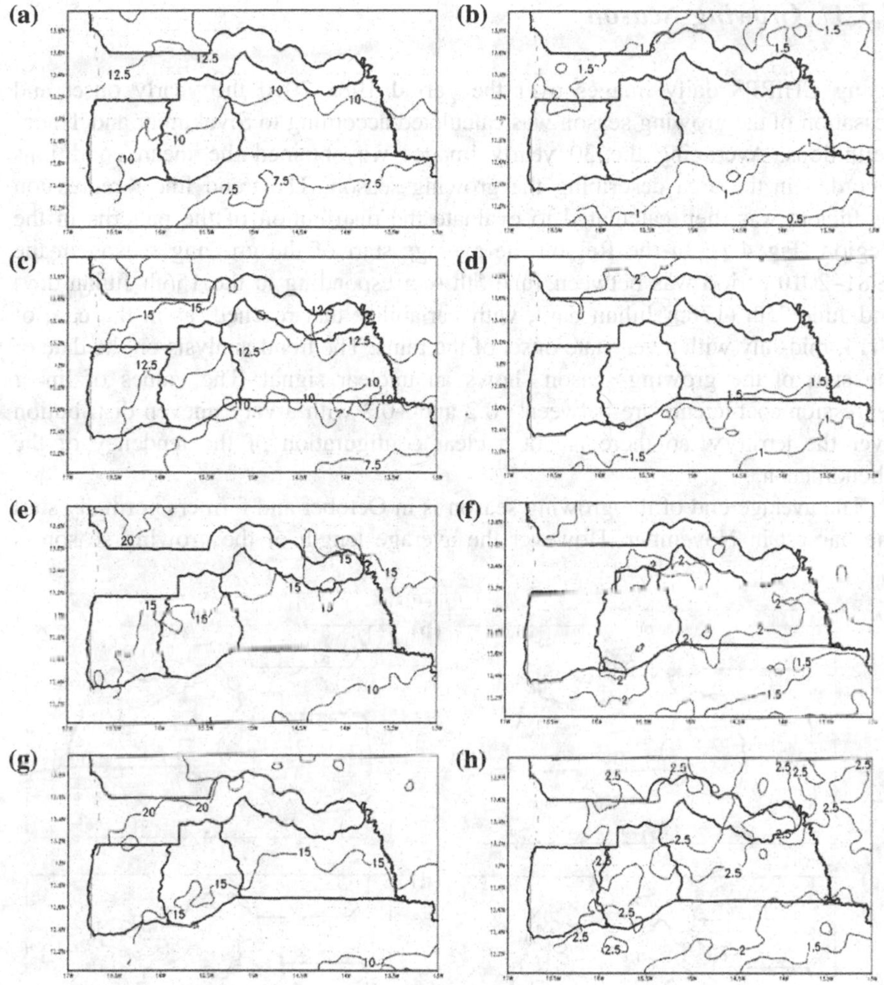

Fig. 4.6 Maximum consecutive dry days in October **a**, **c**, **e**, **g** 2, 5, 10 and 20 years return period respectively; Number of dry periods of more than 5 days **b**, **d**, **f**, **h** 2, 5, 10 and 20 years return period respectively

Episodes of consecutive dry days gives a chance of 9–11 consecutive days as a normal condition, while with a 20-year return period it is possible to arrive at 14–20 days. In both cases the distribution will follow isohyets with a north south gradient from wetter conditions to the driest.

4.3.3 Growing Season

Using CHIRPS daily images over the period 1981–2010 the yearly onset and cessation of the growing season was calculated according to Sivakumar and Traoré definitions. Averaging the 30 yearly images we obtained the mean conditions recorded in the area describing the growing season. The trend linear regression coefficient was then calculated to evaluate the distribution of the patterns in the Region (Fig. 4.7). In the Region the average start of the growing season in the 1981–2010 period was between June 5th (corresponding to the 156th Julian day) and June 25th (176th Julian day), with variability that reached, as in the case of 2014, mid-July with a very late onset of the rains. The trend analysis on the date of the start of the growing season shows an unclear signal. The values of linear regression coefficients are between −0.2 and +0.2 with a very uneven distribution over the territory, so there is not a clear configuration of the tendency of the phenomenon.

The average end of the growing season is in October and farmers normally start the harvest in November. However the average length of the growing season is

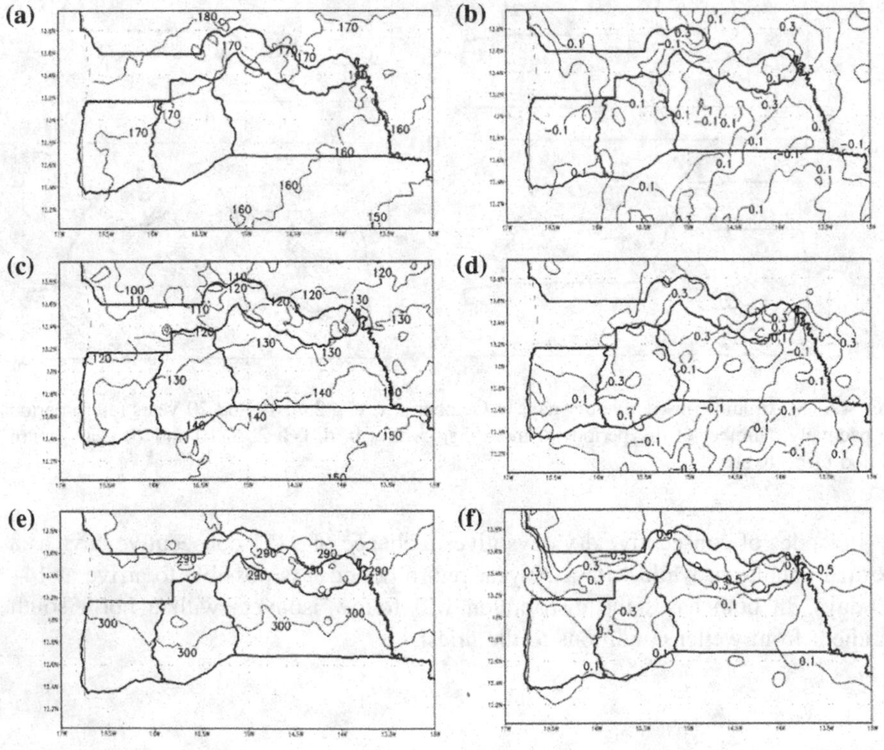

Fig. 4.7 Average start (Julian day), length (days) and end (Julian day) of season, **a**, **c** and **e** respectively with the relative linear regression coefficient, **b**, **d** and **f** respectively, 1981–2010

120–140 days. The result is a consistent pattern to a shift towards November for the end of the cropping season. In general for Casamance there is a slight increasing tendency in the length of the growing season. Especially in Sédhiou Region maximum values are observed arriving at a ratio of increase in length of the growing season by 0.3 days a year.

4.4 Discussion

Rain distribution is one of the main factors that cause problems for rice crops in Mid-Upper Casamance. Poor valleys water management, soil degradation and erosion phenomena, the lack of proper management of water in the plot and inappropriate agricultural practices are all factors creating a risk for crop yield.

To improve land and drainage management in the Region, the measure of climate change risks is necessary to assist this development process. Indeed, where the agricultural system is very vulnerable to natural hazards, adaptation measures are urgently required, such as the reorganization of hydraulic management in the valleys, with an evaluation of alternative cropping systems and/or introduction of new rice varieties and the risk to achieve all of this must be quantified.

The probability of dry spells or extreme events in the most critical stages of rice crop supports alternative strategic choices and it is possible to evaluate the best options in terms of crops and varieties combination for the Casamance Region. In fact, each crop, and consequentially each variety, has a particular sensitivity to drought spells. Using the probability maps of extreme events produced in this study it is possible to select the most appropriate combination of crops and varieties considering future climatic conditions.

We want to stress the fact that using a high resolution daily precipitation gridded dataset it is possible to describe phenomena at a very local scale, giving tailored climatic information also at municipality level.

The study characterized some crucial agro-climatic hazards impacting on the Region rice crop production; this paper presents the most important findings. First of all we found that in the last decade the good rainfall distribution in the initial phase of the growing season determined a low probability of unfavorable conditions for crops. A downward trend in extreme rains in July also reassures us that in the coming years we can expect better average conditions for the emergence of rice crop. This doesn't mean that hazardous events are not expected in future. In fact using the return period of extreme events in the June–July period a map was produced showing that amounts of 80 mm/day and more are expected 1 year every 20, so a yearly probability of 5%. This information is also useful for drainage design and management in the Region.

The dynamics of the growing season show that in the Region there is generally a slight trend in the end of season shifting from early to end November. This aspect is very interesting and reassures farmers and decision makers of the opportunity to complete the crop cycle in the case of late starts to the season. But this shift also has negative effects on farming activities; indeed a continuation of favorable conditions for crop development, with delayed maturation and harvesting, can lead to problems of losses due to straying animals and generate conflicts with livestock farmers.

Instead, the intercepted signal on the start of the growing season is weaker and shows an unclear trend in the Region. This could be linked to the current variability in the onset of the West African Monsoon.

Considering natural climate variability, caution will be maintained for agronomic planning in the Region, but a set of quantitative values is now available to support the decision-making process. This information must be constantly updated in order to give end users the very latest available and accurate information of the climate evolution.

4.5 Conclusions

The agro-meteorological characterization provided in this paper aims to support the decision-making process in agriculture through tailored climatic information for the Middle and Upper Casamance Region. The integration of local knowledge with the most recent gridded model outputs intercepts criticalities for rice crop yield.

A set of probability thresholds for the most important climatic threats is given allowing end users to have a clear picture of possible risk scenarios. These values help decision makers in their choices, indeed it is now possible to define the acceptable risk threshold; in other words they can set a level of potential losses that a farmer or politics considers acceptable, given probable climatic conditions. The extreme events' probability could also support new varietal and agronomic choices, guiding farmers in optimizing the production process. Furthermore, these values could support agronomists in the redesign and sizing of the hydraulic system management in the rice valleys of the Region.

The climate variability recorded in the Region suggests caution on the possible recurrence of flood and drought episodes, requiring an improvement in water management systems to reduce the risk of yield losses.

The rainfall estimation gridded dataset describes phenomena through their natural modulation over the territory; it therefore supports local communities in their adaptation by showing a climate change distribution not measurable by the ground meteorological network. This approach allows us to release information at very local level with detailed maps supporting the usability of the analysis results.

Feedback from end users, such as local associations and decentralized agricultural technical services, could guide future fine tuning of the climatic index analysis. Indeed, dialogue between the scientific community and local users is the key element in the innovation process for the improvement of agronomic techniques and strategies to face climate change. Moreover we recommend a synergy among different levels of end users to achieve a coordinated rural development process that considers the most accurate information available on climate.

This approach could be reproduced with General Circulation Models outputs in order to evaluate some possible future scenario giving a picture of what climatic conditions will be like for rice crops in Casamance.

References

Ati, O.F., C.J. Stigter, and E.O. Oladipo. 2002. A comparison of methods to determine the onset of the growing season in northern Nigeria. *International Journal of Climatology* 22: 731–742. doi:10.1002/joc.712.

Evans, M. 2003. Ni paix ni guerre: The political economy of low-level conflict in the Casamance. Background Research for Humanitarian Policy Group Report, (13). London: Overseas Development Institute.

Fitzpatrick, R.G.J., C.L. Bain, P. Knippertz, J.H. Marsham, and D.J. Parker. 2015. The West African monsoon onset: A concise comparison of definitions. *Journal of Climate*. doi:10.1175/JCLI-D-15-0265.1.

Ficher, B.L. 2007. Statistical error decomposition of regional-scale climatological precipitation estimates from the tropical rainfall measuring mission (TRMM). *Journal of Applied Meteorology and Climatology* 46: 791–813. doi:10.1175/JAM2497.1.

Funk, C., P. Peterson, M. Landsfeld, D. Pedreros, J. Verdin, S. Shukla, G. Husak, J. Rowland, L. Harrison, A. Hoell, and J. Michaelsen. 2015. The climate hazards infrared precipitation with stations—a new environmental record for monitoring extremes. *Scientific Data* 2: 150066. doi:10.1038/sdata.2015.66.

Graef, F., and J. Haigis. 2001. Spatial and temporal rainfall variability in the Sahel and its effects on farmers' management strategies. *Journal of Arid Environments* 48 (2): 221–231.

Gumbel, E.J. 1954. Statistical theory of extreme values and some practical applications. Applied mathematics series 33. U.S. Department of Commerce, National Bureau of Standards.

Maikano, I. 2006. Generate prototype WCA recommendation maps for selected sorghum and millet cultivars based on updated end-of-season dates (PRODEPAM, activity). Rapport de stage. Institut International de Recherche sur les Cultures des Zones tropicales semi-arides (ICRISAT/Bamako).

Manzelli, M., E. Fiorillo, M. Bacci, and V. Tarchiani. 2015a. Lowland rice production in southern Senegal (Middle Casamance): Challenges and prospects for sustaining their restoration and development. *Cahiers Agricultures* 24 (5): 301–312. doi:10.1684/agr.2015.0772.

Manzelli, M., I. Seppoloni, E. Zucchini, M. Bacci, E. Fiorillo, and V. Tarchiani. 2015b. La riziculture traditionnelle de bas-fond en Moyenne Casamance dans un contexte de changements globaux: enjeux et perspectives. In Eaux et sociétés face au changement climatique dans le bassin de la Casamance. Paris: L'Harmattan. ISBN 978-2-343-07690-4.

Roudier, P., B. Sultan, P. Quirion, and A. Berg. 2011. The impact of future climate change on West African crop yields: What does the recent literature say? Symposium on Social theory and the environment in the new world (dis)Order. *Global Environmental Change* 21 (3): 1073–1083. doi:10.1016/j.gloenvcha.2011.04.007.

Sivakumar, M.V.K. 1988. Predicting rainy season potential from the onset of rains in southern Sahelian and Sudanian climatic zones of West Africa. *Agricultural and Forest Meteorology* 42: 295–305.

Sultan, B., C. Baron, M. Dingkuhn, B. Sarr, and S. Janicot. 2005. Agricultural impacts of large-scale variability of the West African monsoon. *Agricultural and Forest Meteorology* 128 (1): 93–110. doi:10.1016/j.agrformet.2004.08.005.

Traoré, S.B., F.-N. Reyniers, M. Vaksmann, B. Koné, A. Sidibé, A. Yorote, K. Yattara, and M. Kouressy. 2000. Adaptation à la sécheresse des ecotypes locaux de sorgho du Mali. *Sécheresse* 11: 227–237.

Chapter 5
A Methodology for the Vulnerability Analysis of the Climate Change in the Oromia Region, Ethiopia

Elena Belcore, Angela Calvo, Carolin Canessa and Alessandro Pezzoli

Abstract Goal of the vulnerability research of the last years is to evaluate which community, region, or nation is more vulnerable in terms of its sensitive to damaging effects of extreme meteorological events like floods or droughts. Ethiopia is a country where it is possible to find the described conditions. Aim of this work was to develop an integrated system of early warning and response, whereas neither landmark data nor vulnerability drought analysis existed in the country. Specifically, a vulnerability index and a capacity to react index of the population of three Woredas in the Oromia Region of Ethiopia were determined and analysed. Input data concerned rainfall, water availability, physical land characteristics, agricultural and livestock dimensions, as well as population and socio-economic indices. Data were collected during a specific NGO project and thanks to a field research funded by the University of Torino. Results were analysed and specific maps were drawn. The mapping of the vulnerability indices revealed that the more isolated Woreda with less communication roads and with less water sources presented the worst data almost on all its territory. Despite not bad vulnerability indices in the other two Woredas, however, population here still encountered difficulty to adapt to sudden climatic changes, as revealed by the other index of capacity to reaction. Beyond the interpretation of each parameter, a more complete reading key was possible using the SPI (Standardized Precipitation Index) beside these

E. Belcore (✉) · A. Calvo · C. Canessa
Department of Agricultural, Forest and Food Sciences, University of Torino, Largo Paolo Braccini 4, Grugliasco (TO) 10095, Italy
e-mail: elena.belcore@gmail.com

A. Calvo
e-mail: angela.calvo@unito.it

C. Canessa
e-mail: carolincanessa14@googlemail.com

A. Pezzoli
DIST-Politecnico and University of Turin, Viale Mattioli 39, 10125 Turin, Italy
e-mail: alessandro.pezzoli@polito.it

M. Tiepolo et al. (eds.), *Renewing Local Planning to Face Climate Change in the Tropics*, Green Energy and Technology,
DOI 10.1007/978-3-319-59096-7_5

indicators. In a normalized scale between 0 and 1, in this study the calculated annual SPI index was 0.83: the area is therefore considerably exposed to the drought risk, caused by an high intensity and frequency of rainfall lack.

Keywords Vulnerability analysis · Climate change · Ethiopia

5.1 Introduction

In 2007 the document of the Intergovernmental Panel on Climate Change (IPCC 2007) confirmed with absolute certainty the man-made phenomenon of climate change manifested by the increasing of the average global temperatures. One of the main consequences is the variation in rainfall patterns and the increasing probability of extreme meteorological events like floods or droughts. Various areas of the world are therefore at natural disaster risk. Nevertheless, the latter is not just a function of climatic events, but also of the system characteristics. As underlined by UNISDR: "There is no such thing as a 'natural' disaster, only natural hazards" (UNISDR 2009).

The probability of calamitous events depends also on the vulnerability of the system, that is the collection of characteristics and conditions of human-environment systems that made it sensitive to damaging effects of extreme events. Ethiopia is a country where it is possible to find the conditions above described. In the Ethiopian territory, in fact, environment, demographic and economic conditions exposed in many occasion the population to droughts and humanitarian catastrophes due to the ENSO phenomenon and the heating of the Indian Ocean. Nowadays the necessity of limiting the increasing of global temperatures with politics of mitigation is clear, but it become fundamental the ex-ante evaluation of the systems vulnerability to extreme events. In order to adopt measures to develop the capacity of adaptation of population and to improve the resilience, it is important to analyse the system vulnerability.

Vulnerability analysis can have a lead role in adaptation policies designed to reduce climate change impacts and extreme events on ecosystem services that are the foundations of the human wellbeing (MEA 2009). Evaluating which individual, community, region, species or nation is more vulnerable is the goal of the vulnerability research of last decades. However vulnerability studies are still a developing field of research. Actually, the conceptual framework is still fragmented and based on different paradigms, adversarial theories, heterogenic empirical studies and a very diversified terminology (Vincent 2004). For example at now it doesn't exist an universal accepted definition of vulnerability, the meanings attributed to the term are often contested and the related concepts depends on finalities and on the studied system (Cutter 2003). It is therefore necessary to find a common conceptual framework that guarantee coherence during research activities, to enhance policy activities that can be applied to the stakeholders (Kasperson et al. 2001).

The issue seems to be so urgent that the General Assembly of United Nation in June 2015 established with a specific resolution[1] the creation of an intergovernmental working group to define indicators and a common terminology for the disaster risk reduction (AG 2015).

According to Füssel there are four elements of vulnerability that should be define ex-ante: system, attribute of concern, temporal reference and hazard (Füssel 2007). System is the potentially vulnerable system, such a population, an economic sector or a geographic area. The attributes of concern are all the components that are treated by a dangerous event, they include infrastructural elements and socio-economic elements, like food security or wellbeing. Temporal reference is the considered period of time for vulnerability evaluation. Finally, the hazard is a potentially damaging influence on the considered system. It is important to consider the element above as able to describe a vulnerability situation, even if it is not possible to give the same weight at each of them.

On the other side, Nelson (Nelson et al. 2009) underlines that the definitions of vulnerability should not be confused with the conceptual framework of vulnerability, and this last should not be confused with the empirical studies about vulnerability. In fact, definitions describe the components of vulnerability while the conceptual framework gives meaning to definitions themselves and allows analysis based on an analytic context in a transparent and repeatable way. The goal of the empirical studies, however, is to propose a methodology based on definitions and conceptual framework used in order to quantify and evaluate the vulnerability. It is important to stress the different approaches of the literature to vulnerability, to natural disasters and to vulnerability of livelihoods. The approach to vulnerability to natural disasters focus on the relation between environment disasters and human dimension and, in this case, the hazard is considered exogenous to the human being. On the other hand, the vulnerability of livelihoods approach considers vulnerability as human conditions caused by lack of entitlements (Alwang et al. 2001).

In the last years a new analysis approach was introduced, focused on vulnerability to climate change (IPCC 2001). This approach is in line with the literature about natural disasters, but it considers the hazard as the result of endogenous and exogenous characteristics of the system. Furthermore, it treats exposure to a dangerous event as part of vulnerability of the system itself, as we can see in the results of this research.

---

[1]United Nations General Assembly, A/RES/69/284, "Establishment of an open-ended intergovernmental expert working group on indicators and terminology relating to disaster risk reduction", June, 3rd, 2015.

5.1.1 Climate Change, Vulnerability and Resilience

It is well known that rural poverty increases when it increases elsewhere and the expected situation is more difficult for the remote rural areas of the South of the world (often already poor) where people lack resources to fall back on. When rural regions impoverished, more people leave rural areas, reinforcing existing migration trends in a downward cycle.

Aim of this work is the analysis of population vulnerability and resilience in the Region of Oromia (Ethiopia), using a data base of the NGO LVIA (created by a questionnaire on food security) investigated by three students of the University of Turin (Department of Agricultural, Forest and Food Sciences, Department of Culture, Politics and Society and Department of Territory of the University of Turin) during a three month stage in Ethiopia in the summer 2015. An innovative methodology is therefore explained, to evaluate the hazard and the effect of the Climate Change on the drought in Ethiopia using the vulnerability analysis, with a deeper study of the system and of the attribute of concern.

5.2 Area of Study

Ethiopia stretches over 1.12 million of km^2 covering a huge part of Horn of Africa. It is a federal country, composed by nine ethnic administrative regions (Fig. 5.1).

The population in 2015 was around 99 million people, with a forecast of 138 million in 2030 (United Nations 2015).

It is the second more populated country in Africa after Nigeria (CSA 2007). The agricultural sector accounts the 45% of Ethiopian GDP. The 85% of population works in agriculture, but only 1% of farmland has an irrigation system. Agriculture therefore depends on rainfalls. Pastoralism is an important source of livelihood and economy and accounts 15% of Ethiopian GDP. Ethiopia is composed by 9 Regions (or Zones): inside each Region there is a certain number of districts (called Woreda) and each Woreda is composed by municipalities (Kebele), the smallest administrative unit of Ethiopia. In the last century Ethiopia unfortunately registered many anomalies concerning precipitations distribution, that leads to frequent and intense periods of drought in many areas of the country. Several studies justify the precipitation variability of Horn of Africa with the phenomenon of Niño and Niña (ENSO—El Niño Southern Oscillation18) (Hulme et al. 2001, Korecha 2006, Segele and Lamb 2005, Seleshi and Demarèe 1995). The main concern is the seasonal cycle of the rain (Segele and Lamb 2005, Viste et al. 2013), because anomalous rainy cycles may lead to huge damages to the agricultural production of the country, increasing the vulnerability of rural people which lives of subsistence (Sadoff 2008). The area of this study stretch over the territory in three Woredas of

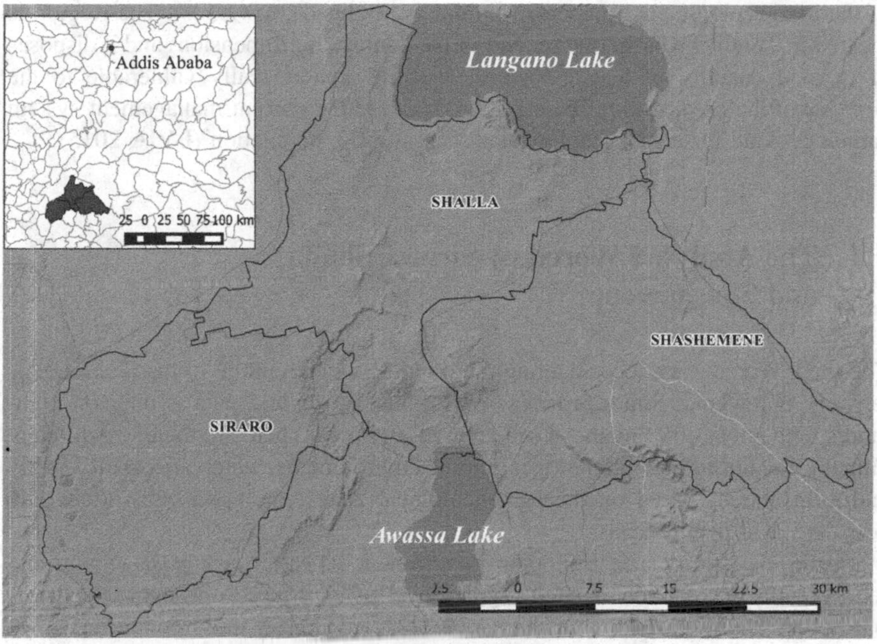

Fig. 5.1 Map of the Siraro, Shalla and Shashame Woredas

the Oromia Region: Siraro, Shalla and Shashamane. These districts are located in the West Arsi zone, along the Rift Valley (Fig. 5.1).

5.3 Geographic Description and Social Conditions of the Oromia Region

The Oromia Region is the biggest of the nine Ethiopian regions with about 33.7 million people (IWMI 2009): the most represented is the Oromo ethnicity (85% of population). Oromia is essentially rural: the 50% of Ethiopian coffee production comes from Oromia. Breeding is also important for the region: it is estimated that at the farmers of Oromia belong about the 45% of the Ethiopian cattle production. The characteristics of the Oromia conditions are similar to the other Ethiopian Regions: there is a strong dependence from the subsistence of the agricultural sector and there are few forms of economic diversification. The cultivated land of the Ethiopian farmers is often lower than 2 hectares and 52% of them has less than 1 hectare (AQUASTAT 2005): in Oromia the average cultivated land is 1.15 hectares per household. On the contrary the average number of cattle per household is here higher than in the other Ethiopian regions (CSA 2007): it must be however considered that this is only an averaged data which greatly varies, especially in function of the agro-ecologic environments (highlands, midlands, lowlands). The Region is located in a drainage basin (the second of the country for dimension) and it disposes of a good quantity of water. Nevertheless the area is really vulnerable to the changes of the precipitations distribution, because the current management system cannot provide for a sustainable water access to the population (LVIA 2015).

5.4 The Analysed Woredas: Siraro, Shalla and Shashamene

The three Woredas are located along the down Rift valley: 80% of the Shashamene territory is lowland, Siraro broadens on highlands, while Shalla is only on highlands, until about 2000 m above sea level (Table 5.1). There are therefore different climatic conditions and different geographic environments, which affect soil fertility and cultivations. Based on the agro-ecological zones, the main cultivations are: corn, teff, barley and beans.

During the last decade Ethiopian government (Disaster risk management and food security sector) documented a great level of exposure to drought and floods (DRMFSS 2016). In addiction there is a socio-economic stress condition which makes the population more sensitive to climatic disturbs.

With an estimated population of about 662,000 inhabitants (Table 5.2), the three Woredas have an high population density (about 310 inhabitants/km^2) compared to the national one (lower than 100) (DRMFSS 2016). Population (mainly of Islamic religion) is very young with a low occupational level outside the agricultural sector: these economic conditions decrease the possibility of economic diversification in order to reduce the population dependence from natural resources.

Table 5.1 Environmental and economic characteristics of the examined Woredas

Woreda	Area km^2	Environment	Economic activities	Crops
Shashamene	779	Lowland (80%) Highland (20%)	Agriculture Livestock Commerce	Teff, barley, millet, sorghum, wheat, corn, beans
Siraro	607	Lowland (68%) Highland (32%)	Agriculture Livestock	Teff, wheat, corn, beans, potatoes
Shalla	749	Highland (1700–2000 m)	Agriculture livestock	Corn

Table 5.2 Population (census 2007 and 2015 forecast)

Shashamene	Siraro	Shalla	Total	2015 Forecast
347,228	145,649	149,804	642,681	661,961

5.5 Method and Materials

Questionnaires and Interviews

Questionnaires and interviews were the instruments used for data collection. It was chiefly used a questionnaire created by LVIA NGO in its Food security project. Furthermore four interviews were submitted: one singular interview to the head of the village and three group of interviews to local water committees, to the women of the villages and to the village community. 180 questionnaires were filled in the three Woredas (15 Kebele per Woreda and 12 households per Kebele) in three months of work: the questionnaire was composed by 200 questions, with a total of 36,000 collected data. More than 40 interviews were therefore submitted to heads of villages and communities.

Food Security Questionnaire

The food security questionnaire was created by LVIA NGO in occasion of a project related to the food security and water's resource in the Woredas of Siraro, Shalla and Shashamane. The questionnaire was composed by 45 questions divided in 5 sections: general information, livelihood, food security, and observation checklist (see Annex). The general information section was made up of 25 questions concerning family composition and water access of the interviewed household: they were usually short and had multi-choice answers. Livelihood was the longest section, containing 30 questions to investigate household economics activities, agricultural systems, yearly income, irrigation types and livestock characteristics. Food security section was composed by 11 questions about the number and the composition of daily meals and diseases related to under-nutrition conditions. The last section, observation checklist, was the shortest and was dedicated to the interviewer, to obtain information on the family composition and characteristics. The questionnaire was created in order to be more understandable as possible and contained short questions easily translatable from English to the local language. It also included a series of control questions. Some questions (20, 21 and 41) were open and asked to the interviewee to discuss about different topics, but people normally did not agree to answer. LVIA cultural mediators helped people to fulfill the questionnaire: they read and translated in local language, explained questions and wrote down the answer of the interviewee. Before conducting and summiting questionnaires, each cultural mediator received the questionnaire and studied the guide line in order to have an homogeneous fulfill. In the questionnaire some key questions were individuated and therefore used to calculate vulnerability and resilience indexes. The identified sub-indicators which composed the vulnerability and resilience indices were 24 and concerned different topics: economy, water, environment, agriculture, livestock, demography, and others. The data elements of each sub-indicators were one or more.

5.5.1 Interviews

The personal interview to the head of the village was useful to acquire general information about the village where the households questionnaire was administered. It was composed by 28 short-answer questions about ethnic and religious composition, inhabitants distribution, quantity of cattle in the village and number and location of the main water sources. Another interview was dedicated to the Water committee (the local branch of the water management Ethiopian office). The interview was submitted to at least two of the seven members of the Committee. The 19 questions concerned citizens' water access, quality and quantity of water and the main problems about water sources management. The group interview to members of the village was structured as a community map. All members of the community were invited to take part at the interview. The consultation was lead outdoor. The first question was to draw on the ground the village boundaries and the main aggregation points (schools, church, mosque and health centers). The community maps are really useful to collect contest information and to create a first relation with the community before starting the submission of the household interviews. A short interview was submitted to a group of women in each village. It was composed by four questions with the aim to generate a discussion between the interviewees about their safeness, empowerment and duties.

The Interview Procedure The interviews and the questionnaires were submitted by one or two interviewers and a cultural mediator. In each village the same procedure was followed. At the beginning there was a short presentation of the project and of the interviewer to the people of the village; therefore all members were invited to take part to the community map, in order to create a first relationship with the population and to find some volunteers for the questionnaires. Community maps were followed by the interviews to the head of the village, to the Water Committee and to a group of women. Therefore the questionnaires were submitted to four households in each village. In order to guarantee the same complete comprehension about the main themes of the project, both questionnaires and interviews were controlled and discuss with the local staff. Interviews and questionnaires took place outdoor, in shared spaces, with the presence of other members of the community due to the necessity of transparency in the research and the spread curiosity of population. Two days in each village were necessary to collect all data.

5.6 Analysed Indicators

5.6.1 Vulnerability

The vulnerability is the capacity of a community to adapt itself when climate changes cause modifications to the environment and condition the life (IPCC 2001). The vulnerability is an indicator, defined as a function of three sub-indicators: exposure (E), sensitivity (S) and adaptive capacity (Ac):

$$V = (E \cdot S)/Ca \tag{1}$$

Both E and S represent the negative effects of the changing conditions, while C is the parameter which may counteract the negative effect of the impact.

5.6.2 Vulnerability and Resilience Macro-Indicators

The vulnerability index is composed by 18 macro-indicators, while the resilience is calculated from 6 macro-indicators (Table 5.3).

Table 5.3 Vulnerability and resilience macro indicators

Indicator	Description	Data source	Questions
E	Population density	Statistic agency of Ethiopia	/
E	Drought analysis	Ethiopian National Meteorological Services Agency	/
S	Cultivated land pro capita	FSQ	4/16
S	Number of members <5 and >65 years old	FSQ	5 + 6
S	Yearly income pro capita	FSQ	(31–32)/4
S	Presence of a second income source	FSQ	19
S	Average litres of water fetch per day	FSQ	(13a + 13b)/2
S	Distance from the main water source	FSQ	12a
S	Average water price	Community map	5
S	Water quality	FSQ, water committee interview	12d
S	Type of soil and soil erosion	Antierosion study	/
S	Average number of meals per day	FSQ	(36a + 36b)/2
S	Number of agriculture varieties	FSQ	23
S	Use of inputs	FSQ	24
Ca	Number of animals per cattle	FSQ	17
Ca	Type of water source in the village	FSQ	12a, 12b, 12c, 12e
Ca	Presence of a warehouse	FSQ	30
Ca	Beneficiary of governmental programs	FSQ	22
Re	Higher educational level in the family	FSQ	9

(continued)

Table 5.3 (continued)

Indicator	Description	Data source	Questions
Re	Participation in associations (cooperative)	FSQ	27
Rc	Internal aid (participation to local associations)	FSQ	34
Re	Access to microcredit	FSQ	33
Re	Beneficiary of food security LVIA project	FSQ	1
Re	Presence of an health center in the village	Community maps	/

FSQ-Food security questionnaire

Exposure
Exposure is defined as each meaningful climatic variation influencing the examined system. Intensity, frequency, duration and physic extension of the hazard are specifically considered. In the disaster risk management, the exposure is a set of elements of the analysed system. There are two types of exposure: physical and social. Physical exposure refers to infrastructures, buildings, ecosystems and cultivations, while social exposure concerns human and animal populations. It is important to understand that an exposed population can be not vulnerable if it presents low sensitivity and a good capacity of adaptation, but a vulnerable population is always exposed. Furthermore, a geographic area can be exposed to hazards, but not vulnerable because uninhabited. In the definition of exposure is necessary to find the social component, or the subject, and the physic component, or the hazard of the system. The following paragraphs describes the procedure for the determination of the physical exposure (E_f) and of the social exposure (E_d) in this work.

Physical Exposure
Physical exposure describes the nature and the level of danger of the context potentially exposed. In Ethiopia the hazard is the drought. As suggested by WMO (2012), in order to analyze physical exposure, the SPI-Standardized Precipitation Index was used. The SPI is based on the cumulative probability of a given rainfall event occurring at a station. The historic rainfall data of the station is fitted to a gamma distribution, as the gamma distribution has been found to fit the precipitation distribution quite well. The process allows the rainfall distribution at the station to be effectively represented by a mathematical cumulative probability function. Therefore, based on the historic rainfall data, it is possible to tell what is the probability of the rainfall being less than or equal to a certain amount. Thus, the probability of rainfall being less than or equal to the average rainfall for that area will be about 0.5, while the probability of rainfall being less than or equal to an

amount much smaller than the average will be also be lower (0.2, 0.1, 0.01 etc., depending on the amount). Therefore if a particular rainfall event gives a low probability on the cumulative probability function, then this is indicative of a likely drought event.

Alternatively, a rainfall event which gives a high probability on the cumulative probability function is an anomalously wet event. (Guttman 1999). In this research the SPI was evaluated using precipitation data provided by the Ethiopian National Meteorological Services Agency (NMSA 2016). Data were aggregated and expressed in millimeters, and referred to four meteorological stations distributed in a circular area with a radius of 50 km. The center of the area was located in Shashamane town, that is in the middle of the three examined Woredas (Fig. 5.2). Data collected between 1994 and 2014 were used. The choice to limit the research at four stations (Fig. 5.2) permits a good data coverage (between 88% and 96%) that guarantees a high result confidence (WMO 2012). The SPI was calculated thanks to an open-source software created by the National Drought Mitigation Center (NDMC) of the University of Nebraska-Lincoln (NDMC 2016). The SPI is a positive value when precipitation data is over the average, while a negative value indicates precipitations under the averaged value. Drought is considered as the time period when the SPI is negative and it is classified on the basis of its duration.

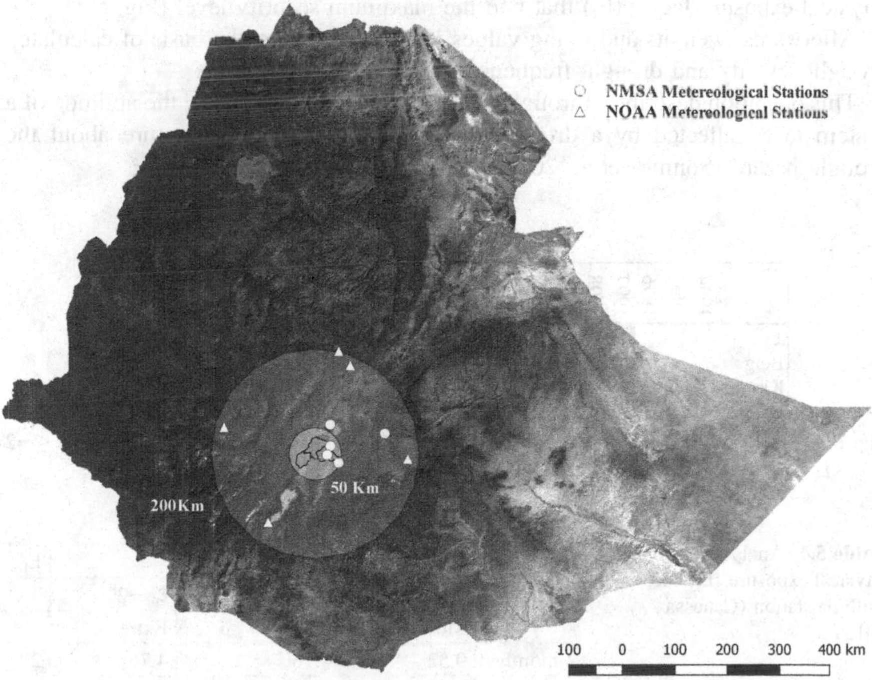

Fig. 5.2 Analyzed area for the evaluation of the physical exposure. Meteorological stations of NMSA (*circlets*), meteorological stations of NOAA (*triangles*)

The meteorological drought is a period of at least one month of negative SPI, the agricultural drought is at least a three months period of negative SPI, while the hydrologic drought has a length of at least six months of negative SPI (Sönmez et al. 2005).

In this research a SPI of four months was used, in order to have a seasonal point of view. Then, to evaluate also the SPI for a long term period, a SPI of twelve months is used considering that it was analyzed a time series of 20 years.

The short term analysis (4 month) is focused on precipitation of the main rainy period, the Belg (February-May) and the Kiremt (June-September) seasons. The annual SPI, instead, allows to obtain a historical analysis about frequency and intensity of long-period droughts in the area.

The value of the physical exposure was achieved starting from SPI through the evaluation, according to a normalized scale of SPI, of the frequency of drought events for every recorded intensity and duration (4 month or 12 month) (WMO 2012). The evaluation considered the data from two meteorological stations: Shashamane and Bulbula. These stations were chosen for their position near the analyzed Woredas, for the good data cover and because their represent two areas at different altitudes. Comparing the results obtained from the analysis of the two stations, Bulbula area shows the most critical condition.

Considering the proximity of the stations and their significance to the analyzed area, it was decided to use the most critical values in order to have a final analysis of physical exposure factor (E_f) that had the maximum security level (Fig. 5.3).

Afterwards, weights and rating values were assigned, on the basis of calculated drought severity and drought frequencies (Table 5.4, 5.5).

This operation describes, through SPI and in a quantitative way, the aptitude of a system to be affected by a drought event according to the literature about the drought hazard (Sönmez et al. 2005, Shahid and Behrawan 2008).

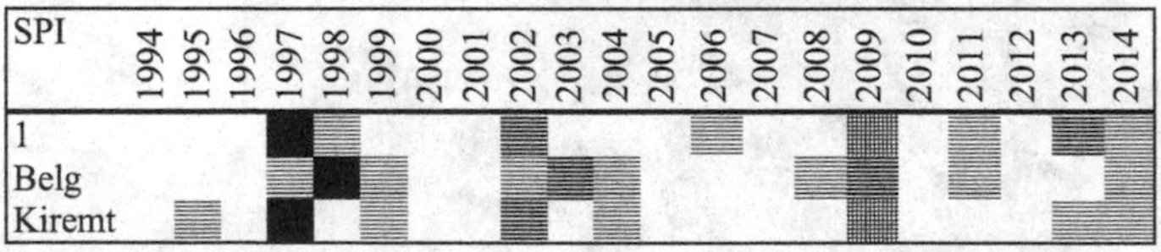

Fig. 5.3 Bulabula SPI 0–0.99 (*clear lines*), −1 to −1.49 (*dark lines*), −1.5 to −1.99 (*grid*), ≤ −2 (*black*)

Table 5.4 Analysis of physical exposure (E_f) for Bulbula station (Canessa 2015)

Drought		SPI		E_f
Intensity	−1 to 1.49	−1.49 to −1.99	≥ −2	
	Moderate %	Severe %	Extreme %	
12 months	9.52	4.76	4.76	20
Belg	9.52	–	4.76	14

Table 5.5 Weight, frequency and rating of E_f

Drought severity	Weight	Occurrence (%)	Rating
Moderate	1	≤9.0	1
		9.1−10.0	2
		10.1−11.0	3
		≥11.1	4
Severe	2	≤3.5	1
		3.6−4.5	2
		4.6−5.5	3
		≥5.6	4
Very severe	3	≤1.5	1
		1.6−2.0	2
		2.1−2.5	3
		≥2.6	4

Physical exposure (E_f) was obtained using the equation:

$$E_f = (SM_v * SM_w) + (SS_v * SS_w) + (SE_v * SE_w) \tag{2}$$

where: *SMv*: attributed rating of moderate drought frequency; *SMw*: weight of the moderate intensity drought; *SSv*: attributed rating of severe drought frequency; *SSw*: weight of the severe intensity drought; *SEv*: attributed rating of extreme drought frequency; *SEw*: weight of the extreme intensity drought.

Finally, averaging the three values (Table 5.4), it was obtained an E_f value equal to 17.66, on a maximum value of 24. The final normalization of the value, in a scale of 0 ÷ 1, reported a final value of 0.73.

5.6.3 *Demographic Exposure*

The demographic exposure indicates the amount of individuals exposed to hazards.

In order to describe the demographical exposure, the population density of the examined Kebeles was considered. In the Kebeles with high values of population density, the availability of resources for inhabitants is low: for this reason the areas that present higher values of population density are more vulnerable than the ones with lower density. The population density was calculated using the data of the Ethiopian population census of 2007 of the Central Statistical Agency of Ethiopia (CSA). All data was updated to the 2015 values considering a grow rate of 3% (World Bank 2016). Four classes corresponding to an increased population density were created, in order to normalize the demographic density values: at each class an E_d (demographic exposure) value between 0.3 and 1 was associated (Table 5.6), corresponding the highest index (1) to the highest density (in this case >600 inhabitants per squared kilometre).

Table 5.6 Demographic exposure (E_d) index

Population density Inhabitants/km^2	E_d
0–199	0.3
200–399	0.5
400–599	0.7
>600	1

Table 5.7 Sensitivity indicator components

Agro-environment	Soil type
	Land degradation
	Quality and quantity of agricultural inputs
	Culture diversification
Water	Availability
	Price
	Distance from the nearest water source
	Quality
Socio-economic	Cultivable land
	Income per capita
	Members of the family economically inactive
	Presence of secondary income sources
	Numbers of daily meals

Sensitivity The sensitivity (*S*) represents the aptitude of populations and infrastructures to be damaged by extreme events (see Eq. 3). Contest conditions and intrinsic characteristics of population itself are the main causes of sensitivity.

$$S = \sum resources\, availability \tag{3}$$

When the hazard has a climatological nature, then an high sensitivity system is extremely sensitive to small variations of climatological conditions (Fellmann 2012).

Analysing the data extrapolated from the Food security survey, twelve indicators that describe the system sensitivity as resources availability were found. This twelve indicators were grouped in three categories: agro-environmental indicators, water indicators and socio-economic indicators (Table 5.7).

Agro-Environment Indicators

Agro-environment indicators concern natural characteristics of the system. The Kebeles population life in based on subsistence agriculture, consequently the ecosystem provides seasonal yield but it is also source of fuel, water and grazing land. Considering the strong relationship between population and its environment, for a correct vulnerability analysis it is essential to find agricultural and environmental indicators, as soil characteristics and agricultural quantity and quality.

Soil type In the areas where the soils are more fertile, the yields are potentially greater and the vulnerability is lower. The soils classification of the three Kebeles was therefore based on their fertility: values between 1 (better situation) and 3 (worst situation) were assigned.

Land degradation The land degradation concerns the loss of soil fertility due to the erosion. Three classes that consider the yearly loss of soil (in tons) were ascribed, based on the USLE (Universal Soil Loss Equation) (Wischmeier and Smith 1978) as used in the Anti-erosion study by LVIA. At each class a value between 1 (better condition) and 3 (worst situation) was assigned (Table 5.8).

Quality and quantity of agricultural inputs In an agricultural subsistence system as the Ethiopian one, the employment of fertilizers, chemicals and improved seeds (even if in small quantity) immediately increases the seasonal yield and guarantees an annual yield. For the determination of the indicator the qualitative data (corresponding to personal judgment of the interviewer on the inputs quality) was converted in a quantitative data (Table 5.9): at different judgments different values were attributed, between 1 (not use or bad quality of fertilizers, seeds and pesticides) and 0 (good quality).

Culture diversification The differentiation of cultivated species increases the agro-biodiversity of the environment. Higher is the number of agricultural varieties and lower is the probability that the yield is totally destroyed by diseases. An higher agricultural diversity represents a minor sensitivity of the system. The value of the indicator was found as the normalization of the number of the cultivated varieties.

Water Indicators

Considering drought as a main hazard, indicators concerning water availability are the key of the sensitivity evaluation. In detail the indicators were built considering all that factors that could negatively affect populations for a poor access to water sources.

Water availability The water availability index was calculated as the average between the litres of water collected during the dry season by the household and the same collected during the rainy season. The average was normalized and translated in a 0–1 scale. The water availability indicator must be interpreted as a vulnerability

Table 5.8 Land degradation index

USLE (Universal Soil Loss Equation)	Value
<7 tons/he/year	1
7–25 tons/he/year	2
>25 tons/he/year	3

Table 5.9 Input quality index of fertilizers, seeds and pesticides

Inputs quality	Value
No use	1
Bad	0.75
Acceptable	0.4
Good	0

shock absorber: the cases of low value of litres of water collected suggest low needs of water consumption, and consequently a minor exposure to drought periods (in this case the index is close to 0). On the contrary, family that showed high consumption of water are more sensitive and the indicator value is near 1.

Water price Water price can also highly influence families sensitivity. In Ethiopia water price is decided by the *Woreda Water Desks,* that are local water committees that manage and monitor water resources on the Woreda territory. The water price is set accordingly to the management costs and to the underground water availability in the area. If the water price increases due to the drought, an higher part of the families yearly income is used to satisfy the water needs, increasing the vulnerability of the system. In order to individuate a numeric value to the water price indicator, four cost classes were created based on the prices collected from the Water Committee interviews. Higher is the index, more the water price improves the household sensitivity (Table 5.10).

Distance from water source The distance between the household and the water source was built considering the time and the human resources dispended by the families for water collection. Usually water sources are kilometres far from the dwellings and, especially during the dry season, people have to wait for hours without the guarantee to collect the water they need. Furthermore, robbery and violence against the women during the trip to the water source are not unusual. Longer is the way to collect the water and higher is the households vulnerability. The distances obtained by the surveys was expressed in kilometres, but the largest part of rural population, especially less instructed members, were not able to quantify distances: for this reason the distances (also in this case normalized) must be intended as perceptive.

Water quality indicator The water quality was referred to the perception of the interviewees about the water drinkable. In this context water for human consumption is rarely controlled or treated and it may contain chemical or physical contaminants. The interviewees gave their opinion about the water quality by mean of a multi-choice answer (Table 5.11).

Table 5.10 Relation between water price and indicator value

Price (ETB) of one jerrican (25 L)	Indicator values
>5	1
4–0.7	0.75
0.6–0.5	0.5
<0.5	0

1 ETB = 0.048 USD (rate of 14 August 2015)

Table 5.11 Drinkable water index

Water interviewee judgment	Indicator value
Bad quality	1
Acceptable quality	0.5
Good quality	0.25
Very good quality	0

Socio-Economic Indicators

Socio-economic indicators are useful to understand if a risk of the population sensitivity exist in relation to both low social position and feeble economic capacity of the families.

Cultivable land Land ownership identify the richness of households: larger is the owned land and higher is the economic return in form of food subsistence. Considering that the examined areas are the most populated of Ethiopia and the largest part of yield is usually destined to self-consume, the land pro-capita was used to create the cultivable land indicator. Lower is its value, higher is the vulnerability.

Income per capita Climatic vulnerability weakens economic capacity of individuals, putting them into a poverty condition that, like a vicious circle, doesn't allow them to go out. This situation is called poverty trap. The richness indicator is described by the pro-capita income calculated as the difference between agricultural incomes and yearly based expenditure divided by the number of the household members. In some cases this indicator resulted negative, probably due to the difficulty to analyse the annual expenditures. In these cases the index was considered equal to zero. All the values were normalized.

Members of the family economically inactive The households with economic inactive members are more vulnerable than the ones where all members works. For the quantification of the indicator were considered the household members younger than five years old and older than sixty-five years old.

Presence of secondary income sources Considering that the population of these areas are mainly occupied in subsistence agricultural activities, the existence of additional income sources indicates the dependence of population to agriculture that hardly suffers drought. The main second income arose from farming, local trade or small activities like apiculture and little works in the nearest town. All these activities were evaluated basing on generated income, work frequency and relation with the agriculture. At each information a value between 0 and 1 was attributed, where 1 represent the worst condition while 0 the better one (Table 5.12).

Number of daily meals A just number of daily meals is index of food self-sufficiency. Assuming that all meals are equally nutrient, higher is the number of meals in a day and higher is the quantity of available food of the family. If the

Table 5.12 Additional income indicator

Additional income	Indicator value
No additional income activity	1
Petty trade	0.5
Seasonal work	0.5
Trade and governmental employment	0
Farming	0

major part of food comes from their own cultivations, than it is possible to use the average number of daily meals as proxy for food self-sufficiency. All values were normalized.

Capacity of Adaptation

The capacity of adaptation is the amount of inherent and context abilities of populations that permits to people to quickly react to an extreme hazard and to adapt to new situations. Therefore, the capacity of adaptation can contribute to mitigate the sensitivity effects and positively respond to physical exposure. The capacity of adaptation is defined by indicators that concern support degree between individuals of the community, government disaster emergency plans and community aid infrastructures. In this work the capacity of adaptation (*Ca*) indicator was calculated as:

$$Ca = \sum \textit{food stock}, \textit{livestock}, \textit{governmental aid}, \textit{water source} \qquad (4)$$

Food stock A warehouse permits to the households to store the seasonal yields until the following agricultural season, instead of selling it after the harvest. A food storage represent a tool for adaptation in drought situation. The value of 1 was assigned to the families that used a private or common warehouse and 0 to the ones that did not have it.

Livestock In Ethiopia livestock is an important economic and social element of people's life: livestock represents a food source, a work force, an income source and a social prestige. Thanks to the sale of part of the cattle, households can cope with natural shocks. The livestock represent the capacity to reply in short time to climatic hazards, even if the economic value of livestock decrease in case on drought because of the market saturation. The cattle number of each household was normalized.

Governmental aid Ethiopian government promotes many aid programs for rural population taking care of different aspects, from food security to education rights. In this case the indicator specifically refers to PSNP-Productive Safety Net Project, a governmental project started in 2005. The project promotes food safety through food distribution, mainly milk powder and palm oil, to families in disadvantaged situations. Households beneficiary can better cope with shocks thanks to external aid.

Water sources A good water resource management associated to collection, storage and pumping systems is the key strategy of adaptation to drought periods. The indicator was calculated using data from different questions of the survey (Table 5.13).

Table 5.13 Water source indicators

Water source in the compound		Water source outside the compound					
Good quality of the water facility	Poor quality of the water facility	Protected water source (physical protection)		Not protected water source			
		No necessity of water treatment	Treated water	No necessity of water treatment, good water quality	No necessity of water treatment, bad water quality	Treated water, good quality	Treated water, poor quality
0	0.2	0.25	0.5	0.5	0.7	0.8	1

5.6.4 Resilience

Resilience is defined as the ability of a potentially exposed system, community or a society, to resist, absorb, accept and recover from disasters effects in a prompt and effective manner, even though the conservation of its essential base structures and functions (UNISDR 2009). In a climate change framework, resilience is strictly interconnected to the capacity of adaptation. It concerns the ability of anticipating and preventing any climatological disaster, but it can also be interpreted as the ability of society of changing and innovating itself thanks to communication and education. In our case the resilience was built as:

$$R = \sum \begin{array}{l} educational\ level,\ intercommunity\ aid,\ health\ centres, \\ cooperative,\ micro-credit,\ NGO support \end{array} \quad (5)$$

Educational level People with an higher educational level has more possibility to cope and go out from insecurity situations. For this reason resilience was also made up of this indicator, defined by the grade of the head of the household. The Ethiopian education system comprises 13 years of school. The data was normalized basing on the maximum grade detect from all interviews.

Intercommunity aid The presence of local institutions, that provide aid to community, permits to people to recover and to evolve in function of climatic hazard impacts. In almost each analysed areas, each family participates to the *iddir* and/or to the *iqub*, which are self-aid community institutions. To be part of an *iddir* or a *iqub* it is necessary to pay a monthly fee and in case of need families receive a monetary aid from the *iddir* or *iqub*. *Iddirs* are very common and they are mainly used to afford funerals and marriages. *Iqubs* are used for other purpose, for example for maintaining the petty sale of milk of women of the village. This intercommunity aid indicator is based on the participation in *iddir* or *iqub*. If interviewee's family took part at *iddir* or *iqub* the indicator value was set at one, otherwise it was set at zero.

Health centres In villages where there is a functional health centre or a health post (this is smaller than the health centre and medical staff is present three days per week), the access to medicines and basic medical care guarantees a rapid recovery post-hazard. Furthermore health posts offers courses about hygiene, children's growth and birth control, that improve community education and preparedness. In villages where there was a health post or centre was attributed the value of 1 to the indicator, if not the value was zero.

Cooperatives participation Cooperatives are able to decrease the social and economic marginality of rural areas thanks to members' training, competiveness and access to market. Agricultural cooperatives give to farmers improved seed, fertilizers and chemicals by credit which can be paid back with the seasonal yield. Furthermore, cooperatives permit the organization between the members to help each other during the tillage and seeding periods providing agricultural knowledge. The agricultural cooperatives system potentially permits to farmers to easily resume the agricultural cycle after a disturb. If people took part to cooperatives the indicator had the value of 1, if not, 0.

Micro-credit The access to credit indicates that the population has economic instruments that permit the adoption of recovery strategy. Microcredit can potentially help vulnerable people to react and cope climatologic hazard, giving them economic resources necessary for adopting structural, physic, environmental and social strategies. The indicator value was 1 in case of access to microcredit and 0 in case of not access.

NGO support NGO projects play an important role in developing countries. Specifically in the analysed areas LVIA NGO is really active in food safety programs that aim to defeat famine helping population to be food independent. In the Kebele of Shashamane, Siraro and Shalla the NGO LVIA built warehouses in the villages in order to permit to the farmers to stock the seasonal yield. Furthermore they distributed seed, agricultural tools, plants and nitrogen-fixing species to improve soil fertility. Communities that receives any NGO aid have more instruments to recovery after hazards. The indicator had the value of 1 in the community which received NGO aid and 0 in the opposite case.

5.6.5 Capacity to React to Vulnerability

The capacity to react to vulnerability represents the relationship between vulnerability and resilience. It allows the detection of the areas that can potentially show more damages and negative consequences due to climatic hazards. The capacity to react to vulnerability identifies the areas with high vulnerability but low resilience value. The areas where the value of the capacity to react to vulnerability is high show critical conditions for population. It can be a useful tools to quickly find the areas that need more improvements. It is the difference between vulnerability and resilience values (Eq. 6),

$$Cr = V - R \tag{6}$$

5.7 Results and Discussion

There is a strong link between vulnerability and resilience, but it is firstly necessary to discuss the result of each indicator in all the examined villages and therefore to see how they relate. Being the number of the household interviews quite high (n = 180), the calculated indicators were grouped in N classes using the Sturges rule (Sturges 1926) and 8 classes were obtained (Eq. 7).

$$N = 1 + 10/3 Logn \tag{7}$$

where: n: number of interviews.

Therefore the class amplitude a was calculated (Eq. 8).

$$a = (I_{max} - I_{min}))/N \tag{8}$$

where: I_{max} = higher indicator value calculated; I_{min} = lower indicator value calculated; N = number of classes

5.8 Vulnerability Classes and Maps

Concerning the vulnerability, data varied from 2.37 to 140.98: it must be however observed that the two higher values, 93.30 and 140.98, were absolutely outliers, because all other data ranged between 2.37 and 37.74. The 60% of the vulnerability values were under 10 (Table 5.14): this fact confirms the World Bank assumption (Sadoff 2008) that the Oromia Region is less vulnerable than the other Ethiopian regions. Nevertheless the mapping of the vulnerability indices (Fig. 5.4) reveals that the Woreda of Siraro presents the worst data almost on all its territory. Siraro is effectively more isolated than the Woredas of Shalla and Shashamene, has less communication roads and also its water sources are scarcer.

Table 5.14 Classes of vulnerability indices

Class	Values	Frequency	Relative frequency (%)	Cumulative frequency (%)
1	2–6	49	27	27
2	6–10	59	33	60
3	10–14	15	8	68
4	14–18	13	7	76
5	18–22	10	6	81
6	22–26	12	7	88
7	26–30	11	6	94
8	>30	11	6	100

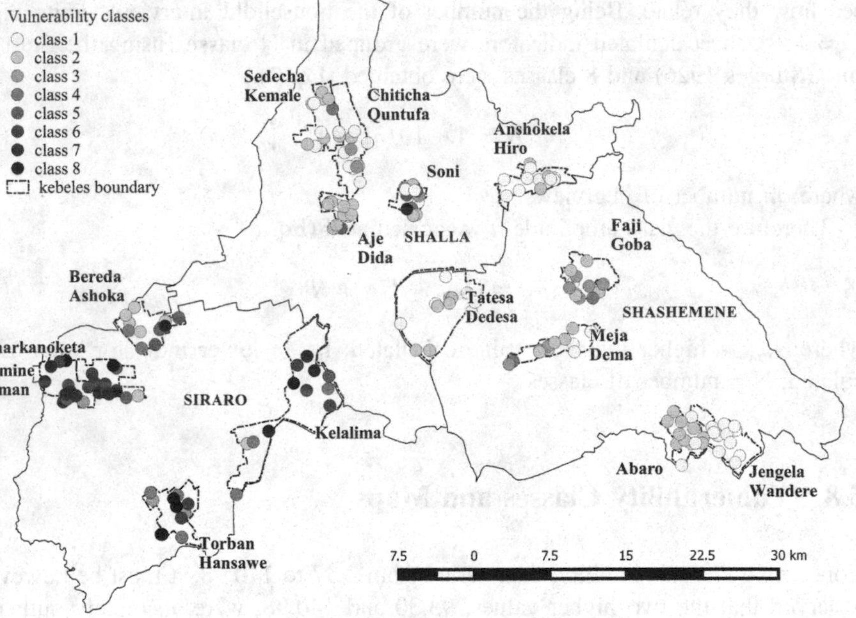

Fig. 5.4 Map of the vulnerability indexes in the three analyzed Woredas

5.9 Resilience Classes and Maps

With a class amplitude of 0.5, the resilience indices ranged between 1.40 and 5.69, and where more evenly distributed than the vulnerability indices, with highest values in the central classes (41% of the data are concentrated between 2.9 and 3.9, Table 5.15). Considering that higher is the resilience, higher is the community ability to adapt to critical situations (scarcity of water, agricultural inputs and

Table 5.15 Classes of resilience indices

Class	Values	Frequency	Relative frequency (%)	Cumulative frequency (%)
1	1.4–1.9	18	10	10
2	1.9–2.4	20	11	21
3	2.4–2.9	19	11	32
4	2.9–3.4	36	20	52
5	3.4–3.9	38	21	73
6	3.9–4.4	16	9	82
7	4.4–4.9	20	11	93
8	>4.9	13	7	100

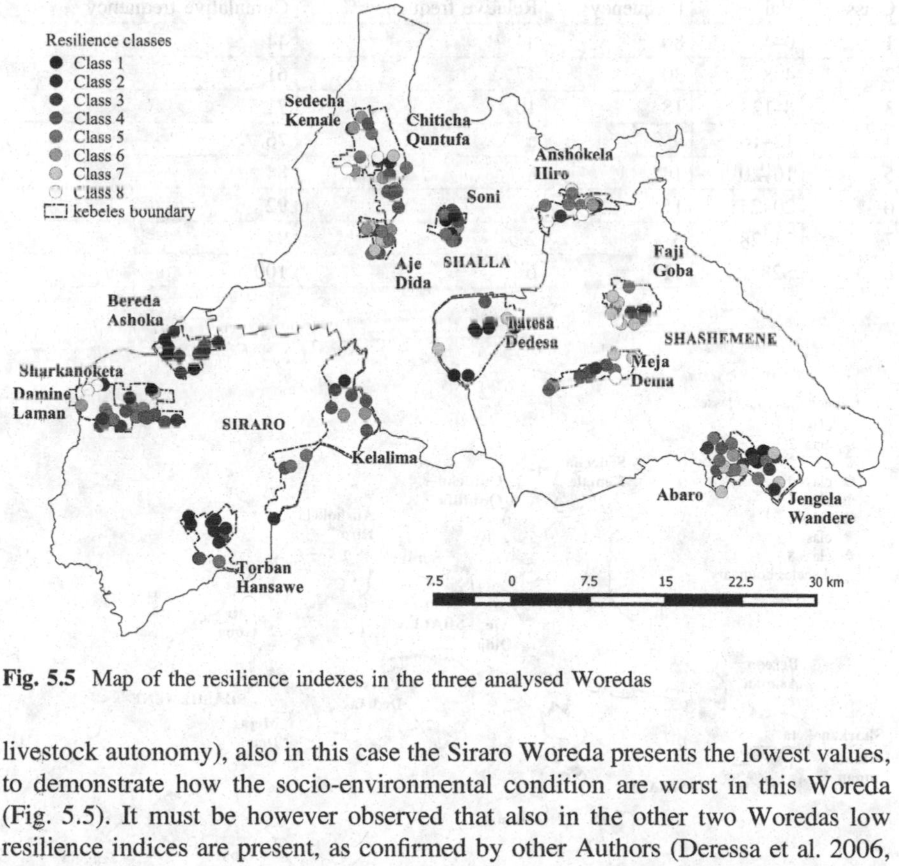

Fig. 5.5 Map of the resilience indexes in the three analysed Woredas

livestock autonomy), also in this case the Siraro Woreda presents the lowest values, to demonstrate how the socio-environmental condition are worst in this Woreda (Fig. 5.5). It must be however observed that also in the other two Woredas low resilience indices are present, as confirmed by other Authors (Deressa et al. 2006, DRMFFS 2016). Despite a not bad vulnerability indices in Shalla and Shashamene, in these Woredas population may still encounter difficulty to adapt to sudden climatic changes. For the Shashamene province this fact may be attributed to an higher population density in its territory and therefore to an higher social exposure that makes this Woreda more vulnerable than the others.

5.10 Capacity to React to the Adversities Classes and Maps

Considering the capacity to react to the adversities, lower it the value, higher is the capacity of the household to react to critical conditions. This index ranged between 0.17 and 137.71, but the last two values were outliers. The class amplitude was 4. The observed capacity of adaptation was quite good (61% of the cases less than 8, Table 5.16), with worst aspects always retrieved in the Woreda of Siraro (Fig. 5.6). Beyond the interpretation of each parameter, a more complete reading key is

Table 5.16 Classes of capacity of adaptation indices

Class	Values	Frequency	Relative frequency %	Cumulative frequency %
1	0–4	80	44	44
2	4–8	30	17	61
3	8–12	18	10	71
4	12–16	9	5	76
5	16–20	14	8	84
6	20–24	15	5	92
7	24–28	3	2	94
8	>28	11	6	100

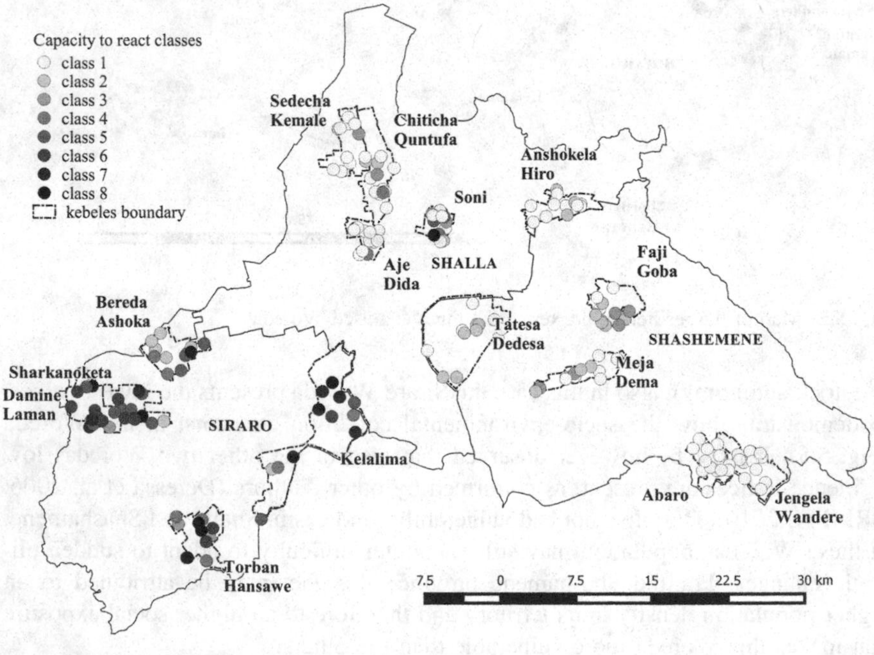

Fig. 5.6 Map of the capacity of adaptation indices in the three analysed Woredas

possible if the SPI index is put beside them because the drought risk is a crucial parameters for all the types of vulnerability and resilience detection. In a normalized scale between 0 and 1, in this study the calculated annual SPI index was 0.83: the area is therefore considerably exposed to the drought risk, caused by an high intensity and frequency of rainfall lack.

5.11 Conclusions

Uncertain precipitations during the two Ethiopian rainy seasons (Belg and Kiremt) and increased temperatures registered in the last years impact the crop production, the livestock management and, as a consequence, the subsistence agricultural production systems that in this country solely depends on rainfall.

Unfortunately at local level there are not specific quantitative studies to afford the problem, also if the Ethiopian Government promoted the Disaster Risk Management and Food Security Sector (DRMFSS) which in 2009 began to qualitative monitor the drought risk bent of the Woredas. The aim of this project is to develop an integrated system of Early Warning and Response, but actually neither landmark data nor vulnerability drought analysis exist in the country.

For this reason the methodology introduced in this study is important, both as environmental and social analysis.

The high risk of drought exposition caused by the increasing of drought's frequency in the three examined Woredas is a useful information for the policy maker to create adaptation plans. At the same time policy makers may use the vulnerability index both to isolate the variables that influence the system's vulnerability and to compare the different situations in the Woredas. Last, but not less important, with the introduction of the index called "capacity to react to the vulnerability", it is possible to pointed out the areas at higher risk for the analysed impact factors.

Annex Food Security Questionnaire

Section 1—General Information

1. Date and time ______________ **2**. Location (woreda, kebele, village): ________

3. Household HH ID ___ **4**. HH size ___ **5**. Members <5 age ___ **6**. Members >65 age ___

7. Sex and age of respondent ______ **8**. Relation interviewer with the HH ________

9. Educational level of HH head ___ **10**. Nb of HH school-age children attending classes regularly ___

11. Do you have any type of water source in the compound? Y/N ___

11A. If yes, specify source type: ☐ tap ☐ birkad ☐ roof catchment ☐ hand-dug well ☐ others ___

11B. Conditions: ☐ very good ☐ good ☐ needs minor improvements ☐ poor

11C. Water quality: ☐ very good ☐ good ☐ acceptable ☐ poor

11D. Water quantity: ☐ more than enough ☐ sufficient ☐ insufficient

11E. In case of more sources available, specify above the main water source and indicate here the others (type, conditions, quality, quantity) ___

12A. If no or limited water source in the compound, specify the distance (km) of the main source used by the HH___

12B. Source type: ☐ protected ☐ unprotected **11B2** Specify ___

12C. Conditions: ☐ very good ☐ good ☐ needs minor improvements ☐ poor

12D. Water quality: ☐ very good ☐ good ☐ acceptable ☐ poor

12E. Water quantity: ☐ enough ☐ sufficient ☐ insufficient

13A. Average liters of water fetched by the HH per day during rainy season _ **13B**. During dry season ___

14A. Is the drinking water used by the HH treated? Y/N ___

14B. If yes specify the method ___

15. Main source of fuel for domestic use ___

Section 2—Livelihood

16. HH land size (ha) ___ **17**. Nb of livestock ___ **18**. Main occupation/productive activity ___

19. Any other additional income generating activity: ___

20. Main 3 factors affecting negatively the agricultural production ___

21. Main 3 factors affecting negatively other productive activities ___

22. Is the HH beneficiary of PSFP or other governmental support programs? Y/N ___

22A. If yes specify which one ___

If the HH is mainly involved in agriculture, fill the following tables and information (22 to 28):

23. Production last year

Crop	Farming practices	Cultivated land (ha)	Average yield/ha (good season)	Average yield/ha (bad season)
% kept	% sold	% lost	Main buyer	Time and reason for sale

24. Inputs: Specify per each type of input, the availability (both in terms of quantity and timing), modalities access (for ex. borrowed, purchased, exchanged, etc.), and the average cost per season.

Input	Availability and quality	Access modalities	Cost

25. Practices: Does the HH make use of compost or any improved organic agronomic practices? Y/N ___

25A. If not, specify why ____

26. How would you rate the support received from agricultural extension services? ☐ Poor ☐ Acceptable ☐ Good

26A. If rated poor, specify main areas for improvement ____

27 Are you part of a cooperative? Y/N ____

27A If yes specify services and benefits you receive from it ____

27B If not, why are you not part? ____

28. Nb of agricultural seasons ____

29. Water source for agricultural activities: ☐ Rain-fed ☐ Irrigation

29A If irrigation, specify access modalities/costs and management body ____

30. Does the HH have a storage space? Y/N ____

30A. If yes, specify capacity, main construction materials, general conditions, what is stored, related average storage time, eventual treatments ____

30B If yes, specify the main use: ☐ later sale ☐ own consumption ☐ others[2] ____

31. Average income of the HH, per year, in ETB:

Main activity			Average income main activity	
2nd activity				Average income 2nd activity
3rd activity		Average income 3rd activity		
Other incomes				

32. Average expenses of the HH, per year, in ETB:

Food	Water	Clothes	Education	HH and hygiene items	Other expenses

33. Did the HH have access to formal credit institutions in the last 3 years? Y/N ____

33A. If yes, specify access modalities, value of loan, interest rate, purpose ____

34. Is the HH part of iddir/iqub? Y/N ____

34A. If yes, specify the main purpose of the iddir/iqub ____

34B. If yes, specify average contribution per month ____

Section 3—Food Security

35. During the last 5 years, did any member suffer from malnutrition? Y/N ____

35A. If yes please specify which member(s), how many, measures undertaken ____

36A. Average nb meals/day during hunger season: ☐☐ 1 ☐ 2 ☐ 3 ☐>3

36B. Post-Harvest Season: ☐ 1 ☐ 2 ☐ 3 ☐ > 3

37. Specify if any difference on nb of meals among HH members ____

[2]Specify.

38A. Composition of main daily meal/hunger season ____

38B. Composition of main daily meal/post-harvest season ___

39. How is the quantity of food available in the HH perceived? ☐ more than enough ☐ enough ☐ insufficient

40A. Main source of food: specify estimated % per each season and each modality indicated by the respondent during hunger season: __% ☐ own production __% ☐ purchased __% ☐ received __% ☐ borrowed

40B. During post-harvest season: __% ☐ own production __% ☐ purchased __ % ☐ received __% ☐ borrowed

41. Discuss any shock during the past 5 year faced by the HH, its impact and main coping mechanisms __

Section 4—Observation Checklist

42. Is there a toilet (Y/N)? __ **42A**. If yes, specify general conditions:☐ poor ☐ acceptable ☐ good

43. Electricity in the house (Y/N) __

44. General conditions of compound and house____

45. Any other relevant observation, comment, information ______________________

Enumerator signature ______________________ Team Leader signature ________

References

AG-General Assembly of United Nation. 2015. Establishment of an open-ended intergovernmental expert working group on indicators and terminology related to disaster risk reduction. New York: A/RES/69/284.

Alwang, J., P. Siegel, S. Jorgensen. 2001. Vulnerability: a view from different disciplines. *Social Protection Discussion Paper* 115.

AQUASTAT. 2005. Irrigation in Africa in figures. AQUASTAT Survey 2005. Rome: FAO.

Canessa, C. 2015. *Cambiamento climatico, siccità e sicurezza alimentare: una metodologia di analisi della vulnerabilità. Il caso delle Woreda Etiopia*. Turin: Thesis University of Turin.

CSA-Central Statistical Agency. 2007. *Population and housing census report*. Addis Abeba: Government of Ethiopia.

CSA-Central Statistical Agency—Government of Ethiopia. 2007. National Statistical Data. Official website: http://www.csa.gov.et/ Last access: 25th August 2016.

Cutter, S.L. 2003. The vulnerability of science and the science of vulnerability. *Annals of the Association of American Geographers* 93–1: 1–12.

Deressa, T., R. Hassan, and C. Ringler. 2006. *Measuring Ethiopian farmers vulnerability to climate change across regional states*. IFPRI Discussion Paper: Washington D.C. 00806.

DRMFSS-Disaster Risk Management and Food Security Sector. 2016. Government of Ethiopia, Official website: http://www.dppc.gov.et/ Last access: 25th August 2016.

Fellmann, T. 2012. The assessment of climate change related vulnerability in the agricultural sector: reviewing conceptual frameworks. FAO/OECD Workshop Building resilience for adaptation to climate change in agricultural sector. Rome: FAO.

Füssel, H.-M. 2007. Vulnerability: a generally applicable conceptual framework for climate change research. *Global Environmental Change* 17(2): 155–167. doi:10.1016/j.gloenvcha.2006.05.002.

Guttman, N.B. 1999. Accepting the Standardized Precipitation Index: a calculation algorithm. *Journal of the American Water Resources Association* 35–2: 311–322.

Hulme, M., R. Doherty, T. Ngara, M. New, and D. Lister. 2001. African climate change: 1900–2100. *Climate Research* 12: 145–168.

IPCC-Intergovernmental Panel on Climate Change. 2001. Climate change 2001: Synthesis report. Contribution of working groups I, II, and III to the Third assessment report of the Intergovernmental Panel on Climate Change ed. R.T. Watson. Cambridge and New York: Cambridge University Press.

IPCC-Intergovernmental Panel on Climate Change. 2007. Climate Change 2007: Synthesis Report. Contribution of Working Groups I, II and III to the Fourth assessment report of the IPCC, ed. R.K. Pachauri and A. Reisinger. Geneva, Switzerland: IPCC.

IWMI-International Water Management Institute. 2009. *Mapping drought patterns and impacts: A global perspective*. Colombo: IWMI Research Report.

Kasperson, J., and R. Kasperson. 2001. *International workshop on vulnerability and global environmental change*, 01–2001. Stockholm: SEI Risk and Vulnerability Programme Report.

Korecha, D.B. 2006. Prredictability of June-September rainfall in Ethiopia. *Monthly Weather Review* 135: 628–650. doi:10.1175/MWR3304.1.

LVIA. 2015. Groundwater potential study and mapping for Arsi Negele, Shashamene, Shala, Siraro. In *Addis*, ed. Oromia Region. Ababa: LVIA and Water Resources Consulting Service.

MEA-Millenium Ecosystem Assessment. 2009. *Ecosystems and human well-being: current state and trends*. Findings of the condition and trends working group. Washington: Island Press.

NDMC-National Drought Mitigation Center. 2016. Program to calculate Standardized Precipitation Index. Web address National Drought Mitigation Centre: http://drought.unl.edu/monitoringtools/downloadablespiprogram.aspx Last access: 4th July 2016.

Nelson, R., P. Kokic, S. Crimp, P. Martin, H. Meinke, and S. Howden. 2009. The vulnerability of Australian rural communities to climate variability and change: Part II—Integrating impacts with adaptive capacity. *Environmental Science & Policy* 13: 18–27. doi:10.1016/j.envsci.2009.09.007.

NMSA-Ethiopian National Meteorological Services Agency. 2016. Official website: http://www.ethiomet.gov.et/ Last access: 25th August 2016.

Sadoff, C. 2008. *Managing water resources to maximize sustainable growth: a World Bank water resources assistance strategy for Ethiopia*. The World Bank: Washington D.C.

Segele, Z., P. Lamb. 2005. Characterization and variability of Kiremt rainy season over Ethiopia. *Meteorology and Atmospheric Physics* 89(1): 153–180. doi:10.1007/s00703-005-0127-x.

Seleshi, Y., and G. Demarèe. 1995. Rainfall variability in the Ethiopian and Eritrean highlands and its links with the Southern Oscillation Index. *Journal of Biogeography* 22: 945–952.

Shahid, S., H. Behrawan. 2008. Drought risk assessment in the western part of Bangladesh. *Natural Hazards* 46(3): 391–413. doi:10.1007/s11069-007-9191-5.

Sönmez, F.K., A.Ü. Kömüscü, A. Erkan, and E. Turgu. 2005. An analysis of spatial and temporal dimension of drought vulnerability in Turkey using Standardized Precipitation Index. *Natural Hazards* 35(2): 243–264. doi:10.1007/s11069-004-5704-7.

Sturges, H.A. 1926. The choice of a class interval. *Journal of the American Statistical Association* 21–153: 65–66.

United Nations. 2015. World population prospects, the 2015 Revision. Key findings and advanced tables. *Working Paper* ESA/P/WP.241.

UNISDR-United Nations International Strategy on Disaster Risk Reduction. 2009. www.unisdr.org Last access: 25th August 2016.

Vincent, K. 2004. Creating an index of social vulnerability to climate change for Africa. *Tyndall Centre Working Paper* 56.

Viste, E., D. Korecha, and A. Sorteberg. 2013. Recent drought and precipitation tendencies in Ethiopia. *Theoretical and Applied Climatology* 112(3): 535–551. doi:10.1007/s00704-012-0746-3.

Wischmeier, W.H., and D.D. Smith. 1978. Predicting rainfall erosion losses—a guide to conservation planning. U.S. Department of Agriculture.

WMO-World Meteorological Organization. 2012. *Standardized precipitation index – User guide*, 1090. Geneva: World Meteorological Organization, Publication No.

World Bank country data. 2016. http://databank.worldbank.org/ Last access: 1th September 2016.

Chapter 6
Tracking Climate Change Vulnerability at Municipal Level in Rural Haiti Using Open Data

Maurizio Tiepolo and Maurizio Bacci

Abstract In Least Developed tropical Countries, the vulnerability assessment to climate change (CC) at local scale follows an indicator-based approach and uses information gathered mainly through household surveys or focus groups. Conceived in this way, the vulnerability assessment is rarely repeatable in time, cannot be compared with those carried out in other contexts and usually has low spatial coverage. The growing availability of open data at municipal level, routinely collected, now allows us to switch to vulnerability tracking (continuous, low cost, consistent with global monitoring systems). The aim of this chapter is to propose and verify the applicability of a VICC-Vulnerability Index to Climate Change on a municipal scale for Haiti. The chapter identifies open data on national, departmental and municipal scale, selects the information on a municipal scale on the basis of quality, identifies the indicators, evaluates the robustness of the index and measures it. The index consists of 10 indicators created using information relating to monthly precipitations, population density, flood prone areas, crop deficit, farmers for self-consumption, rural accessibility, local plans for CC adaptation, irrigated agriculture and cholera incidence. This information is gathered for the 125 mainly rural municipalities of Haiti. The description and discussion of the results is followed by suggestions to improve the index aimed at donors, local authorities and users.

Keywords Climate planning · Local governments · SPI · Vulnerability index

---

M. Tiepolo authored all the sections with the only exception of section 6.2.4 Indicators 1 and 2, authored by M. Bacci.

M. Tiepolo (✉)
DIST, Politecnico and University of Turin, Viale Mattioli 39, Turin 10125, Italy
e-mail: maurizio.tiepolo@polito.it

M. Bacci
National Research Council—Institute of BioMeteorology (IBIMET),
Via Giovanni Caproni 8, 50145 Florence, Italy
e-mail: M.Bacci@ibimet.cnr.it

M. Tiepolo et al. (eds.), *Renewing Local Planning to Face Climate Change in the Tropics*, Green Energy and Technology,
DOI 10.1007/978-3-319-59096-7_6

6.1 Introduction

Vulnerability is "the propensity or predisposition to be adversely affected" (IPCC 2014: 128) and is estimated in various sectors: food security (OCHA, WFP and FEWSNET), environment, agriculture, livelihood. Here we're going to look at vulnerability to climate change (CC), in other words, "to a change in the state of the climate that can be identified by changes in the mean and or the variability of its properties and that persists for an extended period, typically decades or longer" (IPCC 2014: 120). Vulnerability to CC is related to people. Ascertaining it is important for decision making related to adaptation measures and subsequent monitoring and evaluation.

Vulnerability to CC at local scale is usually ascertained with a vulnerability index with a variable number of indicators: from 10 to 33 (Hahn et al. 2009; RdH, MARNDR 2009; Gabetibuo et al. 2010; Sharma and Jangle 2012; Borja-Vega et al. 2013; Etwire et al. 2013; PNUMA 2013; Ahsan and Warner 2014; Bollin et al. 2014). The more indicators we have (namely those created especially to appreciate vulnerability), the harder it is to repeat the assessment in time and replicate it in other contexts (Table 6.1). In actual fact, decision making and monitoring need to track vulnerability more than assess it occasionally.

To satisfy these requirements, we need open data on a small geographic area, routinely collected at low cost (World Bank 2014), of which we know the metadata. The ideal scale is that of municipalities, as these are the administrative jurisdictions closest to the impacts of CC and with specific tasks in the field.

Haiti is the 4th most vulnerable country to CC (Verisk Maplecroft 2016). The projections of the IPCC for the Caribbean region to which Haiti belongs suggest about a 1.2–2.3 °C increase in surface temperature, a decrease in precipitations of about 5%, and a sea level rise of between 0.5 and 0.6 m by 2100 compared to the baseline 1986–2005 (Nurse et al. 2014: 1627–8). If these changes should take place, the economic impacts on export crops (coffee and mango), on the availability of water resources and on health would be tremendous (Borde et al. 2015). Tracking vulnerability to CC on a municipal scale in Haiti is particularly helpful, because the aid that the country benefits to adapt to CC risk being repeatedly channelled to the same areas, without reaching the most vulnerable municipalities. In Haiti today, the vulnerability assessment is practiced on a national and departmental scale. But the departments (from 1268 to 5000 km^2) contain greatly diverse territories. The South East department, for example, runs from the coast to mountain chains than exceed an altitude of 2200 metres in just 2000 km^2. On the contrary, the municipal scale is small enough (from 22 to 639 km^2) for tracking which is also helpful to local projects.

The idea of measuring vulnerability at municipal scale isn't new in Haiti. In 2009, the National council for food security-CNSA (Ministry of agriculture), in collaboration with FEWS NET, produced a vulnerability index on a municipal scale (RdH, CNSA-MARNDR 2009). However, the assessment was carried out only then and we have no knowledge of the metadata of the indicators to be able to continue it.

The MPCE and ONPES Basic statistics program, funded by the European Union, aimed mainly at the national poverty reduction strategy, is expected to generate useful information, but at the moment there is no open data available. Different maps are now freely accessible at http://haitidata.org/ and routinely collected helpful information is available.

The aim of this chapter is to check the applicability of the vulnerability index to CC (VICC) on a municipal scale to Haiti, using routinely collected open data. The choice of the indicators that make up the index derives from the definition of vulnerability (V) as a function of three components: exposure, sensitivity and adaptive capacity.

Exposure (E) refers to "The presence of people, livelihoods, species or ecosystems, environmental functions, services, and resources, infrastructure, or economic, social, or cultural assets in places and settings that could be adversely affected." (IPCC 2014: 1765). Examples are flood or drought prone areas and population density.

Sensitivity (S) is "The degree to which a system or species is affected, either adversely or beneficially, by climate variability or change. The effect may be direct (e.g., a change in crop yield in response to a change in the mean, range, or variability of temperature) or indirect (e.g., damages caused by an increase in the frequency of coastal flooding due to sea level rise)." (IPCC 2014: 1772–73). Examples are anomalies in the prices of food, crop deficit, poverty, damage following hydro-climatic disaster or the number of farmers who grow crops to satisfy their own requirements only.

Adaptive capacity (AC) is "The ability of systems, institutions, humans, and other organisms to adjust to potential damage, to take advantage of opportunities, or to respond to consequences." (IPCC 2014: 1758). Examples are official development aid in the environmental sector, the reduction of illiteracy, forest surface dynamics, the existence of adaptation plans, rural infrastructures (irrigation, roads), safe housing, provided with drinking water, sanitation and hygiene facilities (WASH).

However, the irregularity of information, its availability at departmental level only and not freely accessible, means that most of these vulnerability dimensions are now impossible to survey at municipal level.

The VICC requires recognition of open data which has led to the identification of 10 vulnerability indicators on a municipal scale. The following paragraphs present, (ii) the methodology, which explains the meaning of the indicators selected, their characteristics and how the index is calculated, (iii) the results obtained with the vulnerability map at municipal level, and the advancements with respect to previous knowledge, (iv) the conclusions relating to perennial repetition, the expansion of the information needed to measure the index and the free availability of information.

6.2 Metodology

Vulnerability is a function of Exposure (E), Sensitivity (S) and Adaptive capacity (AC) according to the Equation V = (E + S) + (1 − AC) already used by Heltberg et al. (2010), PNUMA (2013), Tuhladhar et al. (2013), USAID (2014), and UNESCO (2016). Every determinant is described by indicators. The process of indicator development is organised according to the steps suggested by Birkmann (2006), simplified here for operational use.

6.2.1 *Identification of the Open Data*

The open data useful to tracking vulnerability to CC in Haiti drops from 15 to 9 as we move from country to municipal scale. At least one third of the information is not collected with annual frequency, a quarter of the information is supplied by multi-bilateral organisms (CHIRPS, CRED, OECD, FAO, UNDP) (Table 6.2), and two third is freely available (Table 6.3)

6.2.2 *Indicator Performance*

The open data at municipal scale on Haiti clearly grasps the vulnerability of infrastructures, the population and agriculture, but not of health and nutrition, education, natural resources and ecosystems highlighted by Adger et al. (2004), nor the urban vulnerability.

As regards the temporal resolution of certain indicators (monthly precipitations, crop failure, adaptation planning) it is preferable to consider a period of time rather than just one year (for example, the five years following the earthquake of 2010). The drought magnitude and wet extremes have been observed over the past 35 years. In other cases, the indicators refer to a specific year (population density, rural accessibility, irrigated agriculture, farmers for self-consumption, cholera incidence).

6.2.3 *Selection Criteria for the Set of Potential Indicators*

The selection of the indicators must be led by clear criteria which allow the preliminary assessment of each individual indicator (Schiavo-Campo 1999; Hilden and Marx 2013): the pertinence of the indicator with CC, its accurate description (metadata), data availability and regular updating, spatial coverage (to what extent the municipalities are covered), the length of time series and temporal resolution, intelligibility (an indicator which is comprehensible and easy to understand without a deep knowledge of how it has been constructed) (Table 6.4).

6.2.4 Identification of the Indicators

The VICC for the municipalities of Haiti consists of the 10 indicators described below.

Indicator 1-Drought Magnitude

The severity of drought between 2011 and 2015 is a factor of exposure to CC. For various reasons, including the 2010 earthquake, Haiti does not have a rain gauge network on a municipality scale, with at least a thirty-year series of data up to the present day. The stations are managed by various organisations (CNIGS, CNSA, ONEV, SNRE, etc.), which do not provide open data.

An alternative solution, now frequent in climatic studies, is to use datasets of satellite rainfall estimation. Many institutions and research organisations make this information available free of charge or in exchange for payment. The Climate Hazards group with InfraRed Precipitation Stations (CHIRPS) v2 by the U.S. Geological Survey and Earth Resources Observation and Science incorporates 0.05° resolution satellite imagery with in situ station data to create gridded rainfall time series (Funk et al. 2015). CHIRPS is a quasi-global daily dataset spanning 50°S–50°N and a daily resolution over the period 1981–2016 is freely distributed. These characteristics enable the description of rainfall dynamics in each municipality of Haiti with 36 years of data available. Free access offers the chance to update the VICC in future.

We have determined the centroid of every municipality (geometric centre of a two-dimensional figure through the mean position of all its points) using GIS software (Fig. 6.1) to extract from the CHIRPS 125 dataset series of rainfall (one for each municipality) from January 1981 to December 2015.

So we have calculated, according to the recommendations of the World Meteorological Organisation, the Standardised Precipitation Index (SPI) (WMO 2012). The SPI is based only on the probability of precipitation for any time scale. The index is standardized so that an index of zero indicates the average precipitation amount (50% of the historical precipitation amount is below and 50% is above average) (Ci Grasp 2.0 2016). Positive SPI values indicate higher than average precipitation and negative values indicate less than average precipitation. Because the SPI is normalised, wetter and drier climates can be portrayed in the same way; thus, wet periods can also be monitored using the SPI.

The SPI is not linked to the absolute value of rainfall but to its distribution. An SPI of −2 can derive from a rainfall value of just a few millimetres in a semi-arid area or of several hundred millimetres in a very wet area. However, the SPI expresses how this value is more or less frequent than the historical series of

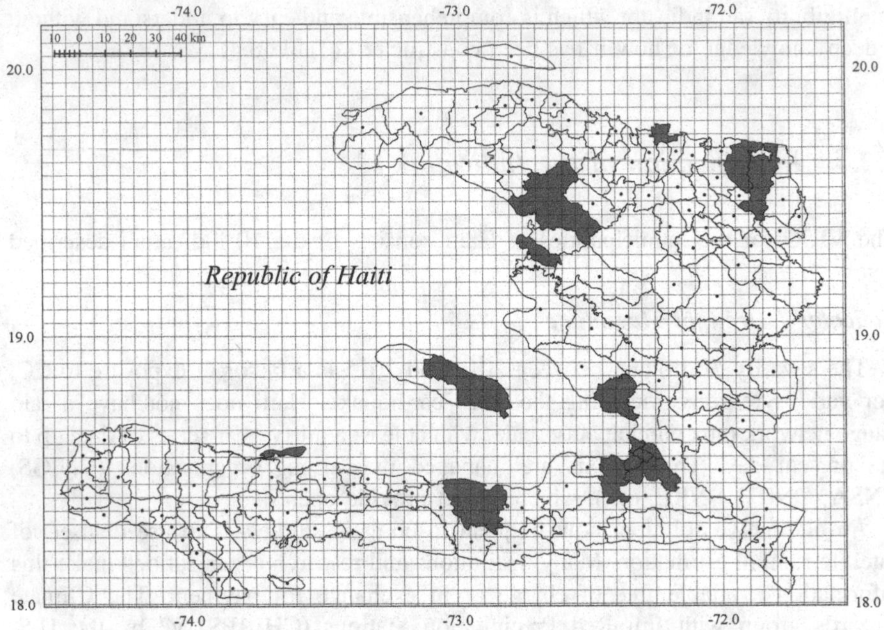

Fig. 6.1 Haiti. The grid of available information on monthly precipitation from CHIRPS showing the centroids (*black* dots) used for climate characterization of each municipality. *Grey* municipalities refers to municipalities with less than 33% of rural population not considered for this chapter

observation. Consequently, the availability of water resources for the population, for vegetation and for farming in the area will probably be jeopardised by phenomena which are very different from the normal distribution of rainfall, because the place has now adapted to a certain pluviometric regime.

A drought event occurs any time the SPI is continuously negative and reaches an intensity of −1.0 or less. The event ends when the SPI becomes positive. The positive sum of the SPI for all the months within a drought event can be referred to as the drought's "magnitude". SPI accumulated values can be used to analyse drought severity.

The SPI can be calculated on different timescales that reflect the impact of drought on the availability of the different water resources. Soil moisture conditions respond to precipitation anomalies on a relatively short scale. Groundwater, streamflow and reservoir storage reflect longer-term precipitation anomalies. In this case, a quarterly SPI was chosen, to intercept relatively prolonged drought phenomena which can jeopardise Haiti's water system and its agricultural crops. The accumulation of rainfall over three months, required to calculate the quarterly SPI, implicates the use of the previous two months to calculate the accumulation. January and February 1981 are missing from the CHIRPS historical series.

Consequently, 1981 has 10 monthly SPI values. Almost all of 1995 is missing, so the SPI for that year was not calculated.

The report for the average value of the drought magnitude 2011–2015 (period of analysis) and its value for 1981–2010 (reference period), was calculated for every municipality. Subsequently, these values have been normalised on a scale of 0–1 on the basis of the minimum and maximum value of the report recorded among all the municipalities (Petite Rivière des Nippes 1.04 = 0, Belle Anse 3.42 = 1). After calculating the drought magnitude, the map was produced (Fig. 6.2) using 4 classes: 0–0.24, 0.25–0.49, 0.50–0.74, 0.75–1.

Indicator 2-Wet Extremes

Extreme precipitations between 2011 and 2015 are an exposure factor of a municipality to CC, in that they can generate flooding. The characterisation of every municipality with respect to extreme precipitations uses, like for drought, the CHIRPS dataset correspondence with the centroid of every municipality. In this case, the options available were to characterise the wet conditions using classic indexes of intense rainfall or using the 1-month SPI:

- R20 mm, number of days' rainfall in excess of 20 mm/year.
- R95p, number of days' rainfall higher than the 95th percentile/year (only during rainy events).

Very wet 1-month SPI (number of months/year with SPI > 1.5) and extremely wet (number of months/year SPI > 2). In this case too, the ratio of the 2011–2015 five-year average and the average for 1981–2010 was calculated for every municipality. Subsequently, these values are normalised on a scale of 0–1, on the basis of the minimum and maximum. Every index presents positive and negative characteristics. R20 mm does not highlight significant changes in the various municipalities so 20 mm is not enough to intercept the dynamics of extreme rains in Haiti.

R95p is more appropriate, in that it is based on the calculation of the distribution of rains in each individual municipality. However, it is a more sophisticated and harder to understand index. Moreover, its distribution in the municipalities of Haiti does not present very significant changes. 1-month SPI extremely wet (SPI > 2) highlight more changes in the municipalities than the 1-month SPI very wet (SPI > 1.5). The 1-month SPI > 2 is relatively easy to explain and understand. This is why the decision was made to use it to characterise the wet extremes (Fig. 6.2) according to 4 classes: 0–0.24, 0.25–0.49, 0.50–0.74, 0.75–1.

Indicator 3-Flood Prone Areas

Haiti has few vast areas exposed to river or marine flooding (North–East, Artibonite, North–East of Port-au-Prince, Cayes). The rest of the territory presents lots of flood prone areas which are small in size. We referred to the freely available mapping from the Centre National de l'Information Géo-Spatiale-CNGIS in 2010, which shows the extent of Haiti flood prone areas, according to field observations,

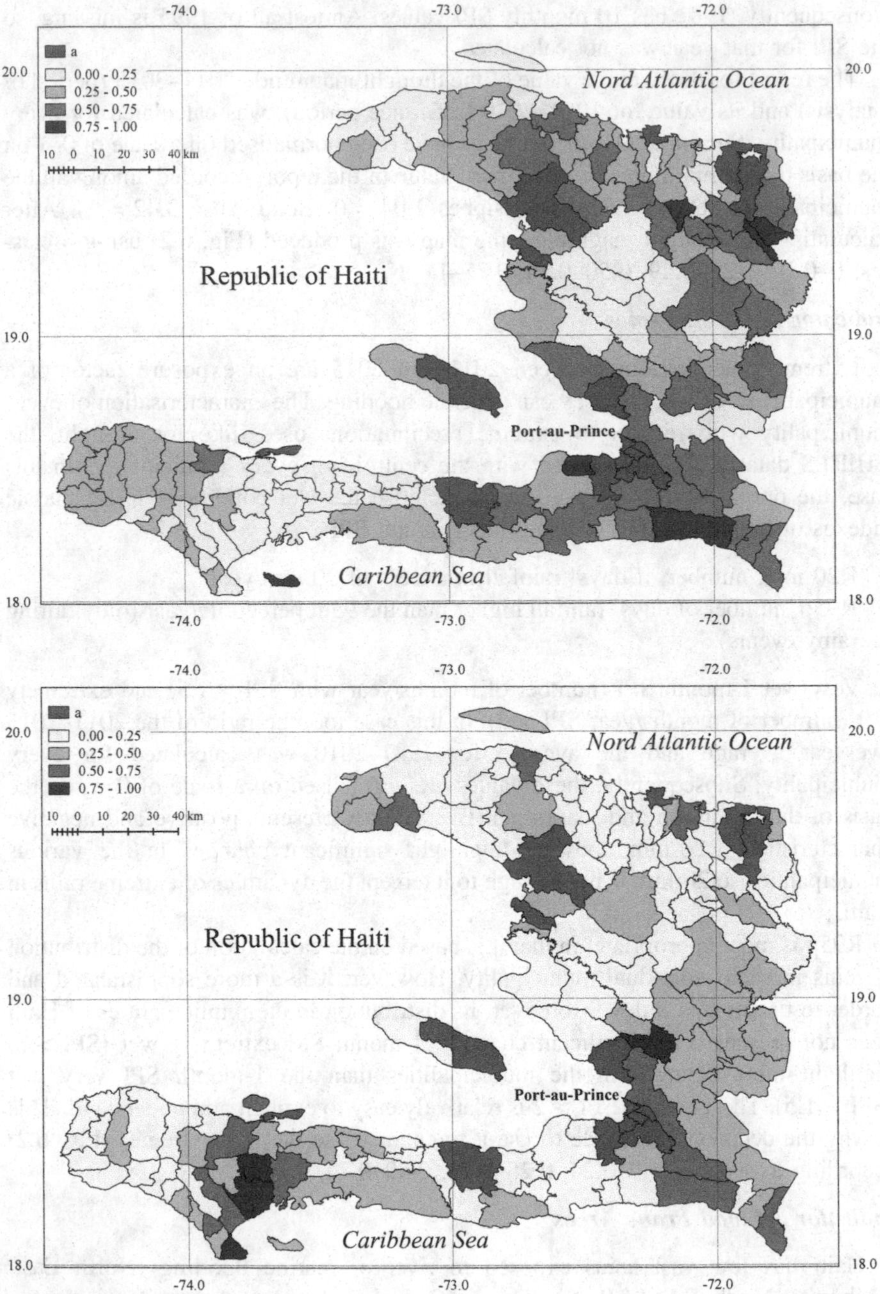

Fig. 6.2 Haiti, municipalities, 2011–2015. Drought magnitude (*top*) and wet extremes (*bottom*): a. urban municipalities

historic and recent records as prepared by NATHAT 1 (May 2010), using Google Maps, and deterministically plotting the largest extent of flooding, according to a digital elevation model of 30 m pixels. The map, overlapping the municipal boundaries, shows that 72% of municipalities have flood prone areas. The flood prone areas have been related to the surface of every municipality and divided into four classes (0–25, 25–50, 50–75, 75–100%) for which the mean value has been chosen (12.5, 37.5, 62.5, 87.5) and then normalised on a scale from 0 to 1 (0.25, 0.50, 0.75, 1.00).

Indicator 4-Population Density

The rise in population density is an exposure factor: with the same territorial surface, a flood causes more damage to a denser population than to a widespread one. It is an indicator proposed also to monitor the Sustainable United Nations' Development Goals-SDGs (UNESC 2015). The density considered is that calculated by the IHSI-Institut Haïtien de Statistique et d'Informatique (RdH, MEF 2012). The value is normalised on a scale of 0–1 depending on whether we start from the less densely populated municipality (Point à Raquette, 74 people per sq km = 0) or the most densely populated one (Saint Louis du Nord, 884 people per sq km = 1).

Indicator 5-Crop Deficit

Crop deficit is a sign of sensitivity to CC. It is recognised as being among the indicators of the 2nd SDG-End hunger, achieve food security and improved nutrition and promote sustainable agriculture (UNESC 2015). The Ministry of agriculture, natural resources and rural development (MARNDR, according to the French acronym) estimates crop production per department using experts, considering the useful farming surface area, sun exposure and other factors with a qualitative-quantitative methodology which is not clearly described. Total production expressed in equivalent cereals and that of the main crops (maize, rice, bananas, beans, potatoes) is the only information supplied in reports on spring crops per municipality (RdH, MARNDR 2014, 2013, 2012a, b and c; FAO-PAM 2010). For the purposes of constructing the VICC, we have used the total production per municipality in equivalent cereals referred to that of 2009 which, according to the MARNDR, was the best year in the last fifteen. For each municipality we assessed the deficit over five years: 2010, 2011, 2012, 2013 and 2014. Production is classified into 5 categories (< 50, 50–70, 70–90, 90–110, 110–130, > 140%) referred to production in 2009. We related these classes to the average value of each of them (40, 60, 80, 100, 120 and 140). Then we calculated the average production of every municipality during the five years. This value varies from 46 to 122%. These values have been normalized on a scale of 0–1 and then inverted, assigning the value of 1 to the lowest (46) and 0 to the highest (122). In this way, the values close to 1 show an average spring crop deficit of around 46% compared to that of 2009. The basic information supplied by the spring campaign reports does not always follow the same standard. Production in 2012 for example was referred to that of the previous year. Therefore, it was necessary to relate it to 2009. Production for 2010 and 2014 was referred to the registration sections (sub-category of the municipality). In this case, when the production values of the sections of a municipality were different, the mean value was calculated.

Indicator 6-Farmers for Self-consumption

Dependence exclusively on farming for subsistence is a sensitivity factor. This information is used by numerous scholars (Adger et al. 2004). The information is supplied by the MARNDR and is part of the Agricultural Atlas of Haiti (RdH, MARNDR 2012a, b and c). The information is provided for five classes which we took to mean values of 80, 60, 40, 22.5 and 7.5 and then normalised on a scale from 0 (7.5) to 1 (80).

Indicator 7-Local Development and Contingency Plans

Among the adaptive capacities the first thing to consider is the capacity of the municipalities to plan the measures for adaptation to CC over a period time. This capacity is among the indicators of the SDG 11-Make cities and human settlements inclusive, safe, resilient and sustainable (indicator 11b1). Haiti's municipalities have basically three types of tools: Local development plans, Contingency plans and, occasionally, Comprehensive plans. The implementation of these plans requires the mobilisation of different players for financing. Freely available plans is important to encourage implementation. Recognition on internet has enabled the identification of 19 plans. We have assigned a value of 0 to municipalities with a plan and a value of 1 to those without a plan.

Indicator 8-Irrigation Systems

The use of irrigation systems in agriculture is an adaptive capacity which is important in a country like Haiti which, in recent years, has been particularly exposed to drought. The monitoring of the SDG 2-End hunger, achieve food security and improved nutrition and promote sustainable agriculture considers it (UNESC 2015). The Agricultural Atlas of Haiti supplies the amount of cultivable land that can be irrigated with respect to the cultivable area. The information is supplied according to five classes (<5, 5–15, 15–30, 30–60, >60%) which we have related to the mean values of every class (2.5, 12.5, 22.5, 45, 75) and normalised on a scale of 0 (75)–1 (2.5).

Indicator 9-Access to All-season Road

Municipalities that are served by all-season roads which allow constant access to services that are not available in the place of residence is an adaptive capacity factor. The Rural Access Index (RAI) proposed by the World Bank (Roberts et al. 2006) and calculated by the MTPTC in May 2015 for every municipal section of Haiti (RdH, MTPTS 2015) is among the indicators proposed by the United Nations to monitor SDG 9-Build resilient infrastructure. The RAI expresses the percentage of population which, in every municipal subdivision (about 600 for the whole of Haiti) lives within 2 km from an all-season road. The information is supplied according to seven classes from 0–25% to 85–100%. We calculated the average value for each class and, when a municipality is made up of sections with a different RAI, the average between the values of every section. Then the values were normalised on a scale of 0 to 1 and rearranged to attribute the maximum value (1) to the municipality with least accessibility (0.13) and the minimum (zero) to the municipality with most accessibility (0.93).

Indicator 10-Cholera Incidence

Access to drinking water, sanitation and medical care expresses an adaptive capacity and forms the 6th SDG-Ensure availability and sustainable management of water and sanitation for all (UNESC 2015). Since there is a lack of open data on access to drinking water and to sanitation at municipal level, we used the cholera incidence as proxy indicator as its persistence "almost 6 years after its appearance, is largely due to the lack of access to safe drinking water and sanitation" (OCHA 2016: 3). This happens, for example, when sewage is dumped straight into the river and the river is used for drinking, irrigation, bathing and washing clothes (Frerichs 2016). Haiti, after the cholera peaks of 1.8 and 3.4‰ in 2010 and 2011 respectively, has still not succeeded in weakening the epidemic, which worsened in 2015 (0.4‰). The Ministry of Human health regularly publishes statistics by municipality. We used those for April–June 2015 (RdH, MSPP 2015), which presents the lowest incidence during the year and considered only municipalities with a cholera incidence higher than 0.1‰. The figures were normalised on a scale of 0–1. Nineteen municipalities do not supply data because they do not have a local centre for the diagnosis and treatment of cholera. We did not think it is fair to proceed in these cases with the imputation of the missing data as the residents from these municipalities simply go to the nearest cholera diagnosis and treatment centres.

6.2.5 Robustness and Applicability of the Index

To ensure that all the indicators are comparable, we normalised the values of each indicator on a scale of 0 (minimum vulnerability) to 1 (maximum vulnerability) (Table 6.5). The indicators 5, 7 and 9 required the overturning of the value. We assigned an equal weight to each indicator, instead of using expert opinions (hard to trace for 125 municipalities) or statistical models like the PCA-Principal component analysis (which require capacities which the users of the index rarely have, whether they are donors or local authorities). The vulnerability index to CC on a municipal scale adds together the values of the single indicators of exposure (E), sensitivity (S) and adaptive capacity (AC) according to the equation $\mathrm{VICC} = (E + S) + (1 - \mathrm{AC}) = (i1 + i2 + i3 + i4 + i5 + i6) + (1 - (i7 + i8 + i9 + i10))$ where $i1$ is drought magnitude, $i2$ is wet extremes, $i3$ is flood prone areas, $i4$ is population density, $i5$ is crop deficit, $i6$ is farmers for self-consumption, $i7$ is local

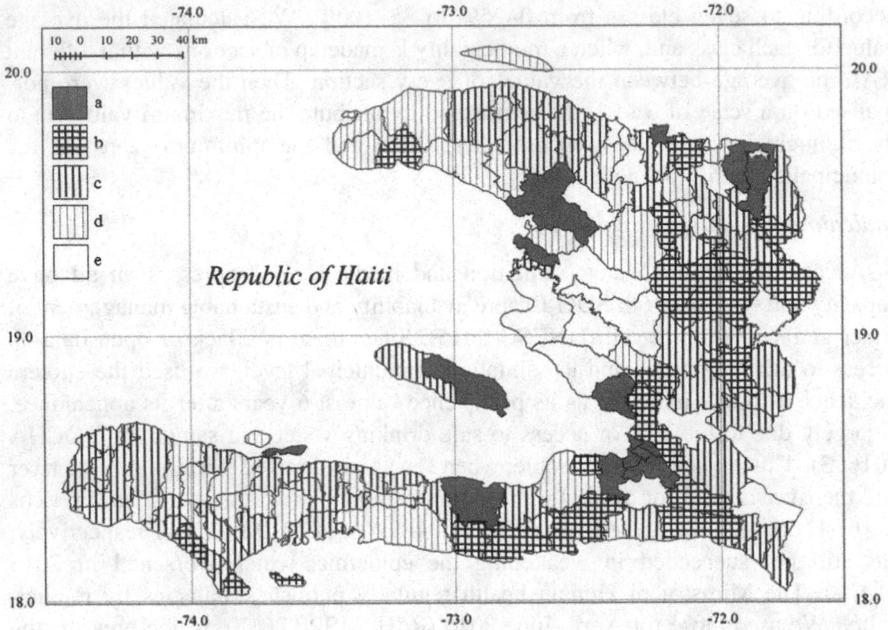

Fig. 6.3 Haiti, vulnerability to CC of the rural municipalities: **a.** urban, **b.** high, **c.** medium, **d.** low, **e.** very low

adaptation or contingency plans, *i8* is rural access, *i9* is irrigation systems, and *i10* is cholera incidence.

Vulnerability can take on a value of 0–10. The real values go from 2.51 (Saint-Marc) to 6.32 (Baie de Henne) (Fig. 6.3; Table 6.8). Considering this distribution, we divided the ranges (3.81 point) into four equal parts corresponding to high, medium, low, and very low vulnerability (Table 6.6).

6.3 Results and Discussion

Vulnerability to CC is calculated for 125 rural municipalities in which 65% of Haiti's population lived in 2012. The remaining 15 municipalities (35% of the population) are exclusively or mainly urban and have not been considered as they require different information to that used for the rural municipalities and, at the moment, this is not freely available. This said, while Port-au-Prince is the second largest Caribbean city, Haiti continues to be 51% rural.

Of the 125 municipalities considered, 24 (1.3 million inhabitants) are highly vulnerable to CC, 58 (2.9 million inhabitants) are medium vulnerable, 31 are low vulnerable (1.5 million inhabitants) and 12 are very low vulnerable to CC (1.2 millions of inhabitants).

Table 6.1 The Haiti open data VICC compared with other conventional surveys to assess vulnerability to CC

Type of information	Type of survey		
	Open data VICC 2nd hand information	Traditional 2nd hand information	Household 1st hand information
Municipalities#	Hundreds	Thousands	Few to dozens
Information used	Census, daily/weekly or annual surveys	Census	Households survey, focus groups
Indicators number	10–13	38	17–31
Indicator weight	No	Yes	Experts, PCA
Processing time	Weeks	Months	Months
Staff	One–two	Unknown	Dozens
Money use	Freed for development activities	Mapping	Data collection and processing
Replicability	Yes	Yes	Rare/No
Refresh rate	Annual-biennial	Decennial	No
Example	This chapter	PNUMA (2013)	Ahsan and Warner (2014)

The municipalities highly vulnerable to CC are concentrated in the Centre department, and around Port-au-Prince. Most of these municipalities stretch across the mountain territories of the central Cordillière, the Montaignes Noires and, above all, the Chaine de la Selle, behind the capital city (Fig. 6.3).

The concentration of highly vulnerable municipalities in the hinterland of the capital city and a few kilometres from the main road linking Port-au-Prince to Cap Haïtien should theoretically simplify the reduction of vulnerability, as the materials needed for the adaptation works do not require movement over long distances, which would be necessary to reach the southwest and northwest areas of the country.

The calculation of the VICC allows further considerations on its single components, starting with exposure. From 2013 to 2015, Haiti was struck by the worst drought in the past 35 years, which peaked in 2014. The drought increased the number of dry months in a year which, in Belle Anse, went from three (1981–2010) to nine (2011–2015), in Bahon from three to eight, in Carice, Gressier and Mont Organisé from three to seven and in Ile à Vache from three to six (Fig. 6.3). This drought affected all the municipalities. In the previous twenty years, there had only been three isolated years of drought (1997, 2000, 2009).

The drought caught the Haitians off-guard, unprepared for an event of such magnitude. The 2013–2015 drought was preceded by two extremely wet years. This however was not an effect of CC, because Haiti frequently has two extremely wet years in a row, due to its position on the hurricane route.

As regards others factors of exposure, the flood prone area exceeds 75% of the municipal surface area in Caracol, Cayes, Ferrier and Grande Saline, and between 50 and 75% of the municipal surface area in Estère, Milot and Saint Raphaël. The

Table 6.2 Haiti 2015. Open data relevant for vulnerability tracking according administrative level

Vulnerability determinants	Open data	Administrative level		
		Country	Department	Municipality
Exposure	Monthly precipitations	●		●
	Flood prone areas/unit area	●	●	●
	Population density	●	●	●
Sensitivity	Anomalies in food prices		●	
	Crop deficit			●
	Farmers cropping for self-consumption			●
	Poverty	●	●	
	Victims of hydro climatic disaster	●		
Adaptive capacity	ODA in environmental sector	●		
	HDI-Human development index	●		
	Maternal mortality	●	●	
	HIV Incidence	●	●	
	Literacy rate	●		
	School enrolment according gender	●	●	
	Forest area	●		
	Local plans for climate adaptation	●	●	●
	Rural accessibility			●
	Irrigated agriculture	●	●	●
	Water and sanitation	●	●	
	Cholera incidence	●	●	●

density of the population peaks in Saint Louis du Nord and is high in Limonade, Quartier Morin, Ferrier (Nord), Grande Saline (Artibonite), Croix des Bouquets (West), Belle Anse (South East) and Ile à Vache (South).

Rural sensitivity consists in two aspects. First, 65% of the municipalities had crop deficits between 2010 and 2014, with very severe deficits in six municipalities (crop yield of less than 46% compared to 2009): Cerca Carvajal, Gressier, La Victoire, La Vallée, Maïssade and Saint Raphaël. Second, 16 municipalities have over 70% of farmers cultivating for self-consumption only.

In relation to adaptive capacity, just nineteen municipalities have a local development or contingency plan. Thirty-eight municipalities have no irrigation systems and only two have irrigated farming on over 60% of the useful farmland (La Victoire and Milot).

Eighty-four municipalities (67%) have very little accessibility to an all-season road. Five municipalities have a very high cholera incidence (> 10/10,000): Baie de Henne, Anse d'Hainault, Croix de Bouquets, Hinche and Quartier Morin. These

Table 6.3 Haiti 2016. Freely available data to build indicators for vulnerability to CC tracking

Vulnerability components	Information	Available at
Exposure	Monthly precipitation	http://chg.geog.ucsb.edu/data/chirps/
	Flood prone areas	http://haitidata.org/layers/cnigs.spatialdata:hti
	Population density	http://www.ihsi.ht/produit_demo_soc.htm
Sensitivity	Crop deficit	http://agriculture.gouv.ht/statistiques_agricoles/Atlas/thematiques_speciphiques.html
	Farmer 4 self-consumption	http://agriculture.gouv.ht/statistiques_agricoles/Atlas/thematiques_speciphiques.html
Adaptive capacity	Plans for CCA	
	Rural accessibility	http://www.mtptc.gouv.ht/media/upload/doc/publications/ROUTE_INDICE-RAI-2015.pdf
	Irrigation systems	http://agriculture.gouv.ht/statistiques_agricoles/Atlas/thematiques_speciphiques.html
	Cholera incidence	http://mspp.gouv.ht/site/downloads/Bulletin%20Trimestriel%20MSPP%20Juillet%202015%20version%20web%20compressed.pdf

four indicators highlight the municipalities with less adaptive capacity (Baie de Henne, Cerca la Source, Mirabelais and Thomonde, Ila à Vache, Anse d'Hainault) and those with more adaptive capacity (Bas Limbé, Saint-Marc, Petite Rivière de l'Artibonite, Kenscoff and Léogane).

The comparison between exposure and adaptive capacity (Fig. 6.4) highlights that there is no relationship between the two determinants. Therefore, the adaptive capacity does not respond to exposure to CC. This awareness alone justifies the need for a VICC for Haiti at municipal level if we need to help decision making with regard to the adaptation of the most exposed municipalities. VICC on a municipal level presents rather different results when compared with the previous maps of the single determinants of vulnerability.

The map of drought magnitude 2011–2015, for example, is completely different from that of drought susceptivity 2010 (RdH 2010: 18). The difference is due to the greater precision of our study, the different methodology, and the fact that the period that we considered was influenced by very severe drought, something which had not occurred in the previous 20 years. Another example is crop deficit.

Table 6.4 Evaluation criteria for vulnerability indicators at municipal level

Indicator	Evaluation criteria					
	Pertinence	Metadata	Availability & updating	Spatial coverage	Length	Intelligibility
Drought magnitude	5	5	5	5	5	4
Wet extremes	5	5	5	5	5	4
Flood prone areas	5	1	2	5	1	5
Population density	5	5	5	5	3	5
Agricultural production	5	5	5	5	4	5
Farmers for self-consumption	5	1	5	5	1	5
Plans for CCA	5	5	5	5	2	5
Rural accessibility	5	3	3	5	1	5
Irrigated agriculture	5	1	5	5	2	5
Cholera incidence	5	1	5	4	2	5

Metadata: 1 weakly documented, 3 methodology documented but not freely available. Availability & updating: 2 routine collection does not exist, indicative information on accuracy, 3 data supposed routinely collected and freely available, 5 data routinely collected and freely available. Spatial coverage: 4 data unavailable on 15% of municipalities. Length: 1 one year available only, 2 two years available, 3 three years available, 4 four years available, 5 over 30 years available without interruptions. Intelligibility: 4 easily understood without expert knowledge

The municipalities that presented the highest food scarcity in 2009 (RdH, MARNDR 2009: 18) are not the same ones that presented the worst crop deficit between 2010 and 2014. The difference depends on the methodology but also on the fact that 2009 was one of the best agricultural years in the last fifteen.

The VICC is useful to all those projects which have such a dimension and duration as to justify a device for the assessment of their actions.

VICC can be also used to direct action. The municipalities with the highest score for crop deficit, local plans, irrigated surface area, RAI and cholera incidence are those that require most support to agriculture, to local development and contingency plans, road accessibility, irrigation systems and WASH.

The VICC can be improved by adding information on sensitivity (food prices, victims of disasters) and adaptive capacity (forested area, school enrolment, HIV incidence) which is currently available at national or departmental level only.

Table 6.5 Indicators value according information organization

Vulnerability indicator	Available information	Indicator range
1 Drought magnitude		1–0
2 Wet extremes		1–0
3 Flood prone areas	<25, 25–50, 50–75, 75–100	1–0
4 Population density	884–74 inhabitants/km^2	1–0
5 Crop deficit	70–90, 50–70, < 50%	0.8, 0.4, 0.2
6 Farmers 4 self-consumption	>70, 50–70, 30–50, 15–30, <15%	1–0
7 Local CCA plans	Y/N	1–0
8 Rural access	0–25, 26–35, 36–45, 46–55, 56–65, 66–75, 75–100	0.87, 0.7, 0.6, 0.5, 0.4, 0.3, 0.12
9 Irrigation systems	<5, 5–15, 15–30, 30–60, >60% SAU	0.3, 0.13, 0.28, 0.45, 0.8
10 Cholera incidence	0,1–2, 2–5, 5–10, >10/10,000 monthly	0, 0.15, 0.35, 0.8, 1

Table 6.6 Robustness of the vulnerability index to CC for Haitian municipalities

Robustness	Choices
Indicators number	10
Indicator exclusion/inclusion	According the evaluation criteria (Table 6.4)
Normalization scheme	Each indicator is normalized in a 0–1 scale
Indicators' weights	Equal weights
Missing data	Indicators 1 and 3: one year (1995) missing Indicator 10: nineteen municipalities missing
Imputation of missing data	No
Aggregation method	Additive

The VICC can be improved also by refining the single indicators. For example, drought magnitude can be calculated as the average of several cells for the bigger municipalities (43%). The RAI can be drastically improved by using the IRA (Tiepolo 2009): an indicator which considers the demographic weight of the towns and their distance from an all-season road, and can be set up after desk work only, without the costs and errors caused by the sample surveys that the indicators used by the World Bank requires (Table 6.7).

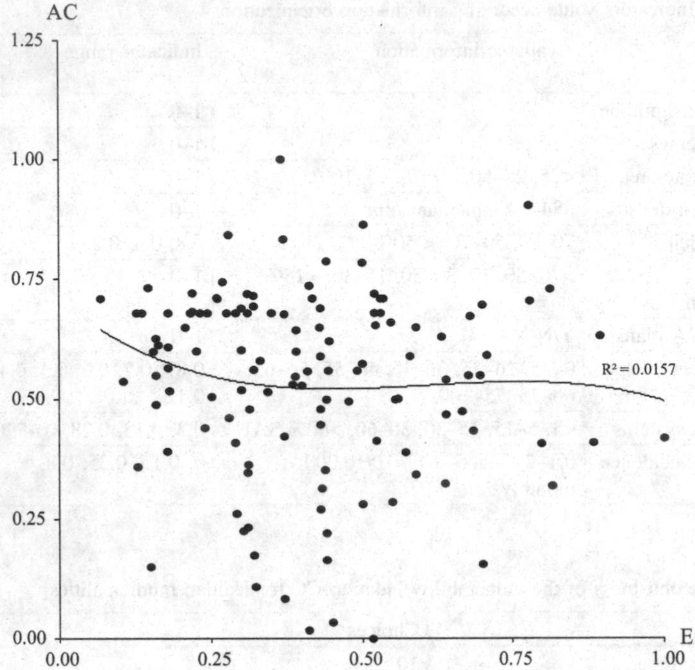

Fig. 6.4 Haiti, municipalities. Exposure (4 indicators) and adaptive capacity (4 indicators)

Table 6.7 Improvement of the indicators of vulnerability to CC

Vulnerability determinants	Country	Department	Municipality	
	Existing	Existing	Existing	Improvement
Exposure			Drought magnitude	Media CHIRPS cells
			Wet extremes	Media CHIRPS cells
			Flood prone areas	FPA/municipal surface
			Population density	P density in FPA
Sensitivity			Crop deficit	
			Farmers for self-consumption	
	Food prices →	→	→	•
	Victims of disaster →	→	→	•
Adaptive capacity			Local CCA plans	
			Rural accessibility →	IRA
			Irrigated agriculture	
	Forest area →	→	→	•
		School enrolment according gender →	→	•
		HIV Incidence →	→	•
			Cholera incidence	

*Estimation

6.4 Conclusions

The VICC at municipal scale for Haiti aggregates 10 indicators consistent with those proposed by the United Nations to monitor the SDGs (UNESC 2015): this is a reduced number compared to other vulnerability indices to CC on a local scale. Kept with just a few indicators, the VICC can easily be updated and allows central and local governments and the Official development aid to orient the measures for adaptation to CC to the most vulnerable sectors, to appreciate the improvements obtained years after year and to assess the impact and sustainability of local projects.

The data needed to measure the 10 indicators to prove the applicability of the VICC were downloaded from the websites of the ministries and those of other authorities, only with regard to information on precipitations that is not freely available in Haiti. This activity would be simplified if just one organisation were to be responsible for open access to information on climate change vulnerability on a municipal scale, such as CNIGS, CNSA, or ONEV-Observatoire National de l'Environnement et de la Vulnerabilité du Ministère de l'Environnement. This organization should update datasets on the vulnerability to CC of Haitian municipalities (useful farming surface area, farmers for self-consumption/Min. Agriculture, monthly cholera incidence/Ministry for Health), supplied always in the same format (maintaining 2009 as the year of reference for agricultural production) and completed (RAI data and not just the map/Ministry of Public works, transportation and communication).

In the future, the organisation in question could provide information on five additional factors of vulnerability to CC available: average monthly prices of food (expanding the number of markets currently monitored by the CNSA), forested (or reforested) area, at five-years intervals at least (CNIGS), victims and damages caused by hydro-meteorological disaster (Ministry of Interior and territorial communities), school enrolment according to gender (Ministry of National education), HIV incidence (Ministry of Health). Further expansions of the number of indicators could prevent the tracking of vulnerability and spatial coverage.

Annexe

See Fig. 6.5 and Table 6.8

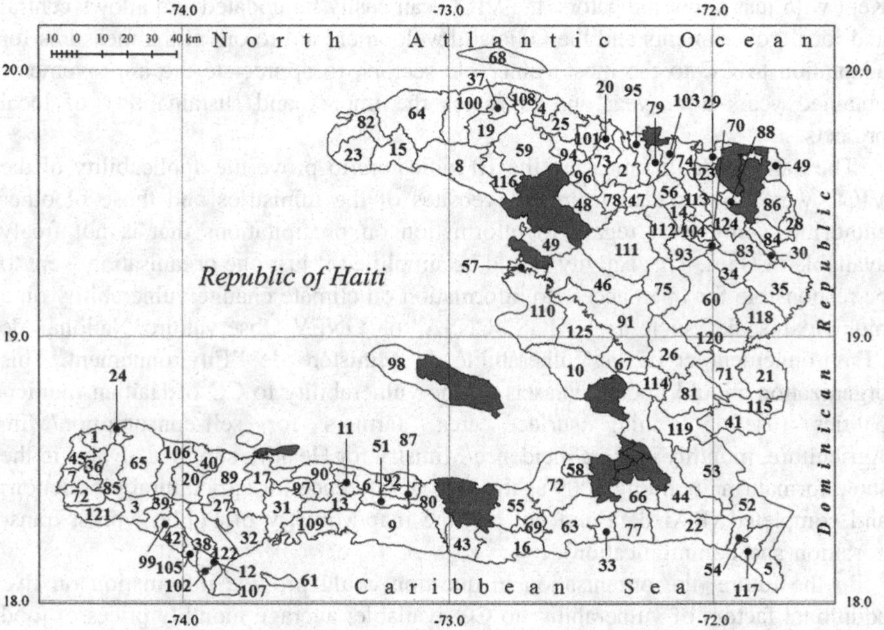

Fig. 6.5 Municipalities of Haiti: urban (*grey*) and rural (*numbers*)

Table 6.8 Haiti 2015. VICC-Vulnerability Index to Climate Change

#	Municipality	Indicator										VICC
		1	2	3	4	5	6	7	8	9	10	
1	Abricots	0.37	0.15	0.25	0.34	0.29	0.72	1.00	0.50	1.00	0.03	4.66
2	Acul du Nord	0.30	0.01	0.00	0.29	0.53	0.45	0.00	0.46	0.93		2.97
3	Anglais	0.12	0.11	0.00	0.21	0.50	0.21	1.00	1.00	0.93	0.15	4.23
4	Anse-à-Foleur	0.04	0.22	0.00	0.48	0.34	0.72	1.00	1.00	1.00	0.12	4.92
5	Anse-à-Pitre	0.60	0.73	0.00	0.10	0.45	0.21	1.00	1.00	0.93	0.00	5.02
6	Anse-à-Veau	0.04	0.34	0.25	0.30	0.29	1.00	1.00	0.38	1.00	0.00	4.61
7	Anse d'Hainault	0.32	0.11	0.25	0.36	0.39	0.45	1.00	0.62	1.00	0.83	5.35
8	Anse Rouge	0.40	0.24	0.25	0.03	0.59	0.00	1.00	1.00	0.72	0.03	4.26
9	Aquin	0.28	0.32	0.50	0.10	0.55	0.45	0.00	0.50	1.00	0.10	3.80
10	Arcahaie	0.51	0.14	0.25	0.28	0.39	0.30	1.00	0.47	0.41		3.46
11	Arnaud	0.11	0.30	0.00	0.22	0.45	0.45	1.00	0.76	1.00	0.03	4.33
12	Arniquet	0.17	0.54	0.00	0.49	0.62	0.21	1.00	0.00	1.00		4.02
13	Asile	0.24	0.17	0.00	0.22	0.25	1.00	1.00	0.53	1.00	0.07	4.47
14	Bahon	0.89	0.01	0.00	0.27	0.40	0.21	1.00	0.38	1.00	0.00	4.16
15	Baie de Henne	0.19	0.54	0.25	0.07	0.63	0.72	1.00	1.00	0.93	1.00	6.32
16	Bainet	0.43	0.27	0.25	0.25	0.71	1.00	1.00	1.00	1.00	0.32	6.22
17	Baradères	0.06	0.28	0.25	0.16	0.38	0.72	1.00	1.00	1.00	0.00	4.86
18	Bas Limbé	0.46	0.03	0.50	0.37	0.48	0.45	0.00	0.00	0.93		3.22
19	Bassin Bleu	0.38	0.20	0.25	0.26	0.36	0.72	1.00	0.88	1.00	0.00	5.05
20	Beaumont	0.27	0.26	0.00	0.15	0.51	0.45	1.00	0.50	1.00	0.00	4.13
21	Belladère	0.55	0.37	0.00	0.25	0.35	0.21	1.00	0.53	1.00	0.18	4.45
22	Belle Anse	1.00	0.51	0.25	0.14	0.44	0.72	1.00	1.00	1.00	0.07	6.13
23	Bombardopolis	0.44	0.37	0.00	0.12	0.63	0.72	0.00	1.00	1.00	0.03	4.32

(continued)

Table 6.8 (continued)

#	Municipality	Indicator										VICC
		1	2	3	4	5	6	7	8	9	10	
24	Bonbon	0.37	0.11	0.25	0.23	0.69	1.00	1.00	0.63	1.00	0.00	5.28
25	Borgne	0.59	0.01	0.25	0.30	0.01	0.72	1,00	0.82	1.00	0.27	4.96
26	Boucan Carré	0.36	0.50	0.25	0.10	0.45	0.45	1.00	0.71	1.00	0.10	4.92
27	Camp-Perrin	0.30	0.41	0.50	0.31	0.42	0.45	1.00	0.25	0.72	0.00	4.35
28	Capotille	0.62	0.15	0.00	0.24	0.49	0.00	1.00	0.88	1.00	0.12	4.50
29	Caracol	0.13	0.22	1.00	0.03	0.65	0.00	1.00	0.18	1.00	0.00	4.21
30	Carice	0.89	0.02	0.25	0.20	0.21	1.00	1.00	1.00	1.00	0.12	5.69
31	Cavaillon	0.13	0.58	0.25	0.17	0.49	0.00	1.00	0.59	0.93	0.10	4.25
32	Cayes	0.12	1.00	1.00	0.03	0.42	0.21	1.00	0.21	0.72	0.23	4.94
33	Cayes -Jacmel	0.50	0.18	0.00	0.51	0.34	1.00	0.00	0.50	0.93	0.00	3.96
34	Cerca Carvajal	0.46	0.20	0.00	0.08	1.00	0.72	1.00	1.00	1.00	0.00	5.47
35	Cerca la Source	0.35	0.17	0.25	0.10	0.60	0.45	1.00	1.00	1.00	0.48	5.40
36	Chambellan	0.49	0.13	0.25	0.33	0.45	0.45	1.00	0.50	1.00	0.12	4.72
37	Chansolme	0.25	0.22	0.00	0.58	0.43	0.45	1.00	0.25	0.93	0.03	4.13
38	Chantal	0.23	0.24	0.00	0.16	0.54	0.21	1.00	1.00	0.41		3.80
39	Chardonnières	0.38	0.14	0.25	0.16	0.58	0.45	1.00	1.00	1.00	0.00	4.96
40	Corail	0.67	0.21	0.00	0.12	0.52	0.72	1.00	0.46	1.00	0.03	4.73
41	Cornillon	0.23	0.34	0.00	0.22	0.62	0.00	1.00	1.00	1.00		4.42
42	Côteaux	0.36	0.32	0.25	0.25	0.42	1.00	1.00	0.80	1.00	0.10	5.49
43	Côtes de Fer	0.26	0.18	0.25	0.27	0.71	1.00	1.00	1.00	1.00	0.10	5.76
44	Croix-des-Bouquets	0.59	0.51	0.50	0.37	0.50	0.21	1.00	0.50	0.72	0.92	5.82
45	Dame Marie	0.33	0.15	0.25	0.36	0.13	0.72	1.00	0.58	1.00	0.17	4.69
46	Dessalines	0.25	0.06	0.50	0.36	0.49	0.00	1.00	0.57	0.00	0.18	3.41

(continued)

Table 6.8 (continued)

#	Municipality	Indicator										VICC
		1	2	3	4	5	6	7	8	9	10	
47	Dondon	0.52	0.04	0.25	0.25	0.44	0.45	1.00	0.50	1.00	0.48	4.93
48	Ennery	0.53	0.20	0.00	0.19	0.47	0.00	0.00	0.62	0.72	0.27	3.00
49	Estère	0.31	0.34	0.75	0.21	0.39	0.00	1.00	0.50	0.41	0.00	3.92
50	Ferrier	0.43	0.40	1.00	0.16	0.74	0.00	1.00	0.50	0.41	0.00	4.64
51	Fonds des Nègres	0.10	0.18	0.00	0.34	0.28	0.72	1.00	0.59	1,00	0.17	4.38
52	Fonds-Verrettes	0.41	0.37	0.00	0.12	0.35	0.00	0.00	1.00	0.72		2.97
53	Ganthier	0.64	0.43	0.25	0.06	0.64	0.21	1.00	0.75	0.41	0.75	5.14
54	Grand Gosier	0.75	0.24	0.25	0.15	0.31	0.21	1.00	1.00	1.00	0.00	4.91
55	Grande-Goâve	0.57	0.05	0.25	0.57	0.74	0.72	1.00	0.50	0.93		5.34
56	Grande Rivière du N.	0.35	0.01	0.00	0.29	0.31	0.45	1.00	0.53	1.00	0.12	4.06
57	Grande Saline	0.34	0.34	1.00	0.44	0.46	0.00	1.00	1.00	0.00	0.18	4.77
58	Gressier	0.82	0.24	0.25	0.37	0.95	0.21	1.00	0.62	1.00	0.35	5.81
59	Gros Morne	0.61	0.27	0.25	0.37	0.54	0.00	1.00	0.62	1.00	0.10	4.75
60	Hinche	0.54	0.40	0.25	0.15	0.53	0.72	1.00	0.63	1.00	0.92	6.14
61	Ile à Vache	0.80	0.30	0.50	0.30	0.45	0.21	1.00	1.00	1.00	0.67	6.23
62	Irois	0.31	0.14	0.00	0.12	0.39	0.72	1.00	1.00	1.00	0.00	4.68
63	Jacmel	0.75	0.20	0.25	0.41	0.69	0.72	1.00	0.47	1.00	0.10	5.60
64	Jean Rabel	0.17	0.24	0.25	0.27	0.46	0.45	0.00	0.57	1.00	0.07	3.48
65	Jérémie	0.28	0.17	0.25	0.28	0.34	0.72	1.00	0.59	1.00	0.10	4.73
66	Kenscoff	0.54	0.27	0.00	0.24	0.62	0.00	0.00	0.46	0.72		2.85
67	La Chapelle	0.28	0.37	0.00	0.17	0.27	0.00	1.00	0.50	0.93	0.03	3.55
68	La Tortue	0.14	0.27	0.00	0.16	0.69	0.48	0.00	1.00	1.00	0.03	3.78
69	La Vallée	0.31	0.17	0.00	0.41	0.76	1.00	1.00	0.11	1.00	0.07	4.82

(continued)

Table 6.8 (continued)

#	Municipality	Indicator										VICC
		1	2	3	4	5	6	7	8	9	10	
70	La Victoire	0.54	0.14	0.50	0.30	0.86	1.00	1.00	0.00	1.00	0.12	5.45
71	Lascahobas	0.47	0.37	0.50	0.26	0.39	0.72	1.00	0.27	1.00	0.57	5.56
72	Léogane	0.49	0.22	0.50	0.52	0.62	0.21	0.00	0.47	0.93		3.96
73	Limbé	0.70	0.02	0.00	0.71	0.67	0.45	0.00	0.53	1.00	0.27	4.35
74	Limonade	0.43	0.54	1.00	0.40	0.53	0.45	1.00	0.20	1.00		5.55
75	Maïssade	0.21	0.22	0.25	0.15	0.79	1.00	1.00	1.00	1.00	0.07	5.69
76	Maniche	0.05	0.77	0.25	0.13	0.39	0.21	1.00	0.50	0.93	0.00	4.24
77	Marigot	0.64	0.24	0.25	0.38	0.48	0.45	1.00	0.88	0.93	0.10	5.35
78	Marmelade	0.31	0.04	0.00	0.32	0.30	0.45	1.00	0.71	1.00	0.07	4.20
79	Milot	0.24	0.18	0.75	0.43	0.46	0.45	1.00	0.00	1.00	0.35	4.87
80	Miragoâne	0.16	0.17	0.00	0.30	0.49	0.72	1.00	0.50	0.93	0.40	4.67
81	Mirebalais	0.65	0.43	0.00	0.26	0.25	0.45	1.00	0.54	0.93	0.83	5.34
82	Môle Saint Nicolas	0.16	0.37	0.25	0.08	0.70	0.45	0.00	1.00	0.93	0.40	4.34
83	Mombin Crochu	0.53	0.30	0.00	0.12	0.57	0.21	1.00	1.00	1.00	0.03	4.77
84	Mont-Organisé	0.78	0.18	0.25	0.17	0.48	0.45	1.00	1.00	1.00	0.07	5.38
85	Moron	0.34	0.28	0.00	0.11	0.52	1.00	1.00	0.50	1.00	0.40	5.14
86	Ouanaminthe	0.72	0.17	0.25	0.54	0.57	0.21	1.00	0.27	1.00	0.10	4.83
87	Paillant	0.05	0.14	0.25	0.23	0.14	0.72	1.00	0.12	1.00	0.00	3.64
88	Perches	0.31	0.02	0.00	0.25	0.63	0.72	1.00	1.00	1.00	0.00	4.93
89	Pestel	0.12	0.58	0.00	0.09	0.44	0.21	1.00	0.20	1.00	0.07	3.71
90	Petit Trou de Nippes	0.07	0.37	0.25	0.14	0.26	0.72	1.00	0.75	1.00	0.33	4.90
91	Petite Rivière Artibonite	0.25	0.17	0.50	0.31	0.38	0.00	0.00	0.47	0.41	0.15	2.63
92	Petite Rivière Nippes	0.01	0.00	0.25	0.27	0.58	1.00	1.00	0.38	0.93	0.25	4.66

(continued)

Table 6.8 (continued)

#	Municipality	Indicator										VICC
		1	2	3	4	5	6	7	8	9	10	
93	Pignon	0.36	0.15	0.50	0.27	0.72	0.21	1.00	0.37	1.00	0.07	4.65
94	Pilate	0.43	0.02	0.25	0.44	0.61	0.21	1.00	0.88	1.00	0.28	5.12
95	Plaine du Nord	0.30	0.02	1.00	0.39	0.54	0.45	1.00	0.32	0.93		4.95
96	Plaisance	0.49	0.01	0.25	0.58	0.39	0.45	1.00	0.57	1.00	0.10	4.85
97	Plaisance du Sud	0.12	0.32	0.25	0.21	0.44	1.00	1.00	0.75	1.00	0.23	5.31
98	Pointe à Raquette	0.28	0.13	0.25	0.00	0.74	0.45	1.00	1.00	1.00		4.85
99	Port-à-Piment	0.10	0.14	0.25	0.28	0.57	0.72	1.00	0.75	1.00	0.00	4.81
100	Port-de-Paix	0.38	0.22	0.25	0.59	0.53	0.45	1.00	0.42	0.93	0.10	4.88
101	Port Margot	0.44	0.02	0.25	0.38	0.59	0.45	1.00	0.59	1.00		4.72
102	Port-Salut	0.37	0.34	0.25	0.37	0.65	0.72	0.00	0.33	1.00	0.45	4.49
103	Quartier Morin	0.36	0.34	1.00	0.44	0.46	0.72	1.00	0.00	0.93	0.92	6.18
104	Ranquitte	0.72	0.15	0.00	0.31	0.47	0.45	1.00	0.29	1.00	0.00	4.39
105	Roche à Bâteau	0.41	0.28	0.25	0.37	0.50	1.00	1.00	0.62	1.00		5.43
106	Roseaux	0.43	0.13	0.25	0.10	0.35	0.45	1.00	0.50	1.00	0.00	4.21
107	Saint Jean du Sud	0.23	0.87	0.25	0.34	0.58	1.00	1.00	0.59	1.00		5.86
108	Saint Louis du Nord	0.13	0.56	0.25	1.00	0.39	0.72	0.00	1.00	1.00	0.17	5.23
109	Saint Louis du Sud	0.20	0.32	0.25	0.32	0.57	0.72	1.00	0.54	1.00		4.92
110	Saint-Marc	0.26	0.15	0.25	0.47	0.39	0.00	0.00	0.47	0.41	0.10	2.51
111	Saint-Michel Attalaye	0.21	0.27	0.25	0.20	0.36	0.00	1.00	1.00	1.00	0.10	4.38
112	Saint Raphaël	0.70	0.03	0.75	0.25	0.77	0.45	1.00	0.75	0.72	0.57	6.00
113	Sainte Suzanne	0.25	0.18	0.00	0.17	0.68	0.72	1.00	1.00	1.00	0.15	5.15
114	Saut d'Eau	0.45	0.87	0.25	0.17	0.39	0.72	1.00	0.53	1.00	0.07	5.44
115	Savanette	0.42	0.34	0.00	0.15	0.29	0.21	1.00	1.00	1.00	0.03	4.45

(continued)

Table 6.8 (continued)

#	Municipality	Indicator										VICC
		1	2	3	4	5	6	7	8	9	10	
116	Terre Neuve	0.44	0.22	0.00	0.12	0.34	0.00	1.00	1.00	1.00	0.00	4.11
117	Thiotte	0.74	0.40	0.00	0.23	0.47	0.45	1.00	1.00	1.00	0.00	5.30
118	Thomassique	0.53	0.43	0.25	0.19	0.82	0.21	1.00	1.00	1.00	0.07	5.50
119	Thomazeau	0.21	0.37	0.25	0.12	0.47	0.21	0.00	0.46	1.00		3.09
120	Thomonde	0.12	0.37	0.25	0.11	0.47	0.72	1.00	0.62	1.00	0.57	5.23
121	Tiburon	0.34	0.18	0.25	0.09	0.62	0.72	1.00	1.00	1.00	0.00	5.21
122	Torbeck	0.25	0.62	0.50	0.38	0.40	0.45	1.00	1.00	0.72		5.33
123	Trou du Nord	0.19	0.17	0.25	0.35	0.63	0.45	0.00	0.33	0.93	0.00	3.29
124	Vallières	0.66	0.18	0.25	0.08	0.50	0.45	1.00	1.00	1.00	0.03	5.16
125	Verrettes	0.13	0.17	0.25	0.39	0.37	0	1.00	0.47	0.41	0.10	3.29

References

Adger, W.N., N. Brooks, G. Bentham, M. Agnew, S. Eriksen. 2004. New indicators of vulnerability and adaptive capacity. Tyndall Centre for Climate Change Research, technical report 7.

Ahsan, N., J. Warner. 2014. The socioeconomic vulnerability index: A pragmatic approach for assessing climate change led risks—A case study in the south-western coastal Bangladesh. *International Journal of Disaster Risk Reduction* 8: 32–49. doi:10.1016/j.ijdrr.2013.12.009.

Birkmann, J. 2006. Indicators and criteria dor measuring vulnerability: Theoretical bases and requirements. In *Measuring vulnerability to natural hazards. Onwards disaster resilient societies*, ed. J. Birkmann, 55–77. Tokyo–New York–Paris: UNU Press.

Borde, A., M. Huber, A. Goburdhun, A. Guidoux, E. Revoyron, E. Nsimba, J.A. Louis, A. Donjia, and J.-L. Kesner. 2015. *Estimation des couts des impacts du changement climatique en Haiti*. New York: PNUD-Ministère de l'économie et des finances.

Bollin, C., K. Fritzsche, S. Ruzima, S. Schneiderbauer, D. Becker, L. Pedoth, and S. Liersch. 2014. *Analyse intégrée de la vulnerabilité au Burundi*, vol. II. GIZ: Analyse de vulnérabilité au niveau local. Bujumbura.

Borja-Vega, C., A. de la Fuente. 2013. Municipal vulnerability to climate change and climate-related events in Mexico. *Policy Research Working Paper* 6417. Washington: The World Bank.

Ci Grasp 2.0. 2016. Glossary. http://www.pik-potsdam.de/cigrasp-2/bg/glossary.html#S. Accessed 15 July 2016.

Etwire, P.M., R.M. Al-Hassan, J.K.M. Kuwornu, and Y. Osei-Owusu. 2013. Application of livelihood vulnerability index in assessing vulnerability to climate change and variability in Northern Ghana. *Journal of Environment and Earth Science* 3 (2): 157–170.

FAO, PAM. 2010. Mission FAO/PAM d'évaluation de la récolte et de la sécurité alimentaire en Haiti. Rome: FAO.

Frerichs, R.R. 2016. *Deadly river. cholera and cover-un post-earthquake Haiti*. Ithaca, NY: ILR Press.

Funk, C., P. Peterson, M. Landsfeld, D. Pedreros, J. Verdin, S. Shukla, G. Husak, J. Rowland, L. Harrison, A. Hoell, and J. Michaelsen. 2015. The climate hazards infrared precipitation with stations—A new environmental record for monitoring extremes. *Scientific Data* 2: 150066. doi:10.1038/sdata.2015.66.

Gbetibouo, G.A., C. Ringler, R. Hassan. 2010. Vulnerability of the South African farming sector to climate change and variability: An indicator approach. *Natural Resources Forum* 34: 175–187. doi:10.1111/j.1477-8947.2010.01302.x.

Hahn, M.B., A.M. Riederer, S.O. Foster. 2009. The livelihood vulnerability index: A pragmatic approach to assessing risks from climate varianbility and change—A case study in Mozambique. *Global Environmental Change* 19: 74–88. http://dx.doi.org/10.1016/j.gloenvcha.2008.11.002.

Heltberg, R., and M. Bonch-Osmolovskiy. 2010. *Mapping vulnerability to climate change*. Washington: The World Bank.

Hilden, M., A. Marx. 2013. Evaluation of climate change state, impact and vulnerability indicators, ETC CCA Technical paper 02/2013.

IPCC-International Panel on Climate Change. 2014. Annex II: Glossary. In *Climate change 2014: synthesis report. Contribution of Working Groups I, II and III to the Fifth Assessment Report of the IPCC*, eds. R.K. Pachauri, and L.A. Meyers, 117–130. Geneva: IPCC.

Nurse, L.A., R.F. McLean, J. Agard, L.P. Briguglio, V. Duvat-Magnan, N. Pelesikoti, E. Tompkins, and A. Webb. 2014. Small islands. In *Climate Change 2014: Impacts, Adaptation, and Vulnerability. Part B: Regional Aspects. Contribution of Working Group II to the Fifth Assessment Report of the Intergovernmental Panel on Climate Change* ed. V.R. Barros, C.B. Field, D.J. Dokken, M.D. Mastrandrea, K.J. Mach, T.E. Bilir, M. Chatterjee, K.L. Ebi, Y.O. Estrada, R.C. Genova, B. Girma, E.S. Kissel, A.N. Levy, S. MacCracken, P.R. Mastrandrea, and L.L. White, 1613–1654. Cambridge, United Kingdom and New York, USA: Cambridge University Press.

OCHA-United Nations Office for the Coordination of Humanitarian Affairs. 2016. Cholera remains an emergency and a priority fo the UN and its partners. *Humanitarian Bulletin Haiti* 57: 3–4.

PNUMA-ORLAC-Programa de las Naciones Unidas para el Medio Ambiente, Oficina Regional para América Latina y el Caribe. 2013. Análisis de vulnerabilidad e identificación de opciones de adaptación frente al cambio climático en el sector agropecuario y de recursos hídricos en Mesoamérica. PNUMA-ORLAC.

RdH-République d'Haiti. 2010. Analysis of Multiple Hazard in Haiti. Port-au-Prince, March.

RdH, MARNDR. 2012a. Atlas agricole d'Haïti. Enquête exploitation 2009. http://agriculture.gouv.ht/statistiques_agricoles/Atlas/preface.html. Accessed 16 Jun 2016.

RdH, MARNDR. 2012b. Politique d'irrigation du MARNDR 2012. Juin.

RdH, MARNDR. 2012c. Evaluation de la campagne agricole de printemps 2012. Septembre.

RdH, MARNDR. 2013. Evaluation previsionnelle de la performance des recoltes de la campagne agricole de printemps 2013.

RdH, MARNDR. 2014. Evaluation prévisionnelle de la performance des récoltes de la campagne agricole de printemps 2014. Octobre.

RdH, MARNDR CSA FEWS NET. 2009. Haiti: Cartographie de vulnerabilité multirisque juillet/aout 2009.

RdH, MEF, IHIS-Ministère de l'Economie et des Finances, Institut Haïtien de Statistique et d'Informatique. 2012. Population totale, population de 18 ans et plus ménages et densités éstimés en 2012. Port-au-Prince: DSDS.

RdH, MSPP-Ministère de la Santé Publique et de la Population. 2015. *Infosanté. Bulletin d'information* 1, 9, juillet.

RdH, MTPTC-Ministère des Travaux Publics, Transports et Communicartion. 2015. RAI (Rural Access Index) Haïti 2015. Accessibilité des sections communales au réseau routier, 6 mai http://www.mtptc.gouv.ht/media/upload/doc/publications/ROUTE_INDICE-RAI-2015.pdf. Accessed 16 Jun 2016.

Roberts, P., K.C. Shyam, C. Rastogi. 2006. Rural access index: A key development indicator. *Transport Papers* 10, March. The World Bank.

Schiavo-Campo, S. 1999. Strengthening "performance" in public expenditure management. *Asian Review of Public Administration* 11 (2): 23–44.

Sharma, B., N. Jangle. 2012. Assessing vulnerability to climate risk in Ahmednagar (Maharashtra) and Vaishah (Bihar) using GIS technique. In *13th Esri India User Conference 2012*.

Tiepolo, M. 2009. *Lo sviluppo delle aree rurali remote. Petrolio, uranio e governance locale in Niger*. Milan: Franco Angeli.

Tuladhar, F.M. 2013. Climate change vulnerability index mapping of Nepal using NASA EOS Data. In *14th Esri India User Conference 2013*.

UNESC-United Nations Economic and Social Council. 2015. Report of the inter agency and expert group on sustainable development goals indicators. https://unstats.un.org/unsd/statcom/47th-session/documents/2016-2-IAEG-SDGs-E.pdf. Accessed 16 Jun 1916.

UNESCO-IHE-Institute for Water Education. 2016. Flood vulnerability indices. http://unescoihefvi.free.fr/vulnerability.php. Accessed 2 July 2016.

USAID. 2014. *Design and use of composite indices in assessments of climate change vulnerability and resilience*. Washington: USAID.

Verisk Maplecroft. 2016. Vulnerability Index. https://maplecroft.com/about/news/ccvi.html. Accessed 2 July 2016.

WMO-World Meteorological Organization, M. Svoboda, M. Hayes, and D. Wood. 2012. *Standardized precipitation index user guide*. Geneva: WMO.

World Bank. 2014. *Open data for resilience initiative*. World Bank: Field guide. Washington.

Chapter 7
Visualize and Communicate Extreme Weather Risk to Improve Urban Resilience in Malawi

Alessandro Demarchi, Elena Isotta Cristofori and Anna Facello

Abstract Since the last century, an unprecedented settlement expansion, mainly generated by an extraordinary world population growth, has made urban communities always more exposed to disasters. Casualties and economic impacts due to hydro-meteorological hazards are dramatically increasing, especially in developing countries. Although the scientific community is currently able to provide innovative technologies to accurately forecast severe weather events, scientific products are often not easily comprehensible for local stakeholders and more generally for decision makers. On the other hand the integration of different layers, such as hazard, exposed assets or vulnerability through GIS, facilitates the risk assessment and the comprehension of risk analysis. This work presents a methodology to enhance urban resilience through the integration into a GIS of satellite-derived precipitation data and geospatial reference datasets. Through timely and meaningful hydro-meteorological risk information, this methodology enables local government personnel and decision makers to quickly respond and monitor natural phenomena that could impact on the local community. This methodology is applied to the January 2015 Malawi Flood case study and result will be discussed along with further recommended developments.

Keywords Flood · Urban resilience · Weather risk

A. Demarchi (✉) · E.I. Cristofori · A. Facello
TriM—Translate into Meaning, Corso Sommeiller 24, 10128 Turin, Italy
e-mail: alessandro.demarchi@trimweb.it

E.I. Cristofori
e-mail: elena.cristofori@trimweb.it

A. Facello
e-mail: facelloanna@gmail.com

M. Tiepolo et al. (eds.), *Renewing Local Planning to Face Climate Change in the Tropics*, Green Energy and Technology,
DOI 10.1007/978-3-319-59096-7_7

7.1 Introduction

The course of the XX century has been characterised by two unprecedented demographic phenomena. Firstly, an exceptional growth of world population which is currently more than four times compared to the beginning of the last century (UNDESA 2011). Secondly, the rise of the urban population which in 2010 has exceeded the rural population for the first time in the history (UNDESA 2011).

Combined together, these two demographic trends have determined a rapid expansion of either planned or informal urban areas. All over the world, to meet this insatiable demand for land, communities were settled also in dangerous areas prone to natural disasters. Indeed, new settlements and infrastructures have been built on active faults, under landslides or mudslides, in flooding areas, in areas of volcanic eruptions, etc. This unprecedented urban sprawling phenomenon has involved both developed and developing countries. On the other hand, increases in frequency and intensity of extreme weather events, mainly due to variation either in the mean state of the climate or in its variability, are posing an always more serious risk to worldwide urban communities (Teegavarapu 2012; IPCC 2012). Indeed, even if there is still some uncertainty about the rates of yearly change and it is difficult to distinguish between extreme events exacerbated by climate change from those that represent part of the current natural variability of climate, the Intergovernmental Panel on Climate Change (IPCC) identifies climate impacts as one of the most relevant environmental issues over the 21st century (Cardona et al. 2012; IPCC 2014).

As a consequence, the increased likelihood of urban communities to be affected by natural events, along with climate change effects, is always more frequently causing large-scale disasters. Indeed, over the last century, the number of reported calamities due to natural phenomena has dramatically risen (CRED and UNISDR 2016; Tiepolo and Cristofori 2016b).

Among all natural phenomena, the number of disasters related to hydro-meteorological events is predominant. According to CRED and UNISDR (2016), the overwhelming majority of natural disasters is indeed due to floods, storms and other weather-related events. These phenomena are clearly affecting the largest number of people and they are causing the largest amount of economic damage among all the different typologies of phenomena. Only over the last twenty years, EM-DAT recorded 6457 weather-related disasters which claimed 606,000 lives and 4.1 billion people injured (CRED and UNISDR 2016).

Considering the high level of exposure and the high level of vulnerability due to unprepared human communities, hydro-meteorological phenomena are particularity dangerous in developing countries where these events are causing always more massive impacts in terms of causalities, economic and environment damages (Tiepolo and Cristofori 2016b). According to United Nations Office for Disaster Risk Reduction (UNISDR), Disaster Risk Reduction (DRR) aims to reduce the damage caused by natural hazards like floods, droughts, cyclones, etc., through an ethic of prevention. DRR framework includes on the one hand, long-term actions, such as a more adequate urban and territorial planning to guarantee a sustainable development and to make the local communities safer. On the other hand, DRR includes also short and medium-term actions, such as undertaking prevention and preparedness actions in order to reduce damage due to approaching hazardous phenomena (Eiser et al. 2012).

Nowadays, scientific community is able to provide innovative technologies to accurately forecast hazard weather phenomena. However, results coming from weather forecast must be translated into other languages, to be operationally usable by decision makers. As a matter of fact, within this translation a critical role is mainly played by two actors. The first actor is the scientific community, who produces numerical results, offers their interpretation and gives an assessment of the uncertainty. The second actor is represented by the users who have to understand the scientifically based information and use them to make decisions. Therefore the representation of the final risk information in an understandable, meaningful and usable manner is a key issue in order to implement a proper DRR strategy. This issue is particularly related to the temporal and spatial scale at which the phenomena is studied and to the end-users needs or competencies.

With the aim of implementing effective actions to enhance adaptation and resilience, the increase of awareness about severe hydro-meteorological events through an improved access to technologies is considered essential (UNISDR 2015). In this framework, Geospatial Information Technologies offer a broad assortment of innovative solutions in many different sectors and activities such as: Disaster Risk Management (DRM) and DRR (Altan et al. 2013; Miyazaki et al. 2015; Cristofori et al. 2016a), assessment of environmental issues (Melesse et al. 2007; Chigbu and Onukaogu 2013; Bonetto et al. 2016) and climate change analyses (Donoghue 2002; Sundaresan et al. 2014; Tiepolo and Cristofori 2016a). In particular, Geographic Information Systems (GIS) offer an essential environment for the effective collection, management, representation and multidisciplinary analysis of miscellaneous geographic data (Tomaszewski 2014). The integration and visualization of various data derived from different sources into one system, permits to provide immediately understandable cartographic products that can address a variety of issues such as potential impacts due to hydro-meteorological phenomena on land use, health care facilities, transportation infrastructures or households (Ferrandino 2014; Duncan et al. 2014; Cristofori et al. 2016b). In addition, Remote Sensing (RS), providing timely and reliable information about the Earth's surface and atmosphere, is a primary method for observing weather and climate dynamics (Cristofori et al. 2015), while technical and non-technical users may accurately define the position of relevant environmental and geographic features through Global Positioning System (GPS) tools and applications. Finally, geospatial information technologies allow to effectively reduce costs, both in term of time and economic resources required for carrying out in situ spatial surveys (Cay et al. 2004). These technologies indeed permit to access a big amount of information characterized by a large spatial and temporal coverage and, at the same time, they provide the possibility to easily update the data for monitoring activities (Pandey and Pathak 2014).

The main objective of this paper is therefore to present a multidisciplinary approach to strengthen urban resilience by creating a GIS in which integrate, manage and visualize hazard information along with general reference data coming from different sources. Through an enhanced accuracy and usability of extreme precipitation warnings, this methodology aims to enable local government personnel and decision makers to quickly respond and monitor natural phenomena that could cause an impact for the local community.

In the following paragraphs will be presented, firstly the data used and the methodology, secondly the description of the case study (the 2015 Malawi flood event), thirdly the obtained results, and finally the conclusions and some future developments.

7.2 Data and Method

The proposed methodology is an operative development of a previously published work (Cristofori et al. 2015). One of the principal aims was to properly monitor and forecast extreme rainfall and the potential related physical impacts on populated places. A methodology to easily process satellite-derived precipitation data for the assessment of extreme precipitation hazard was presented along with the application of the methodology to a past event (2013 Malawi Flood). Therefore data used to perform the monitoring and forecasting of extreme precipitation within this paper are the same used in the work presented by Cristofori et al. (2015).

Since the possibility to immediately and easily identify lifelines, infrastructure and facilities affected may effectively supports decision-makers, emergency and recovery teams, additional datasets concerning exposure information have been considered in this work. In this framework, the development of a tailored GIS represents an effective tool to easily and quickly visualize, analyse and integrate into the same system, hazard information with general reference data. The flexibility of the GIS environment allows indeed to easily updating and adding many typologies of data coming from different sources, such as raster satellite-derived precipitation and numerical weather prediction—Hazard data—and geospatial reference vector datasets—Exposure data.

Thus, this methodology allows end-users to adapt the warning messages to the profile and to the specific characteristics of the geographic features (e.g. populated places, infrastructures, facilities, cultivated and grazing areas). In this way, both before and in the aftermath of a hydro-meteorological event, the areas and the resources affected may be identified more precisely and more effective warning messages may be generated. The possibility to issue different and tailored warning messages represents one of the principal innovative development presented in this work.

Moreover, since this methodology aims to be economically sustainable, it has been generated using only open-source data in an open-source environment (QGIS software, released version 2.8.3). Data used for the hazard assessment are indeed two open-source rainfall products derived from satellite precipitation estimates and from numerical weather prediction model: the near-real time version of the Tropical Rainfall Measuring Mission (TRMM) 3B42RT and the Global Forecast System (GFS). TRMM (period of record 1997–2015), was a joint satellite mission of NASA in cooperation with the Japan Aerospace Exploration Agency (JAXA) which covered the latitude band between 50° North and 50° South; it was designed to study tropical rainfall for weather and climate research. Data are available at 0.25° × 0.25° spatial resolution, with 3 h revisit time (Huffman et al. 2007, 2010). TRMM was mainly used to monitor the evolution and magnitude of extreme precipitation events in real-time. Actually, a new joint mission, the Global Precipitation Measurement (GPM), developed by the same agencies and based on the notable successes of the TRMM, is operative.

GFS is a weather forecast model produced by the National Centers for Environmental Prediction (NCEP) of United States. The dataset covers the entire world and it contains several atmospheric and land-soil variables such as temperatures, winds, and precipitation to soil moisture. GFS may be used to operationally forecast an event up to 16 days in advance. Data are available at 0.5° × 0.5° spatial resolution, with 3 h revisit time.

TRMM is mainly used to monitor the evolution and magnitude of extreme precipitation events in real-time while GFS supports the prediction of possible hazardous weather phenomena. Therefore a complementary use of these two products represents a fundamental tool for local governmental personnel and decision makers because it allows to predict in advance and to monitor the temporal and spatial evolution of both short-term heavy rainfall and medium-term persistent rainfall events.

Data used for exposure assessment have been acquired from open-source web portals, such as OpenStreetMap (OSM), Geonode, Global Administrative Areas Database (GADM), FAO Geonetwork. Therefore, depending also on the specific end-user needs, several data have been downloaded, processed and integrated into the system. Nonetheless, if the availability of geospatial datasets, needed to assess areas, proprieties or population most at risk are only partial, not-updated or even denied, extraction techniques may be performed to acquire further information from earth observation data (Ehrlich and Tenerelli 2012; Bonetto et al. 2016).

The main phases of the proposed methodology are hazard assessment, integration of exposure information, and generation of risk warnings.

7.2.1 *Hazard Assessment*

The first step to assess the magnitude of hazardous hydro-meteorological events is the calculation of accumulated precipitations over the area of interest. Several time frames may be chosen to provide hazard information concerning the typology of event (i.e. 6–24 h for short-term heavy rainfall and 2–7 days for persistent rainfall).

This operation is performed in GIS environment using a specific in-house tool which allows to cumulate for each pixel the relative values of the 3-h TRMM precipitation products covering the time span considered.

Once calculated the cumulated precipitation, the second step is the determination of extreme precipitation thresholds, which may be defined as the amount of rainfall that may trigger a "disaster" (such as flooding, landslides, etc.). Consequently, once a given threshold is exceeded, a warning is issued by the alert system. The values of thresholds have to be defined in advance in order to deliver a warning message when a hazardous hydro-meteorological event is approaching and/or affecting a specific geographic area. Different approaches exist to calculate adequate thresholds and the scientific community is still conducting investigations on this topic. In this methodology, a statistical meteorological approach has been followed in order to derive national and regional percentile values. Processing the after-real-time TRMM historical dataset (from 1998 to 2015), the 95th, 97th and 99th percentile have been calculated for different accumulated precipitation timeframes and have been chosen as low, medium and high alert levels.

7.2.2 *Integration of Exposure Information*

Hazard information derived from satellite precipitation estimates or from numerical weather prediction models are often represented using images that give only a rough idea of the potential impact of a flood. Being expressed using raster or gridded format, the rainfall values are indeed returned on pixel scale and the understanding of such information can be particularly difficult for generic and non-scientific users (Cristofori et al. 2015). Moreover, in order to shift from "pure" hazard information to a more "comprehensive" risk conception, the assessment of exposed features has to be as well considered and integrated in the system.

In order to identify the potentially most affected features due to an approaching event in the near-real-time after a disaster, the overlapping between the hazard information assessed and the exposure layers is performed in the created GIS. In particular, the following georeferenced layers are considered:

- administrative subdivision, to better contextualize the affected area;
- census data expressed on administrative area, to assess the number of population potentially affected;
- built-up areas and villages, to identify which human settlements are affected;
- infrastructures (i.e. road network, rail network, train stations, airports and harbours), to identify the best routes for emergency and recovery teams avoiding possible network interruption;
- facilities (i.e. school, religious, government and health-care buildings), to address emergency and recovery teams, means, resources and eventually rescued people avoiding facilities at-risk and/or already affected;

- land-use/land-cover characterization, to know the typology of affected areas;
- natural resources (i.e. lakes and rivers), to identify further hazardous sources which may contribute to generate cascading effects related to hydro-meteorological events.

7.2.3 Generation of Risk Warnings

The third and last phase of the proposed methodology regards the generation of risk warning messages. As previously described, once a given threshold is exceeded, a warning is issued by the alert system. Depending on the magnitude of hydro-meteorological phenomenon, the warning message may be differentiated in three levels of alert. A set of thresholds, composed by different crescent values corresponding to each level of resulting warning, has been defined. In this way, when the value of rainfall forecasted or monitored is exceeding the thresholds based on TRMM time series, the corresponding alert level is issued as following:

- Low alert. Precipitation >95th percentile
- Medium alert. Precipitation >97th percentile
- High alert. Precipitation >99th percentile

Applying this 3-level alert system to administrative areas, human settlements, infrastructure and facilities, several typologies of information may be generated in the form of easy to understand final risk maps. These cartographic products allow to meaningful and immediate visualize the areas and the assets affected or at the risk according to the level of risk warning issued.

7.3 Case Study

As previously stated, this methodology has been conceived as a low-cost tool sustainable also for developing countries which are very often characterized by a shortage both in term of money and personnel. Therefore, in order to test and validate the methodology developed, the 2015 Malawi flood has been chosen as case study. Since December 2014, exceptional heavy rains affected Malawi causing rivers to overflow and floods. More than 400 mm and more than 500 mm were recorded respectively on one and three days of cumulative rainfall which represent the highest measurements since the establishment of meteorological stations in the highlands area of the county (i.e. the Chichiri station in Blantyre) (Fig. 7.1).

Between the 10th and the 12th of January 2015, 15 Malawi districts (out of 28), and in particular the southern districts such as Nsanje, Chikwawa, and Phalombe,

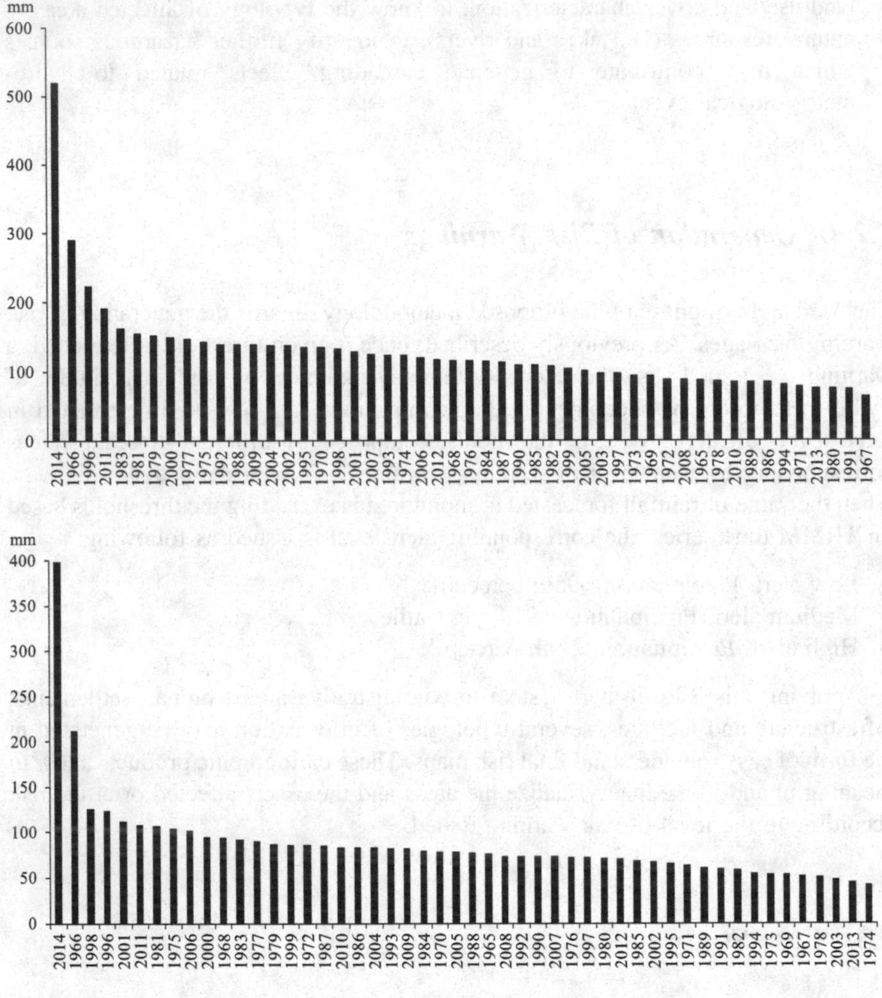

Fig. 7.1 Cumulative rainfall recorded in the Chichiri meteorological station since its establishment until December 2014; max 3 day rainfall (*top*) and max 1 day rainfall (*bottom*) in order of magnitude

were hit by severe flooding, affecting more than 1 million people and killing 106 person. Extensive damages to houses, infrastructures (i.e. roads, bridges, water and power supply) and facilities (i.e. schools) were recorded. Considering that several crops of maize were destroyed and livestock killed, the economic system of Malawi, essentially based on subsistence farming, was severely impacted (Government of Malawi 2015).

7.4 Results

In this case study a set of monitoring activities, which could have been performed by government personnel and decision makers both during and in the aftermath of the extreme event, have been simulated. The results of this simulation are presented and discussed in this section.

Following the methodology workflow described in the paragraph 7.2, the first operation performed has regarded the calculation of the accumulated precipitations recorded by TRMM satellite sensor over the selected 72 h (from 00:00UTC 10/01/2016 to 00:00UTC 13/01/2016) resulting as mainly characterized by heavy rainfalls. The south-eastern part of the country appears to have been particularly affected since several pixels show values greater than 200 mm of precipitation over the selected period (Fig. 7.2a). Considering that satellite measurements are expressed on a square matrix of 0.25° × 0.25°, the precipitation estimates are provided as mean values equally distributed over the area of each pixel. For this reason, these values may differ from single ground measurements recorded on-site by the meteorological stations.

Comparing accumulates precipitations with thresholds values, gridded alerts have been derived (Fig. 7.2b). It is possible to observe that almost all the southern areas of the country would have been interested by a general state of alert ranging

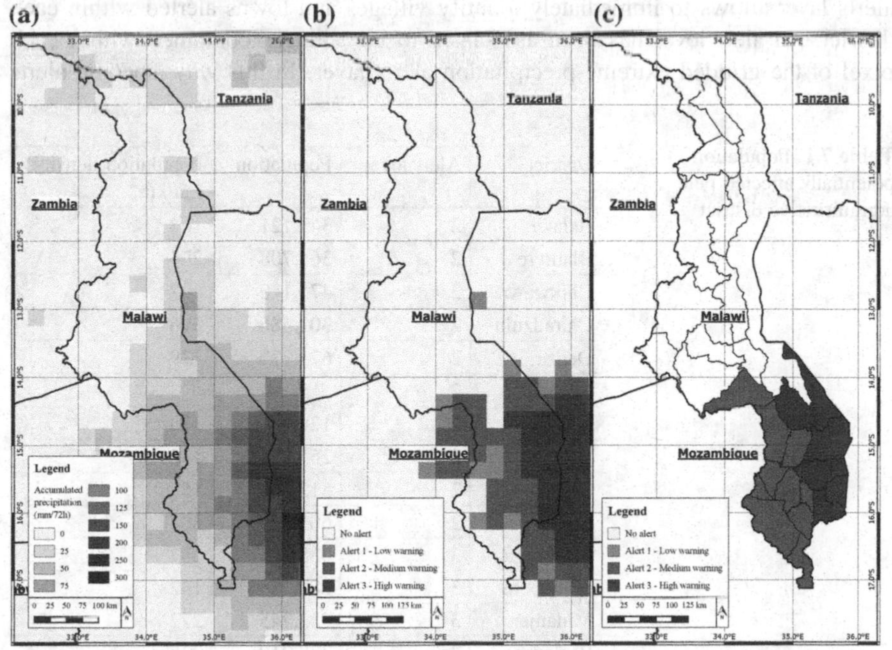

Fig. 7.2 72 h accumulated precipitation based on TRMM-3B42RT issued the 12th January 2015 (**a**), extreme precipitation alerts (**b**) and administrative districts alerted (**c**)

from low alert to high alert messages. On the one hand, this cartographic product could be very useful especially for depicting a first overall description of the areas potentially most affected and for describing the dynamic and the magnitude of the hydro-meteorological event.

On the other hand, an adequate geographic contextualization with an assessment of geographic features, people or infrastructures potentially mostly affected would help to drive decisions during both prevention and response phases of a disaster. Indeed the integration of administrative borders in the system allows to immediately pinpoint the districts mainly affected (Fig. 7.2c). Moreover the GIS application enables an immediate quantification of possible affected population (Table 7.1).

The second step of the methodology is devoted to enhance the understandability and usability of extreme precipitation warnings by highlighting the places and assets potentially most at risk. In the framework of the Shire River Basin Management Program, the Government of Malawi together with the World Bank developed on top of the GeoNode application the Malawi Spatial Data Platform (MASDAP) (Balbo et al. 2013). Therefore several data of potentially affected features such as populated places (i.e. villages and towns), the road network, the health-care facilities (i.e. hospital) and the educational buildings (i.e. schools and universities) have been downloaded from the MASDAP platform. The downloaded layers have been then uploaded into the GIS and integrated with the extreme precipitation alerts.

The intersection of the populated places layer with the extreme precipitation alerts layer allows to immediately identify villages and towns alerted within each district. An alert level has been associated to the villages contained within each pixel of the gridded extreme precipitation alerts layer. In this way, specific alerts

Table 7.1 Population potentially affected per administrative district

District	Alert level	Population	Population density Inh/km^2
Balaka	2	349,121	164
Blantyre	2	364,708	204
Chikwawa	2	475,140	97
Chiradzulu	2	301,586	396
Dedza	2	671,137	179
Mwanza	2	–	–
Neno	2	124,430	79
Nsanje	2	255,995	132
Ntcheu	2	513,865	158
Thyolo	2	603,129	361
Machinga	3	538,345	137
Mangochi	3	885,355	132
Mulanje	3	543,745	272
Phalombe	3	338,219	238
Zomba	3	614,268	197

have been addressed to each village in relation to the amount of precipitation measured in the surrounding of their location which roughly corresponds to a square area of 750 km^2. A map produced for the southern Malawi highlights the alerts calculated considering the populated places (Fig. 7.3b). The analysis performed so far takes into account only the hazard layer as a criterion to calculate the extreme precipitation warnings. On the other hand the calculation of exposure allows the identification of assets most at risk and a most effective prioritization of emergency actions. Therefore the proximity to the main river network has been considered as an exposure criterion to perform the third step of the proposed methodology. This criterion aims to:

- highlight the risk of those settlements lying within 1 km from the riverbanks—in this case the alert level has been incremented by one level without trespassing the maximum level (villages characterized by a high alert did not change the level)
- conceal the risk of those populated places located far away from the riverbanks—in this case the alert level has been decremented by two levels

The application of this exposure criterion allows emergency managers and local government personnel to better manage their resources and means. Through the

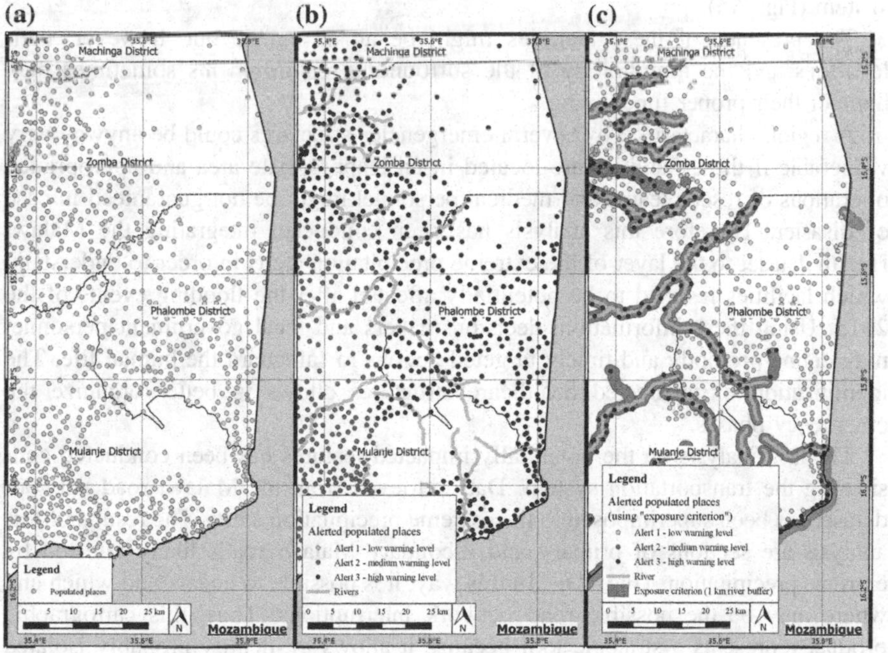

Fig. 7.3 Populated places in Phalombe district, Malawi (**a**), alerted settlements issued for 12th January 2015 on the basis of extreme precipitation alerts in Phalombe district (**b**) and alerted settlements issued for 12th January 2015 on the basis of the "exposure criterion" (**c**)

identification of most affected areas, they may prioritize their actions directly to those villages characterized by the greater alerts (Fig. 7.3c).

Further analyses have been then performed in order to identify potential impacts on urban and local facilities due to the hydro-meteorological event of 2015. Similarly to the analysis performed using the populated places layer, the facilities datasets have been integrated in the system and overlaid to the hazard information.

Another layer that has been used is the Malawi educational buildings dataset. It contains the geographic location of schools and universities which may represent a fundamental asset in case of flooding event for essentially two reasons. Firstly, educational buildings for most of the day are populated by children and young students, which usually are the most affected population by flood. Secondly these structures can be used as gathering points for the displaced populations. Therefore an analysis of facilities potentially at risk allows decision makers to implement specific monitoring actions and to define promptly and timely targeted activities to safeguard the life of students. Moreover, an immediate identification of the structures suitable as collection points may eventually accelerate the rescue activities. Educational buildings affected, such as those located in the south/south-eastern part of the country, where several facilities resulted characterized by a high level of alert, are recognized (Fig. 7.4). Since emergency facilities, such as health-care structures, are particularly important during and after a disaster, the dataset containing their location has been downloaded from MASDAP and integrated in the system (Fig. 7.5).

The presence of these facilities might be an advantage but, however, their locations and the morphology of the surrounding environments sometimes may hamper their proper functioning.

A region characterized by several emergencies structures could be anyway very vulnerable if these facilities are located in an inappropriate area and appropriated operations of rescue teams and medical personnel might be not guaranteed in case of disaster. Therefore this analysis has been conducted integrating the Malawi Health dataset to the layer of the extreme precipitation alerts in order to understand which facilities resulted to be potentially affected after the flooding events of the 2015. Using these information, decision makers and local government personnel may define promptly and timely targeted actions to safeguard the human life. The identification of unaffected healthcare structures allows to better organize the emergency phases.

The last analysis on the potentially impacted features has been conducted considering the transportation system. Data retrieved from the Malawi road network dataset has been superimposed to the extreme precipitation alerts. The results of this analysis are sections of primary and secondary Malawi roads likely affected by extreme precipitations (Fig. 7.6). In this way, it is possible to understand which and where may be the possible road network interruptions. Thus, this cartographic product represents a strategic tool because it allows to identify probably isolated and unreachable settlements.

Moreover, the possibility to hold such information in near-real time and in the immediate aftermath of disastrous event allows to greatly improve the emergency

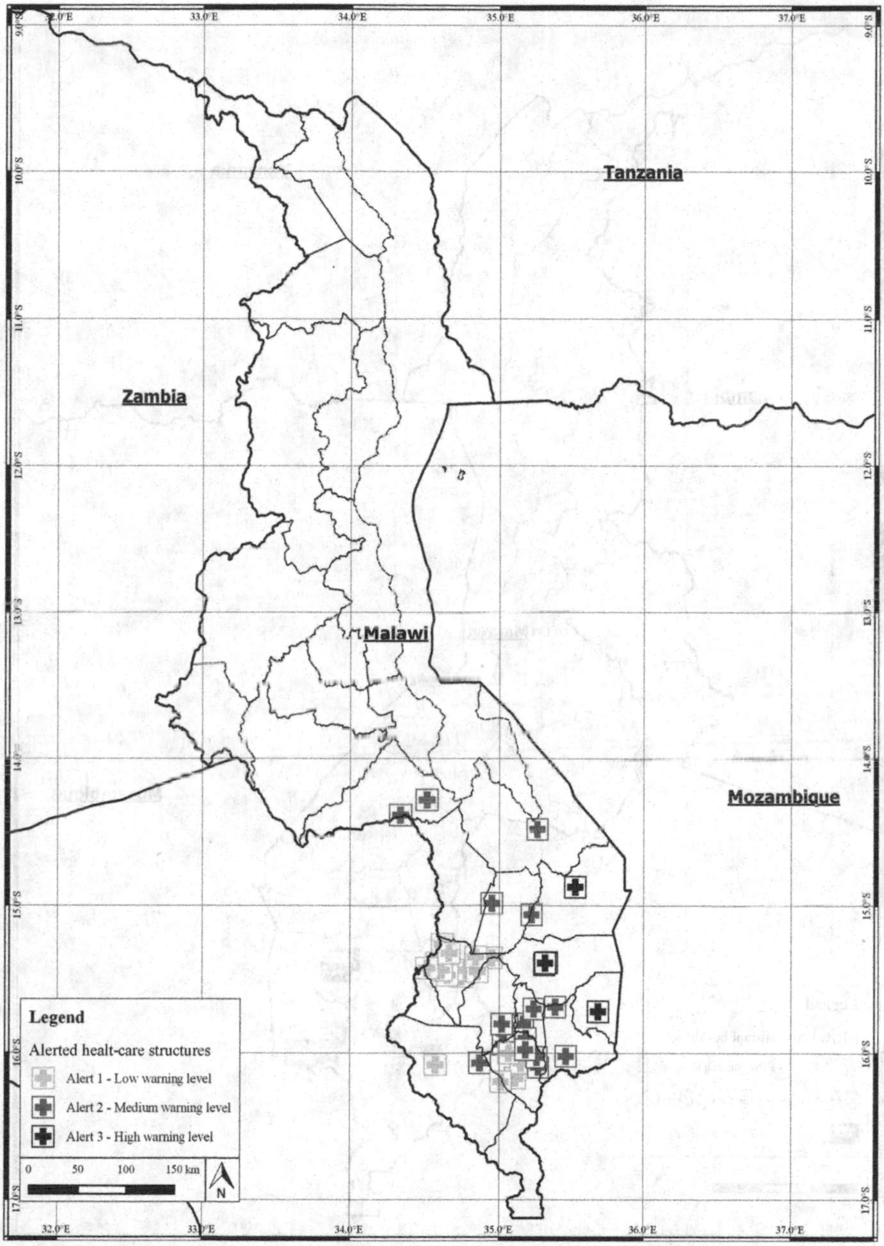

Fig. 7.4 Malawi educational buildings alerted on the basis of extreme precipitation alerts issued for the 12th January 2015

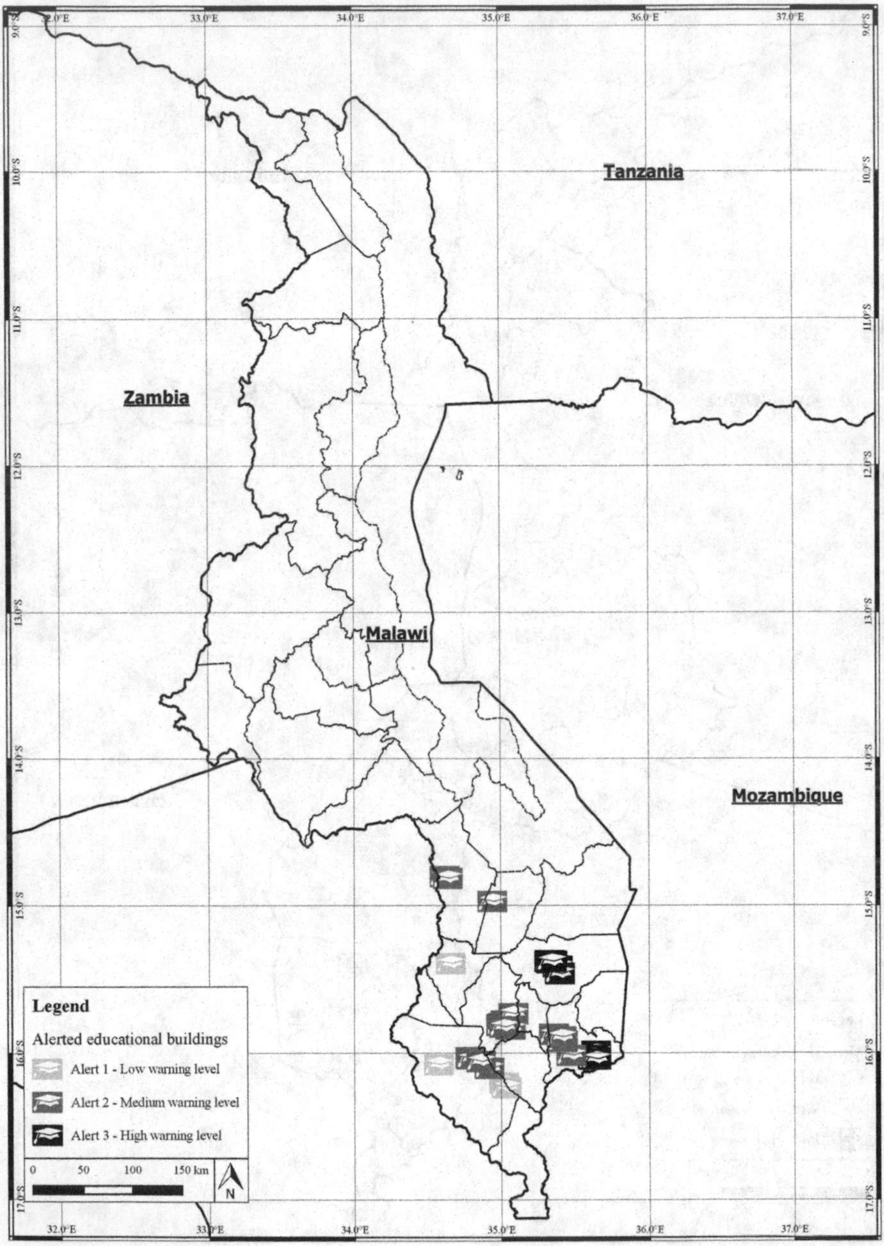

Fig. 7.5 Malawi health-care facilities alerted on the basis of extreme precipitation alerts issued for 12th January 2015

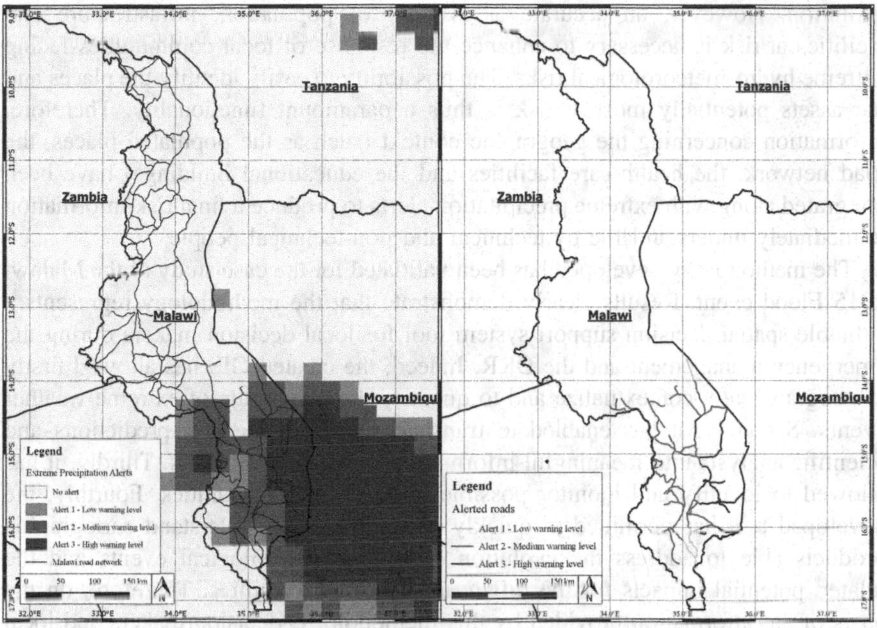

Fig. 7.6 Extreme precipitation alerts issued for the 12th January 2015 and the Malawi road network (on the *left*) and alerted section of the Malawi road network in relation to the amount of precipitation measured (on the *right*)

phases in terms of results and response time. Indeed, the possibility to plan in advance the routes to be used by rescue teams or to locate and pinpoint back-up network systems certainly allows to save great amount of resources, both in terms of time and money.

7.5 Conclusions

The methodology developed in this work is aimed at enhancing the adaptation and resilience to hydro-meteorological risks of local and urban communities. For this purpose, this paper presents a multidisciplinary approach to produce immediately understandable and easy to use information about the potential impacts generated by extreme rainfall events.

The methodology relies on the development of a tailored GIS that allows the integration, management and visualization of different types of information commonly used to perform the risk assessment (i.e. satellite derived precipitation estimates and geospatial datasets concerning the geographic features of the context analysed).

Simple hazard representations (i.e. near-real time accumulated precipitations) are very useful tools for a first delineation of the event in term of magnitude and spatial evolution. However, an accurate assessment of population, infrastructure and facilities at risk is necessary to enhance the response of local communities facing extreme hydro-meteorological risks. The possibility to easily identify the places and the assets potentially most at risk is thus a paramount functionality. Therefore, information concerning the geographic context (such as the populated places, the road network, the health-care facilities and the educational buildings) have been integrated along with extreme precipitation alerts to produce a final risk information immediately understandable by technical and non-technical people.

The methodology developed has been validated for the case study of the Malawi 2015 Flood event. Results clearly demonstrate that the methodology represents a valuable spatial decision support system tool for local decision makers during the emergency management and the DRR. Indeed, the created GIS has allowed firstly to geographically contextualize and to quantify the magnitude of extreme weather events. Secondly, it has enabled to translate results of weather predictions and scientific analysis into meaningful information for decision makers. Thirdly, it has allowed to identify and monitor possible impacts on communities. Fourthly, the developed tool has permitted to quickly produce easy-to-understand cartographic products able to address the evolution of hydro-meteorological events and the related potential impacts for the urban and local communities. Therefore, on the basis of the information provided by this methodology, decision makers and local government personnel may evaluate promptly and timely targeted actions, implementing all necessary measures to safeguard the life and property in areas at risk.

Moreover, the flexibility of the GIS environment enables the development of a repeatable and standardisable procedure that may be adapted to support a variety of actors involved in resilience and adaptation projects also in other geographic areas.

It should be noticed that this methodology relies on the use of open-source data in an open-source environment. Open-source geospatial information technologies allow to effectively reduce costs, both in term of time and economic resources, required for carrying out in situ spatial surveys. However, the accuracy of the alerts issued by this tool is strictly related on the resolution of the satellite imageries used. Using open-source data characterized by medium-low resolutions, this tool is suitable only for analyses at national and regional scale. Therefore, further developments of the research should consider the implementation of analyses at a higher resolution to cover local scale events. Moreover, including topographic information (i.e. DTM-Digital Terrain Model and DEM-Digital Elevation Model) and layers of historical floods, can significantly improve the reliability and accuracy of the methodology.

References

Altan, O., R. Backhaus, P. Boccardo, F. Giulio Tonolo, J. Trinder, N. Van Manen, and S. Zlatanova. 2013. *The Value of Geoinformation for Disaster and Risk Management (VALID): Benefit Analysis and Stakeholder Assessment*. Joint Board of Geospatial Information Societies (JBGIS): Copenhagen.

Balbo, S., P. Boccardo, S. Dalmasso, and P. Pasquali. 2013. A public platform for geospatial data sharing for disaster risk management. In *International Society for Photogrammetry and Remote Sensing (ISPRS) Archives*, ed. F. Pirotti, A. Guarnieri, and A. Vettore: 189–195, Padua, Italy. XL-5/W3, W (The role of geomatics in hydrogeological risk, 27–28 February 2013.

Bonetto, S., A. Facello, E.I. Cristofori, W. Camaro, and A. Demarchi. 2016. An approach to use earth observation data as support to water management issues in the Ethiopian rift. In: *Symposium on climate change adaptation in Africa 2016—Addis Ababa Ethiopia*. 2016.

Cardona, O.D., M.K. van Aalst, J. Birkmann, M. Fordham, G. McGregor, R. Perez, R.S. Pulwarty, E.L.F. Schipper, and B.T. Sinh. 2012. Determinants of risk: exposure and vulnerability. In *Managing the risks of extreme events and disasters to advance climate change adaptation. A special report of working groups I and II of the Intergovernmental Panel on Climate Change (IPCC)*, ed. C.B. Field, V. Barros, T.F. Stocker, D. Qin, D.J. Dokken, K.L. Ebi, M. D. Mastrandrea, K.J. Mach, G.K. Plattner, S.K. Allen, M. Tignor, and P.M. Midgley, 65–108. Cambridge, UK, and New York, USA: Cambridge University Press.

Cay, T., F. Iscan, and S.S. Durduran. 2004. The cost analysis of satellite images for using in GIS by the pert. In *XX ISPRS Congress, Com. IV*, 12–23 July 2004, XXXV(B4). 2004, Istanbul, pp. 358–363.

CRED—Centre for Research on the Epidemiology of Disasters and UNISDR-United Nations Office for Disaster Risk Reduction. 2016. *The Human Cost of Weather-related Disasters 1995–2015*. http://www.emdat.be/publications.

Chigbu, N. and D. Onukaogu. 2013. Role of geospatial technology in environmental sustainability in Nigeria-An overview. In *FIG working week 2013 environment for sustainability*. Abuja, Nigeria, 6–10 May, 2013.

Cristofori, E.I., A. Albanese, and P. Boccardo. 2016a. Early warning systems & geomatics: Value-added information in the absence of high resolution data. In *Planning to cope with tropical and subtropical climate change*. ed. M. Tiepolo, E. Ponte, E. Cristofori, 141–153. doi:0.1515/9783110480795-009.

Cristofori, E.I., S. Balbo, W. Camaro, P. Pasquali, P. Boccardo, and A. Demarchi. 2015. Flood risk web-mapping for decision makers: A service proposal based on satellite-derived precipitation analysis and geonode. *In IEEE International Geoscience and Remote Sensing Symposium, IGARSS* 2015 (2015): 1389–1392.

Cristofori, E.I., A. Facello, A. Demarchi, W. Camaro, M. Fascendini, and A. Villanucci. 2016b. A geographic information system as support to the healthcare services of nomadic community, the Filtu Woreda case study. In: *Symposium on climate change adaptation in Africa 2016—Addis Ababa Ethiopia*. 2016.

Donoghue, D.N.M. 2002. Remote sensing: Environmental change. *Physical Geography* 26 (1): 144–151.

Duncan, M., K. Crowley, R. Cornforth, S. Edwards, R. Ewbank, P. Karbassi, C. McLaren, J.L. Penya, A. Obrecht, S. Sargeant, and E. Visman. 2014. *Integrating science into humanitarian and development planning and practice to enhance community resilience*. Full guidelines.

Ehrlich, D., and P. Tenerelli. 2012. Optical satellite imagery for quantifying spatio-temporal dimension of physical exposure in disaster risk assessments. *Natural Hazards* 68 (3): 1271–1289. doi:10.1007/s11069-012-0372-5.

Eiser, J.R. and White et al. 2012. Risk interpretation and action: a conceptual framework for responses to natural hazards. *International Journal of Disaster Risk Reduction* 1: 5–16. doi:10.1016/j.ijdrr.2012.05.002.

Ferrandino, J. 2014. Incorporating GIS as an interdisciplinary pedagogical tool throughout an MPA program. *Journal of Public Affairs Education* 20 (4): 529–544.

Government of Malawi. 2015. *Malawi 2015 floods post disaster needs assessment report Malawi 2015 floods*.

Huffman, G.J., R.F. Adler, D.T. Bolvin, G. Gu, E.J. Nelkin, K.P. Bowman, Y. Hong, E.F. Stocker, and D.B. Wolff. 2007. The TRMM multisatellite precipitation analysis (TMPA): Quasi-global, multiyear, combined-sensor precipitation estimates at fine scales. *Journal of Hydrometeorology* 8 (1): 38–55.

Huffman, G.J., R.F. Adler, D.T. Bolvin, and E.J. Nelkin. 2010. The TRMM multi-satellite precipitation analysis (TMPA). In *Satellite rainfall applications for surface hydrology,* ed. F. Hossain, and M. Gebremichael, 3–22. Springer.

IPCC-Intergovernmental Panel on Climate Change. 2014. *Contribution of Working Group II to the Fifth Assessment Report on Climate Change 2013: The Physical Science Basis*. Cambridge: United Kingdom.

IPCC. 2012. Managing the risks of extreme events and disasters to advance climate change adaptation. *A special report of working groups I and II of the Intergovernmental Panel on Climate Change*. Cambridge, United Kingdom.

Melesse, A.M., Q. Weng, P.S. Thenkabail, and G.B. Senay. 2007. Remote sensing sensors and applications in environmental resources mapping and modelling. *Sensors* 7 (12): 3209–3241. doi:10.3390/s7123209.

Miyazaki, H., M. Nagai, and R. Shibasaki. 2015. Reviews of geospatial information technology and collaborative data delivery for disaster risk management. *ISPRS International Journal of Geo-Information* 4 (4): 1936–1964.

Pandey, J., and D. Pathak. 2014. *Geographic Information System*. New Delhi: The Energy and Resources Institute, TERI.

Sundaresan, J., K.M. Santosh, A. Déri, R. Roggema, and R. Singh. 2014. *Geospatial Technologies and Climate Change*. Springer International Publishing.

Teegavarapu, R.S.V. 2012. *Floods in a changing climate*. Extreme precipitation. New York: Cambridge University Press.

Tiepolo, M., and E.I. Cristofori. 2016a. Climate change characterisation and planning in large tropical and subtropical cities. In *Planning to cope with tropical and subtropical climate change*, ed. M. Tiepolo, E. Ponte, and E. Cristofori, 6–41. Berlin: De Gruyter Open.

Tiepolo, M., and E. Cristofori. 2016b. Planning the adaptation to climate change in cities: an introduction. In *Planning to Cope with Tropical and Subtropical Climate Change*, ed. M. Tiepolo, E. Ponte, and E. Cristofori, 1–5. Berlin: De Gruyter Open.

Tomaszewski, B. 2014. *Geographic Information Systems (GIS) for Disaster Management*. Boca Raton: CRC Press.

UNISDR. 2015. Making development sustainable: the future of disaster risk management. *Global assessment report on disaster risk reduction.* Geneva, Switzerland: United Nations Office for Disaster Risk Reduction.

UNDESA-United Nations Department of Economic and Social Affairs, Population Division. 2011. *World Population Prospects: The 2010 Revision, Highlights and Advance Tables*.

Chapter 8
Building Resilience to Drought in the Sahel by Early Risk Identification and Advices

Patrizio Vignaroli

Abstract Agriculture in the Sahel region is characterized by traditional techniques and is strongly dependent on climatic conditions and rainfall. As a consequence of recurrent droughts in East and West Africa, integrated famine Early Warning Systems (EWS) have been established in order to produce and disseminate coherent and integrated information for prevention and management of food crises. Since the early 1990s, analysis tools and simulation models based on satellite meteorological data have been developed to support Multidisciplinary Working Groups (MWG) in cropping season monitoring. However, many of these tools are now obsolete from the IT perspective or ineffective due to unavailability of regularly updated meteorological data. To ensure a long-term sustainability of operational systems for drought risk zones identification CNR-IBIMET has developed 4Crop, a coherent Open Source web-based Spatial Data Infrastructure to treat all input and output data in an interoperable, platform-independent and uniform way. The 4Crop system has been conceived as a multipurpose tool in order to provide different categories of stakeholder, from farmers to policy makers, with effective and useful information for climate risk reduction and drought resilience improvement. Advice to farmers is a fundamental component of prevention that can allow a better adaptation of traditional cropping systems to climatic variability. Past experiences show that agro-meteorological information and weather forecasts can play a key role for food security, reducing the vulnerability of farmers, strengthening rural production systems and increasing crop yields. A participatory and cross-disciplinary approach is essential to build farmers' trust and to develop effective integration between scientific and local knowledge to increase the rate of dissemination and utilization of advice.

Keywords Drought · Early warning systems · Risk management · Adaptation

P. Vignaroli (✉)
National Research Council—Institute of Biometeorology (IBIMET),
Via Giovanni Caproni 8, 50154 Florence, Italy
e-mail: p.vignaroli@ibimet.cnr.it

M. Tiepolo et al. (eds.), *Renewing Local Planning to Face Climate Change in the Tropics*, Green Energy and Technology,
DOI 10.1007/978-3-319-59096-7_8

8.1 Introduction

Agriculture in the Sahel region is characterized by traditional techniques and is strongly dependent on climatic conditions and rainfall. Moreover, the vulnerability of rainfed crop production systems to drought is worsened by soil degradation, low resilience levels due to poverty and inability to implement effective risk mitigation strategies (FAO 2011).

Food security of the rural population largely depends on environmental conditions and on spatial and temporal climate variability (Jouve 1991). In general, low rainfall during the growing season can lead to lower crop yields and, sometimes, to food crises (Sultan et al. 2005).

Crop yields may suffer significantly with either a late onset or early end of the rainy season, as well as a high frequency of damaging dry spells (Mugalavai et al. 2008). Early rains at the beginning of the season are often followed by dry spells that can last a week or longer. As the amount of water stored in the soil at this time of year is negligible, crops planted early can suffer water shortage stresses during a prolonged dry spell (Marteau et al. 2011). The choice of sowing date is therefore of paramount importance to farmers. Indeed, a successful early planting can produce an earlier crop that may have a higher yield (Sivakumar 1988). Later planting reduces the risk of early crop failures, but yields can decline because of the shorter growing period remaining before the end of the rains (Jones 1976).

Concerning dry spells, there are two extremely critical periods when water shortage can cause wilting of the plants and yield reductions. The first occurs immediately after sowing when plants are germinating. A dry spell at this time can halt germination and cause a sowing failure, requiring a second sowing with consequent loss of seeds and growing days. The second critical period occurs during flowering when a water deficit can severely reduce crop yields (Benoit 1977). The ability to effectively estimate the onset of the season and potentially dangerous dry spells thus becomes vital for planning crop practices aiming to minimize risks and maximize yields. Increasing yearly variability due to climate change has made farmers' traditional knowledge on the planning of sowing and selection of crop type and variety rather ineffective (Waha et al. 2013).

This paper describes the activities of CNR-IBIMET to set up a web-based and open access crop monitoring system for drought risk reduction and resilience improvement.

8.2 The Area of Interest

The geographical area coincides with the western part of the Sahel region. The area extends between Senegal and Chad and is bounded to the north and south by the Saharan and Sudanese climatic zones (Fig. 8.1).

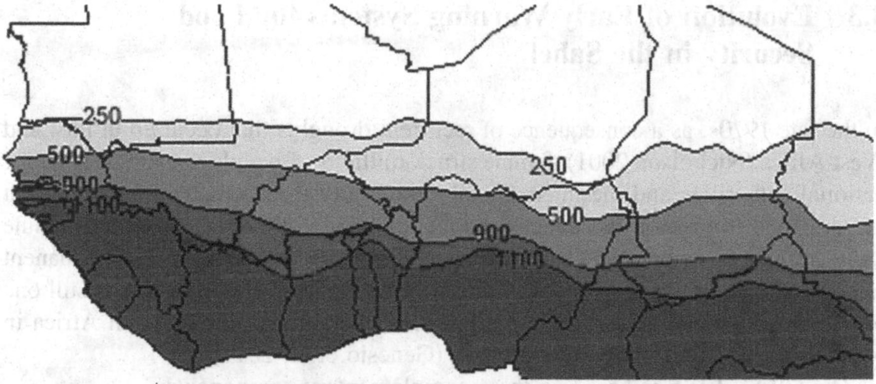

Fig. 8.1 Climatic zones in Western Sahel based on mean annual rainfall 1961–1990: Sahelian (250–500 mm), Soudanese-Sahelian (500–900 mm), Soudanese (900–1,100 mm), Guinean > 1,100 mm (SDRN-FAO)

The wet period mostly occurs in the summer, following the African monsoon. The onset of the rainy season essentially depends on latitude but also has high inter-annual variability (Sivakumar 1988). The rainfall pattern has a unimodal distribution and the annual cumulated rainfall shows a gradient from north to south, going from 150 to 600 mm (Wickens 1997).

With some exceptions, the Sahel landscape is composed of semi-desert grassland, scrub and wooded grasslands, where Acacia species (*Acacia spp.*) play a dominant role (Nicolini et al. 2010). Although the area has been inhabited for at least 9000 years, in the last few decades, over-exploitation of land, closely related to population growth, has led to soil depletion and lower productivity (Di Vecchia et al. 2006).

Human pressure and climate change are leading to a rapid and often irreversible degradation of natural resources (Wezel and Rath 2002).

Millet and sorghum are the primary food crops. In the northern part of the region, which is very sensitive to climatic fluctuations, the most widespread crop is millet, with a cycle length of between 70 and 90 days, depending on the variety. Its water requirement varies from 300 to 500 mm (Bacci et al. 1992). Even if the drought resistance of this crop is naturally high, its sensitivity to water stress greatly increases in the period just after sowing: an interruption of rainfall following the onset of the rainy season can cause the death of seedlings (Kleschenko et al. 2004). Crop failure leads to loss of seeds as well as greater labor demands as farmers must replant their fields. Delayed sowing also entails risks of missing the nitrogen that becomes available through mineralization following the first rains, which is especially important when no fertilizer is applied. (Bacci et al. 1999).

In the southern Sahel areas, the better climate conditions allow farmers to implement different strategies of intensification and diversification of cropping systems. Cereals are often associated with cash crops, such as groundnut and sesame, although these crops are more vulnerable to drought.

8.3 Evolution of Early Warning Systems for Food Security in the Sahel

In the late 1970s, as a consequence of recurrent droughts that occurred in East and West Africa (Nicholson 2001), famine struck millions of people, creating a need for national authorities and the international community to coordinate efforts to limit the dramatic impact on the affected population. Since then, an integrated famine early warning system (EWS) has been established in West Africa by the Permanent Inter-State Committee for Drought Control in the Sahel (CILSS), which is still one of the most effective and reliable mechanisms operating in sub-Saharan Africa in order to prevent and manage food crises (Genesio et al. 2011).

The CILSS EWS is based on three complementary components:

- an information network operating at country level and involving national technical services and international organizations in order to produce and disseminate coherent and integrated information on population's food security status;
- a regional-national coordination, led by CILSS, to facilitate dialogue, consultation and coordination between partners;
- a relief body, to manage emergency activities at national level.

In the early 2000s, the EWS activities were still mainly focused on food availability by monitoring agricultural seasons and agro-climatic events. The main information product concerned the national cereal balance, often coupled with a qualitative assessment of food risk zones provided by agro-meteorological models or composite indicators (Tefft et al. 2006).

However, this approach has proven to be outdated and ineffective in identifying complex multi-source crises such as that of Niger in 2005 (Cornia and Deotti 2008; Vallebona et al. 2008).

A conceptual model to integrate multi-disciplinary data and for the harmonization of information flow within the existing EWS has been provided by the Food Crises Forecasting Timetable (CPCA), developed by AGRHYMET regional Centre of Niamey, Niger, in collaboration with the World Meteorological Organization. The CPCA was designed to respond to decision makers' need to be alerted at an appropriate time depending on the extent of the food crisis (local, national or regional), the coping capacity of the affected population and amount of required food aid (Vignaroli et al. 2009).

This approach was then transferred and implemented in an operational way in the Harmonized Framework (*Cadre Harmonisé*) for the analysis and identification of areas at risk and vulnerable groups in the Sahel established by CILSS in collaboration with national and international organizations involved in the Food Crisis Prevention Network in the Sahel (RPCA). The *Cadre Harmonisé* is a process aiming to achieve technical consensus and to enhance information generated by existing information systems in order to evaluate food and nutrition insecurity on an objective and consensual basis (CILSS 2014). Since 2013, Economic Community of West African States (ECOWAS) has selected the *Cadre Harmonisé* as the main reference framework for food security analysis in 16 countries of West Africa and provides tools for the classification, analysis and reporting of food insecurity, as well as joint approaches for undertaking monitoring, assessments, data collection, and database management.

The Multidisciplinary Working Groups (MWGs) led by National Meteorological Services (NMSs) play a strategic role in the operation of national EWSs especially during the rainy season. MWGs are in charge of providing timely and reliable information on rainfed crops status and agro-hydro-meteorological conditions in support of decision-making processes, from farmers to national and international stakeholders.

8.4 An Open Source Geoprocessing Tool for Drought Monitoring

Early warning systems operate on two levels: prevision and prevention. Prevision is mainly based on tools and models utilizing numeric forecasts and satellite data to predict and monitor the growing season. This approach is centered on the early identification of risks and the production of information within the time prescribed for decision-making (Vignaroli et al. 2009). Agrometeorological models have a central role in this chain, as they can transform meteorological data into levels of risk for agriculture (Di Vecchia et al. 2002), identifying the geographical areas exposed to food insecurity risk due to cereal production shortage in relation to local demand. Advice to farmers is thus a fundamental component of prevention, allowing a proactive response through a better adaptation of the traditional crop calendar to climatic variability (Di Vecchia et al. 2006; Roudier et al. 2014).

Since the early 1990s, crop monitoring analysis tools and simulation models based on meteorological satellite data have been developed within different international cooperation programs to allow monitoring of the cropping season in CILSS countries (Samba 1998; Traore et al. 2011). Software was usually a stand-alone application, transferred to NMSs but without continuous user support or updates. Furthermore, the scarcity of funds for hardware and software maintenance, besides the unavailability of regularly updated and timely meteorological data, led to the failure or ineffectiveness of many of these tools.

In order to meet the needs of the CILSS meteorological services and to ensure a long-term sustainability of operational systems for drought risk zones identification CNR-IBIMET has developed 4Crop (Vignaroli et al. 2016), a coherent Open Source web-based Spatial Data Infrastructure to treat all input and output data in an interoperable, platform-independent and uniform way (Fig. 8.2). This activity is supported by the Global Facility for Disaster Reduction and Recovery (GFDRR)—The Challenge Fund.

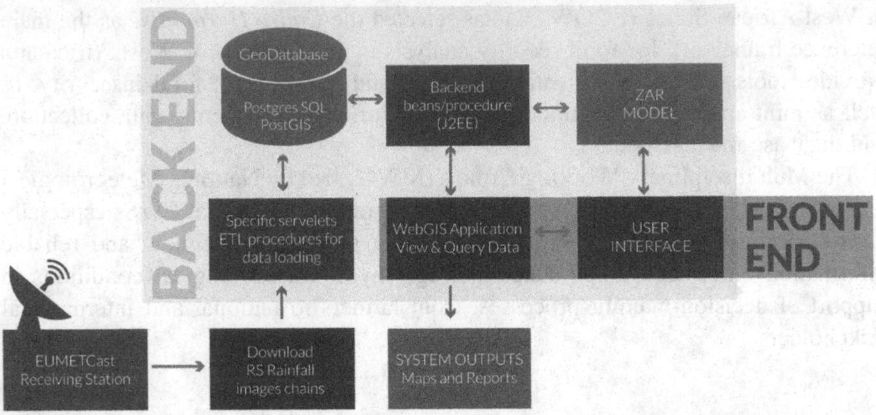

Fig. 8.2 System architecture (Vignaroli et al. 2016)

This web application, targeting Niger and Mali National Meteorological Services, is based on the Crop Risk Zone (CRZ) model (Vignaroli et al. 2016), the updated version of *Zone à Risque* (ZAR) model (CNR-IBIMET 2006) distributed as stand-alone software to the NMSs of CILSS countries within the framework of SVS (Vulnerability Monitoring in the Sahel) Projects, implemented with the World Meteorological Organization at the AGRHYMET Regional Center of Niamey,[1] Niger. The ZAR model accuracy was validated with field data collected in 2006 and 2007 in Mali, Niger and Senegal within the framework of the AMMA project[2] with the support of National Meteorological Services (Bacci et al. 2009). An operational test was also performed on the 2009 season in Burkina Faso, Chad, Mauritania, Mali, Niger and Senegal in collaboration with the NMSs (Bacci et al. 2010).

The CRZ model performs a soil water balance to evaluate the satisfaction of crop water requirements in each phenological stage during the growing period. The model is initialized by a rain threshold (10, 15 or 20 mm); this threshold depends on the crop calendar and the varieties traditionally used by farmers in different agro-ecological zones (Bacci et al. 2009). To best adapt simulation to the real behavior of various cropping systems, the CRZ model allows users to customize some parameters: type of crop and variety, sowing conditions (rain threshold and start of simulation period) and extent of analysis area. So far the model has been tested on the following four crops: pearl millet (85 and 130 days), cowpea (75 days), groundnut (100 and 140 days), sorghum (110 days).

[1]AGRHYMET (Agriculture, Hydrology, Meteorology Regional Center) is a specialized agency of the Permanent inter-state committee against drought in the Sahel (CILSS).

[2]African Monsoon Multidisciplinary Analyses (AMMA) is an international project to improve knowledge and understanding of the West African monsoon (WAM) and its variability.

The model data input can be summarized as follows:

- Gridded daily cumulated rainfall estimate images.
- Gridded daily cumulated precipitation forecast (0–240 h).
- Gridded average daily PET from MOD16 Global Terrestrial Evapotranspiration data set.
- Average start of growing season (computed on the last 10 years).
- Average end of growing season (computed on the last 10 years).
- Gridded soil water storage capacity from Harmonized World Soil Database (FAO/IIASA/ISRIC/ISS-CAS/JRC 2009).
- Phenological phase lengths and crop coefficient Kc (Allen et al. 1998) for each simulated crop.

Due to the lack of a dense weather station network in Africa and the availability and consistency of long-term rainfall data for the Sahel Region, open satellite-derived data sets have been used to provide the CRZ model with the input data required for analysis. NCEP/NOAA Global Forecast System (GFS) is the reference data source for forecast images at 0.25 ° resolution; the Climate Prediction Center (CPC) Rainfall Estimator supplies daily Rainfall Estimates (RFE) at 0.1 ° resolution and EUMETSAT Earth Observation Portal makes available historical series of Multi-Sensor Precipitation Estimate (MPE) at 3 km resolution.

Chains for automatic data downloading have been implemented to ensure a regular update of each data set, while specific procedures and services have been built up to handle model input data flow.

The CRZ model is composed of three modules of analysis (Fig. 8.3). The first two use rainfall estimate images as input data and operate in diagnostic mode. The third module employs precipitation forecast images and works in predictive mode.

Installation Module

It provides an overview of the dates of successful seeding, showing areas where sowing failures may have occurred due to water stress. Zones where a crop was installed later than normal—because of the late onset of the rain season or due to a first sowing failure—are also highlighted. The following outputs are generated:

- image of crop installation dates (where sowing and crop installation conditions occurred);
- image of sowing failure dates (where sowing conditions occurred but not installation conditions);
- image of re-sowing dates (where, after a sowing failure, sowing conditions occurred again);
- image of crop installation anomalies (comparison between actual and average crop installation dates).

Monitoring Module

It performs a diagnostic of crop condition after its installation. The algorithm assesses the satisfaction of water requirements and shows the areas where a water stress happened. The model also provides an estimate of the potential crop yield as

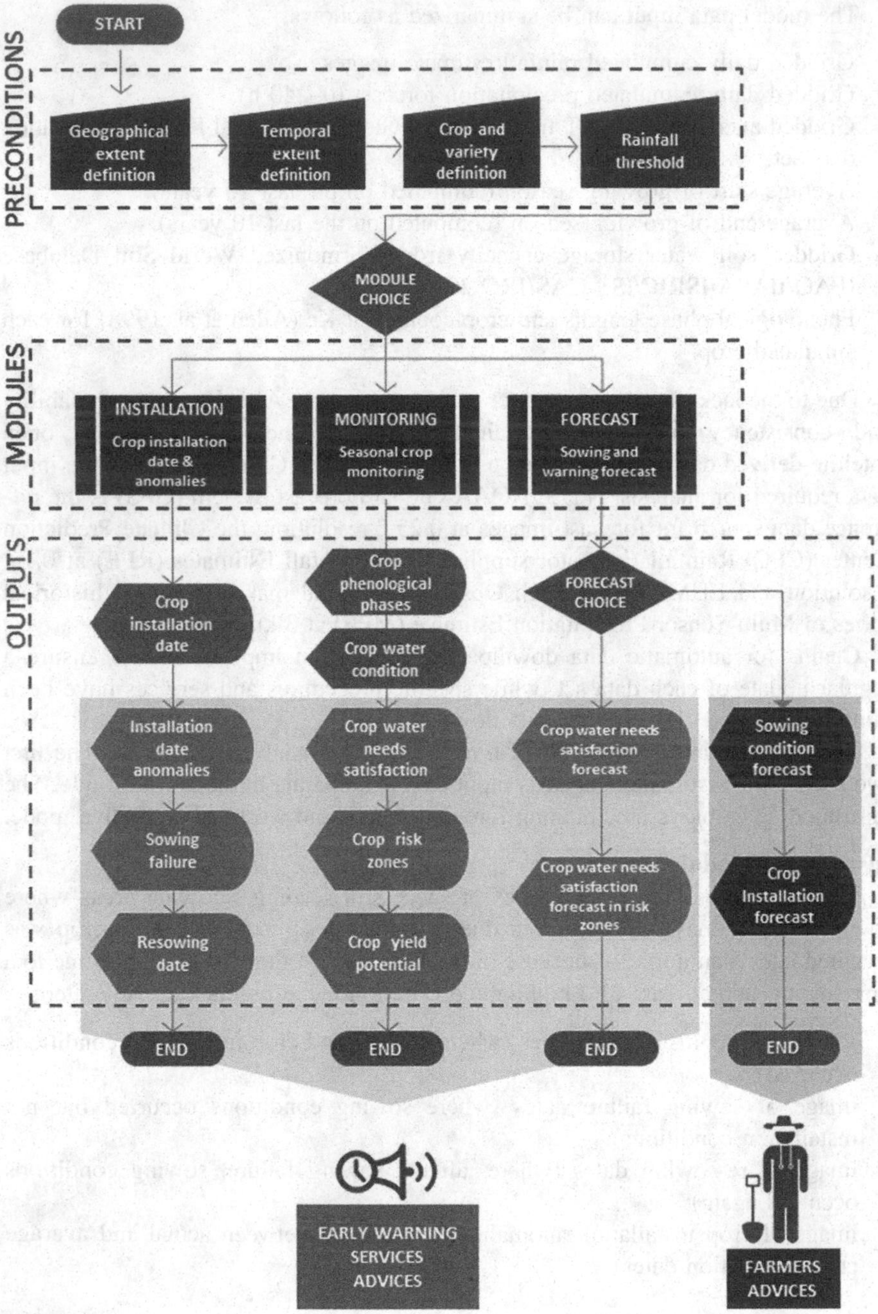

Fig. 8.3 CRZ model flow chart

a consequence of the intensity of water stress and the phenological stage in which it occurred. This module produces the following outputs:

- image of actual phenological phases image (where crop installation occurred);
- crop water requirements satisfaction (water stress level in areas where crop installation occurred);
- available soil water (water actually available in the soil for areas where crop installation occurred).

Forecast module
It is composed of two sub-routines. The first, Sowing Forecast, provides the following outputs on the occurrence of favorable conditions for sowing and the subsequent crop installation:

- forecast date of sowing conditions (where forecasted rainfall satisfies sowing conditions);
- forecast date of crop installation (where the last ten-days estimated rainfall satisfied sowing conditions and the forecasted rainfall allows crop installation conditions to be satisfied).

The second, called Warning Forecast, performs a prognosis on the possible occurrence of a water stress situation for crops already sown. This module produces the following outputs:

- forecast of crop water requirements satisfaction (in areas where crop installation occurred, forecasted satisfaction level of crop water requirements is shown);
- forecast of crop water requirements satisfaction in risk zones (in areas where crop stress conditions have already been identified, crop water requirements are forecasted).

Graphic Users' Interfaces (GUI) and specific WEB GIS applications have been implemented to allow users initializing of analysis modules and to perform model outputs management and visualization.

8.5 Results and Discussion

The challenge of the 4Crop web application is to provide open access to CRZ model output and results. The goal is to support CILSS early warning system and any other local users in decision-making to foster climatic risk management and resilience to drought. In order to avoid a language barrier that could prevent a wider use of the web application, 4Crop is available in French, the official language of the target countries[3] (Fig. 8.4).

[3]In the first phase of GFDRR project, activities have been focused on Mali and Niger. The second phase will concern Burkina Faso, Mauritania and Senegal.

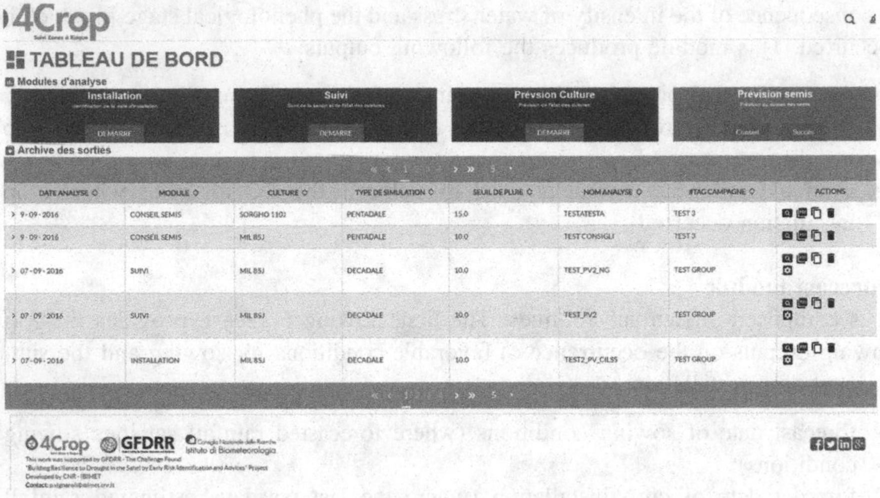

Fig. 8.4 System users' interface

Although Internet connections in most African countries may be a limitation because of poor efficiency of communications networks, the decision to adopt a web-based system allows problems due to the continuous need for software updating and maintenance to be overcome and it also ensures a timely and regular update of input satellite data. In addition, a stand-alone version would still need a large amount of data to be downloaded via the Internet to update the database of satellite images, while the loading of system user interface, if well designed and implemented, requires significantly less resources. Finally, Internet services are reaching good quality in the capital cities of most Sahel countries (Nyirenda-Jere and Biru 2015) where the system users (NMSs) are located.

Moreover, the proposed approach is meant to encourage the integration and sharing of interoperable and open source solutions, thus contributing to the setting up of distributed climate services in developing countries (WMO 2014).

The involvement of national Meteorological Services, as primary users of the system, during all the phases of its development allowed 4Crop to be made easier and more intuitive to use by improving the ergonomics of the Graphic Users' Interface through a co-design approach. This has also contributed to better tailoring the system outputs in terms of both content and formats, to best meet the needs of different users.

In this sense, 4Crop system has been conceived as a multipurpose tool (Fig. 8.5) in order to provide different categories of stakeholders, from farmers to political decision makers, with effective and useful information for implementing specific actions at national and local level for climate risk reduction and drought resilience improvement.

The outputs provided by the 4Crop system in diagnostic mode are specifically addressed to support drought monitoring activities performed by the national

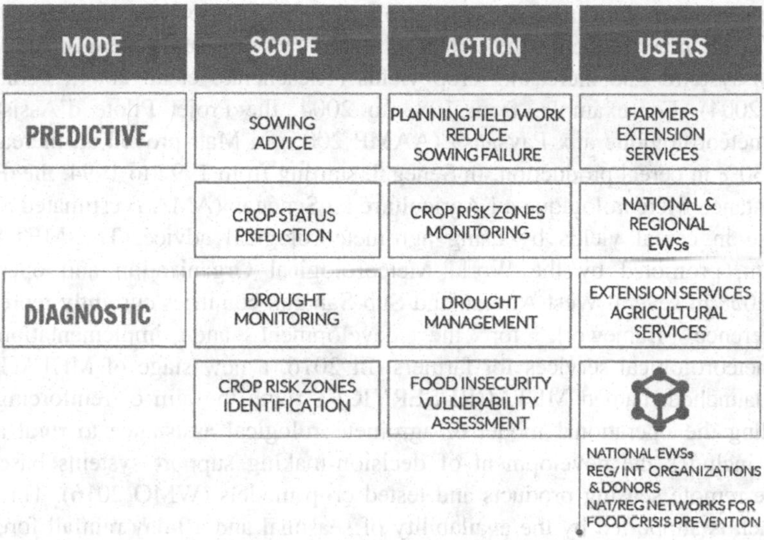

MODE
SCOPE
ACTION
USERS
PREDICTIVE
SOWING ADVICE
PLANNING FIELD WORK REDUCE SOWING FAILURE
FARMERS EXTENSION SERVICES
CROP STATUS PREDICTION
CROP RISK ZONES MONITORING
NATIONAL & REGIONAL EWSs
DIAGNOSTIC
DROUGHT MONITORING
DROUGHT MANAGEMENT
EXTENSION SERVICES AGRICULTURAL SERVICES
CROP RISK ZONES IDENTIFICATION
FOOD INSECURITY VULNERABILITY ASSESSMENT
NATIONAL EWSs
REG/ INT ORGANIZATIONS
& DONORS
NAT/REG NETWORKS FOR FOOD CRISIS PREVENTION

Fig. 8.5 "4Crop" multipurpose tool

Multidisciplinary Working Groups during the agro-pastoral season. An early identification of crop risk areas represents a key step in providing a preliminary evaluation on the dimension of expected food crises (local, national, regional) in order to alert national and international institutions responsible for their prevention and management (Vignaroli et al. 2009). Moreover, monitoring of crop risk areas throughout the cropping season contributes to the final assessment of the real food insecurity status of the population concerned within the framework of national and regional EWS set up in West Africa (CILSS 2014).

The 4Crop products generated in predictive mode, based on GFS 10-days precipitation forecast, are mainly conceived for "advisory-support" activities to farmers, in order to implement appropriate, short-term strategies for minimizing drought risk in crops (i.e. identification of the optimal sowing period, adoption of suitable cropping practices to improve soil fertility and water use efficiency) and contribute to building farmers' resilience.

Field observations in Mali since 1983 show that when farmers use agrometeorological information to plan the sowing date and choose the varieties to be used, yields are higher than with traditional choices (Hellmuth et al. 2007). In addition, some simulation studies performed in Cameroon indicate that crop yields of maize and groundnut with an optimal planting date are usually higher than yields obtained using traditional ones under climate change (Laux et al. 2010; Tingem et al. 2009). This is in agreement with Waha et al. (2013), as the adaptation of sowing dates in this latter study usually resulted in higher crop productivity in most regions and cropping systems of sub-Saharan Africa.

Advice to farmers is indeed a fundamental component of prevention that allows a better adaptation of the traditional crop calendar to climatic variability.

Past experiences show that agro-meteorological information can play a key role for food security, reducing the vulnerability of farmers, strengthening rural production systems and increasing crop yields (Kleschenko et al. 2004,; Pérarnaud et al. 2004). For example, from 1983 to 2004, the«Projet Pilote d'Assistance Agro-météorologique aux Paysans» (AAMP 2005) in Mali proved an increase of about 30% in cereal production. In Senegal, starting from 1992 to 1994, the«Projet d'Assistance Météorologique à l'Agriculture au Sénégal»(AMAS) estimated a 20% increase in cereal yields by using agrometeorological advice. The METAGRI program, promoted by the World Meteorological Organization and operating since 2008 in several West African and Sub-Saharan countries currently represents a reference framework for the development and implementation of agro-meteorological services for farmers. In 2016, a new stage of METAGRI is being launched, named METAGRI SERVICES, with the aim of reinforcing and extending the operational model of agrometeorological assistance to rural areas, which includes the development of decision-making support systems based on reliable remote sensing products and tested crop models (WMO 2016). This new approach is supported by the availability of seasonal and 10-day rainfall forecasts (Njau 2010) and the possibility for farmers to access IC technologies (mobile phones) in order to disseminate information in rural areas (Aker and Fafchamps 2015).

Empirical studies among African farmers have shown that climate forecasts can help reduce their vulnerability to drought and climate extremes, while also allowing them to maximize opportunities when favorable rainfall conditions are predicted (Roudier et al. 2014; Martinez et al. 2012). However, it should be underlined that weather forecasts in situations of climate uncertainty generally don't modify the farmer's decision, but reinforces what has already been decided based on his/her own experience and observations (Roncoli et al. 2009). In fact, farmers combine information from different sources and multiple timeframes (Orlove et al. 2010). So they have some expectation about the coming season based on prior experience, empirical observations, and local knowledge, and then adjust their strategies when shorter-term forecasts and real-time information become available (Orlove et al. 2010).

In this regard, the analysis of the behavior of different groups of smallholder farmers in two agro-ecological areas of Senegal provides examples of how farmers could use predictive climate information at different timescales, as seasonal and 10-day rainfall forecasts. (Roudier et al. 2014). In most cases, farmers identify at least one strategy they could use in response to forecasts. Most adaptation strategies pertain to 10-day rather than seasonal forecasts, though the latter are used especially where farmers have a wider range of options in planning preparatory activities, as well as the choice of crops and/or varieties to sow, the amount of mineral or organic fertilizer to distribute before sowing, extent of the crop surfaces.

Adjusting sowing date is the most common change in traditional practices adopted by farmers in response to 10-day rainfall forecasts; farmers decide to sow earlier or later than originally planned depending both on the configuration of rainfall in the forecasted 10 days (wet or dry) and in the preceding 10 days.

This information is particularly important at this time, which is marked by high uncertainty and anxiety for farmers (Roudier et al. 2014). On the same bases, farmers could also decide to intensify weeding practices—to control weed proliferation and facilitate infiltration to limit runoff and soil erosion—as well as to anticipate harvest to prevent germination of grains or avoid damage by pests and diseases that thrive in humid conditions (Roudier et al. 2014).

Overall, although farmers put their trust in local knowledge, they are open to receiving and integrating scientific knowledge in their decisional process. This encourages extending the use of rainfall forecasts in producing and disseminating advice as a fundamental part of climate services to farmers in order to foster a proactive reaction to climate change and to cope with drought. In this sense, the adoption of a participatory, cross-disciplinary approach is essential to build farmers' trust and develop a fruitful collaboration between scientific and traditional knowledge. Indeed, this interaction has been recognized as fundamental for the adoption of innovation in climate change adaptation because the enhancement of local knowledge facilitates effective communication and increases the rate of dissemination and use of advice (Nyong et al. 2007).

8.6 Conclusions

4Crop is a multipurpose system implemented in collaboration with the National Meteorological Services of Niger and Mali to monitor the cropping season and provide an early identification of crop risk zones in order to establish effective responses for drought risk management at different levels. Within the perspective of the setting-up of distributed climate services, the overall 4Crop approach could represent an enabling factor to allow a switch from generic information to advice tailored to specific user needs. Indeed, 4Crop outputs have been designed as elements of information chains targeted on two main user categories under the principle of information-action.

The first group are decision makers at national/regional level who use information to implement actions and policies to prevent and manage food crises and strengthen food security of the population. In this case, the critical elements are represented by the sustainability of early warning systems and the ability to generate timely and appropriate information for decision makers. Strengthening of the capacity of national institutions—particularly weather services—as well as the setting up of distributed climate services that allow stakeholders at different levels to access and share information products through web services protocols and standards, can thus represent elements that facilitate decision-making and improve its effectiveness.

The second group of users are decision makers at local level, essentially farmers, who have to decide which crops and how to grow them. This requires not only finding a suitable solution for translating information products into services that are simple but effective for farmers, but also establishing a framework for communication and collaboration between scientific institutions, national technical services and rural communities for an effective dissemination and adoption of the advice.

To be effective and sustainable, climate services for farmers must be integrated into a concerted and participatory strategy (including content and format of advice, communication channel, partnership with other local stakeholders and training), which enables final users to adopt scientific innovation in ways that are socially acceptable and environmentally sound.

In terms of beneficiaries, the approach will also pay special attention to gender: in many cases, women are the backbone of the farming community, however they are at a disadvantage in their access to climate and weather information as well as access to training programs.

References

AAMP. 2005. Assistance Météorologique Opérationnelle au Monde Rural au Mali. Développement—Acquis—Perspectives. Coopération Suisse.

Aker, J.C., and M. Fafchamps. 2015. Mobile phone coverage and producer markets: evidence from West Africa. *The World Bank Economic Review* 29 (2): 262–292. Doi:doi.org/10.1093/wber/lhu006.

Allen, R.G., L.S. Pereira, D. Raes, and M. Smith. 1998. *Crop evapotranspiration—Guidelines for computing crop water requirements—FAO irrigation and drainage paper 56*. Rome: FAO.

Bacci, L., G. Maracchi, and B. Senni. 1992. *Les stratégies agrométéorologiques pour les pays Sahéliens*. Florence: IATA-CeSIA.

Bacci, L., C. Cantini, F. Pierini, G. Maracchi, and F.N. Reyniers. 1999. Effects of sowing date and nitrogen fertilization on growth, development and yield of a short day cultivar of millet (Pennisetum glaucum L.) in Mali. *European Journal of Agronomy* 10 (1): 9–21. doi:10.1016/S1161-0301(98)00046-X.

Bacci, M., V. Tarchiani, P. Vignaroli, L. Genesio, and A. Di Vecchia. 2009a. *Identification et suivi des zones à risque agro-météorologique au Sahel*. Rome: Aracne Editrice.

Bacci, M., A. Di Vecchia, L. Genesio, I. Hassan, M. Ndiaye, V. Tarchiani, and P. Vignaroli. 2009. D 5.2.e: AMMA information products tested within the CPCA for EW at national and farmer's scale. African Monsoon Multidisciplinary Analysis (AMMA) project.

Bacci, M., A. Di Vecchia, L. Genesio, V. Tarchiani, and P. Vignaroli. 2010. Drought impact detection on crops in the Sahel: a case study for the 2009 campaign. In proceeding of the GI4DM 2010 Conference Geomatics for crisis management, Feb 2–4, Turin, Italy.

Benoit, P. 1977. The start of the growing season in Northern Nigeria. *Agricultural Meteorology* 18: 91–99.

CILSS-Comité Permanent Inter-Etats de Lutte contre la Sécheresse dans le Sahel. 2014. Identification et analyse des zones à risque et des populations en insécurité alimentaire et nutritionnelle au Sahel et en Afrique de l'Ouest. Manuel. Version 1.0 http://www.agrhymet.ne/PDF/Manuel%20CH_version%20finale.pdf. Accessed 8 July 2016.

CNR-IBIMET. 2006. SPM Suivi des pluies par Meteosat & ZAR Modèle Zone à Risque. Version 2.0. Centre Regional AGRHYMET. ISBN 88-900502-1-7.

Cornia, G.A., and L. Deotti. 2008. Niger's 2005 food crisis: extent, causes and nutritional impact. EUDN/WP, 15. https://www.researchgate.net/profile/Giovanni_Cornia/publication/237773711_NIGER'S'_2005_FOOD_CRISIS_EXTENT_CAUSES_AND_NUTRITIONAL_IMPACT/links/54aab7920cf2bce6aa1d58f5.pdf. Accessed 5 July 2016.

Di Vecchia, A., P. Vignaroli, and B. Djaby. 2002. Les crises alimentaires et les systèmes de prévision au Sahel. https://www.researchgate.net/publication/267374263_Les_crises_alimentaires_et_les_systemes_de_prevision_au_Sahel. Accessed 12 July 2016.

Di Vecchia, A., M. Bacci, G. Pini, V. Tarchiani, and P. Vignaroli. 2006. Meteorological forecasts and agrometeorological models integration: a new approach concerning early warning for food security in the Sahel. In AARSE 2006: Proceeding of the 6th AARSE international conference on earth observation and geoinformation sciences in support of Africa's development, 30 Oct–2 Nov 2006, Cairo, Egypt: The National Authority for Remote Sensing and Space Science (NARSS). ISBN 1-920-01710-0.

FAO/IIASA/ISRIC/ISS-CAS/JRC. 2009. Harmonized world soil database (version 1.1). FAO, Rome, Italy and IIASA, Laxenburg, Austria. http://www.fao.org/docrep/018/aq361e/aq361e.pdf. Accessed 13 July 2016.

FAO. 2011. *The state of the world's land and water resources for food and agriculture (SOLAW)—Managing systems at risk*. Rome-London: Food and Agriculture Organization of the United Nations and Earthscan.

Genesio, L., M. Bacci, C. Baron, B. Diarra, A. Di Vecchia, A. Alhassane, I. Hassane, M. Ndiaye, N. Philippon, V. Tarchiani, and S. Traoré. 2011. Early warning systems for food security in West Africa: evolution, achievements and challenges. *Atmospheric Science Letters* 12 (1): 142–148. doi:10.1002/asl.332.

Hellmuth, M.E., A. Moorhead, M.C. Thomson, and J. Williams. 2007. *Climate risk management in Africa: learning from practice 01-261*. Columbia University, International Research Institute for Climate and Society.

Jones, M.J. 1976. Planting time studies on maize at Samaru, Nigeria, 1970–1973. Samaru Miscellaneous Paper (Nigeria) no. 57.

Jouve, P. 1991. Sécheresse au Sahel et stratégies paysannes. *Science et changements planétaires/Sécheresse* 2 (1): 61–69.

Kleschenko, A., L. Grom, M. Ndiaye, and R. Stefanski. 2004. The impact of agrometeorological applications for sustainable management of farming, forestry and livestock systems. Report CagM WMO/TD No. 1175. World Meteorological Organization.

Laux, P., G. Jäckel, R.M. Tingem, and H. Kunstmann. 2010. Impact of climate change on agricultural productivity under rainfed conditions in Cameroon-A method to improve attainable crop yields by planting date adaptations. *Agricultural and Forest Meteorology* 150 (9): 1258–1271. doi:j.agrformet.2010.05.008.

Marteau, R., B. Sultan, V. Moron, A. Alhassane, C. Baron, and S.B. Traoré. 2011. The onset of the rainy season and farmers' sowing strategy for pearl millet cultivation in Southwest Niger. *Agricultural and Forest Meteorology* 151 (10): 1356–1369. doi:10.1016/j.agrformet.2011.05.018.

Martinez, R., D. Hemming, L. Malone, N. Bermudez, G. Cockfield, A. Diongue, J. Hansen, A. Hildebrand, K. Ingram, G. Jakeman, M. Kadi, G.R. McGregor, S. Mushtaq, P. Rao, R. Pulwarty, O. Ndiaye, G. Srinivasan, E. Seck, N. White, and R. Zougmore. 2012. Improving climate risk management at local level—Techniques, case studies, good practices and guidelines for World Meteorological Organization members. In *Risk management—Current issues and challenges*, ed. N. Banaitiene. Europe: InTech. doi:10.5772/51554.

Mugalavai, E.M., E.C. Kipkorir, D. Raes, and M.S. Rao. 2008. Analysis of rainfall onset, cessation and length of growing season for western Kenya. *Agricultural and Forest Meteorology* 148 (6–7): 1123–1135. doi:j.agrformet.2008.02.013.

Njau, L.N. 2010. Seasonal-to-interannual climate variability in the context of development and delivery of science-based climate prediction and information services worldwide for the benefit of Society. *Procedia Environmental Sciences* 1: 411–420. doi:10.1016/j.proenv.2010.09.029.

Nicholson, S.E. 2001. Climatic and environmental change in Africa during the last two centuries. *Climate Research* 17: 123–144. doi:10.3354/cr017123.

Nicolini, G., V. Tarchiani, M. Saurer, and P. Cherubini. 2010. Wood-growth zones in Acacia seyal Delile in the Keita Valley, Niger: Is there any climatic signal? *Journal of Arid Environments* 74: 355–359. doi:10.1016/j.jaridenv.2009.08.017.

Nyirenda-Jere, T., and T. Biru. 2015. Internet development and internet governance in Africa. ISOC Report. http://www.internetsociety.org/sites/default/files/Internet%20development%20and%20Internet%20governance%20in%20Africa.pdf. Accessed 22 July 2016.

Nyong, A., F. Adesina, and B.O. Elasha. 2007. The value of indigenous knowledge in climate change mitigation and adaptation strategies in the African Sahel. *Mitigation and Adaptation Strategies for Global Change* 12 (5): 787–797. doi:10.1007/s11027-007-9099-0.

Orlove, B., C. Roncoli, M. Kabugo, and A. Majugu. 2010. Indigenous climate knowledge in southern Uganda: the multiple components of a dynamic regional system. *Climatic Change* 100 (2): 243–265. doi:10.1007/s10584-009-9586-2.

Pérarnaud, V., A. Bootsma, P. Isabirye, and B.L. Lee. 2004. Communication of agrometeorological information. Report CagM WMO/TD No. 1254. Geneva: World Meteorological Organisation.

Roncoli, C., C. Jost, P. Kirshen, M. Sanon, K.T. Ingram, M. Woodin, L. Somé, F. Ouatara, B. J. Sanfo, C. Sia, and P. Yaka. 2009. From accessing to assessing forecasts: an end-to-end study of participatory climate forecast dissemination in Burkina Faso (West Africa). *Climatic Change* 92 (3–4): 433–460. doi:10.1007/s10584-008-9445-6.

Roudier, P., B. Muller, P. D'Aquino, C. Roncoli, M.A. Soumaré, L. Batté, and B. Sultan. 2014. The role of climate forecasts in smallholder agriculture: lessons from participatory research in two communities in Senegal. *Climate Risk Management* 2: 42–55. doi:10.1016/j.crm.2014.02.001.

Samba, A. 1998. Les logiciels DHC de diagnostic hydrique des cultures. Prévision des rendements du mil en zones soudano-sahéliennes de l'Afrique de l'Ouest. *Sécheresse* 9 (4): 281–288.

Sivakumar, M.V.K. 1988. Predicting rainy season potential from the onset of rains in Southern Sahelian and Sudanian climatic zones of West Africa. *Agricultural and Forest Meteorology* 42: 295–305.

Sultan, B., C. Baron, M. Dingkuhn, B. Sarr, and S. Janicot. 2005. La variabilité climatique en Afrique de l'Ouest aux échelles saisonnière et intra-saisonnière. II: applications à la sensibilité des rendements agricoles au Sahel. *Sécheresse* 16 (1): 23–33.

Tefft, J., M. McGuire and N. Maunder. 2006. Planning on the future. An assessment of food security early warning systems in sub-Saharan Africa. Synthesis Report. Rome: FAO.

Tingem, M., M. Rivington, and G, Bellocchi. 2009. Adaptation assessment for crop production in response to climate change in Cameroon. *Agronomy for Sustainable Development* 29 (2): 247–256. doi:10.1051/agro:2008053.

Traoré, S.B., A. Alhassane, B. Muller, M. Kouressy, L. Somé, B. Sultan, P. Oettli, A.C. Siéné Laopé, S. Sangaré, M. Vaksmann, M. Diop, M. Dingkhun, and C. Baron. 2011. Caracterizing and modelling the diversity of cropping situations under climatic constraints in West Africa. *Atmospheric Science Letters* 12 (1): 89–95. doi:10.1002/asl.295.

Vallebona, C., L. Genesio, A. Crisci, M. Pasqui, A. Di Vecchia, and G. Maracchi. 2008. Large scale climatic patterns forcing desert locust upsurges in West Africa. *Climate Research* 37: 35–41. ISSN: 1616–1572. doi:10.3354/cr00744.

Vignaroli, P., R. Rocchi, T. De Filippis, V. Tarchiani, M. Bacci, P. Toscano, M. Pasqui, and E. Rapisardi. 2016. The crop risk zones monitoring system for resilience to drought in the Sahel. In *Geophysical Research Abstracts* 18, EGU2016-16616-3, 2016. EGU General Assembly 2016.

Vignaroli, P., V. Tarchiani, M. Bacci, and A. Di Vecchia. 2009. La prévention des crises alimentaires au Sahel. In *Changements globaux et développement durable en Afrique*. Rome: Aracne Editrice.

Waha, K., C. Müller, A. Bondeau, J.P. Dietrich, P. Kurukulasuriya, J. Heinke, and H. Lotze-Campen. 2013. Adaptation to climate change through the choice of cropping system and sowing date in sub-Saharan Africa. *Global Environmental Change* 23 (1): 130–143. doi:10.1016/j.gloenvcha.2012.11.001.

Wezel, A., and T. Rath. 2002. Resource conservation strategies in agro-ecosystems of semi-arid West Africa. *Journal of Arid Environments* 51 (3): 383–400. doi:10.1006/jare.2001.0968.

Wickens, G.E. 1997. Has the Sahel a future? *Journal of Arid Environments* 37 (4): 649–663. doi:10.1006/jare.1997.0303.

WMO-World Meteorological Organization. 2014. Implementation Plan of the Global Framework for Climate Services. WMO, GFCS http://www.wmo.int/gfcs/sites/default/files/implementation-plan//GFCS-IMPLEMENTATION-PLAN-FINAL-14211_en.pdf. Accessed 20 July 2016.

Chapter 9
Rethinking Water Resources Management Under a Climate Change Perspective: From National to Local Level. The Case of Thailand

Francesca Franzetti, Alessandro Pezzoli and Marco Bagliani

Abstract Likewise many other countries in Southeast Asia region, Thailand has historically enjoyed relatively abundant water resources. Nevertheless, recently, this flood-prone country's attention has shifted to drought, as evidence of a globally changing climate. In order to gain better insights of Thai water resources and disaster management, a review of institutions involved and policies promulgated at national level has been conducted. What comes up from this review is that, on paper, Thailand does present a very complex and sophisticated disaster management devise which, apparently, does not seem to be linked in any way to ordinary water resources management, and what is more important is that a gap emerges when it comes to translate a national-level master plan into lower administrative levels (namely at regional, provincial, district and local administrative organization levels). Poor communication, overlapping roles and responsibilities amongst concerned agencies, lack of budget availability and no long-term vision plans are only few of the shortcomings hindering an effective implementation of disaster prevention and mitigation plans. Hence, this chapter seeks to rethink water-related disaster management in Thailand by (re-)shaping the institutional and policy landscapes, envisaging more holistic coordination mechanisms and information flow which would engage all administrative levels (from national level to local level) and concerned stakeholders.

Keywords Disaster management · Water resources · Thailand

F. Franzetti (✉) · A. Pezzoli
Interuniversity Department of Regional and Urban Studies and Planning, Politecnico and University of Turin, viale Mattioli 39, 10125 Turin, Italy
e-mail: francesca.franzetti@edu.unito.it

A. Pezzoli
e-mail: alessandro.pezzoli@polito.it

M. Bagliani
Department of Economy and Statistics Cognetti de Martiis, University of Turin, CLE-Lungo Dora Siena 100, 10100 Turin, Italy
e-mail: marco.bagliani@unito.it

M. Tiepolo et al. (eds.), *Renewing Local Planning to Face Climate Change in the Tropics*, Green Energy and Technology,
DOI 10.1007/978-3-319-59096-7_9

9.1 Introduction

9.1.1 The Socio-economical Analysis

The interest that lies behind this research can be expressed quoting the World Economic Forum' Global Risk Report 2015 "Global water crises—from drought in the world's most productive farmlands to the hundreds of millions of people without access to safe drinking water—are the biggest threat facing the planet over the next decade" (Ganter 2015).

Water resources are one of the key elements necessary to support global economic and social development. However, nowadays evidence shows that their exploitation at an unsustainable rate, exacerbated by population growth, rapid industrialization and non-effective natural resources management, combined with the increasing threat of the global climate change are making this precious resource even more scarce and finite than ever experienced before by the humankind.

What several regions across the world are currently experiencing suggests the need to take immediate action in rethinking water resources management, introducing a more climate-sensitive approach.

Southeast Asia, in general, and Thailand, in particular, are no exception to these global threats. The region has seen, and will very likely see in the near future, a sharp increase of water-related extreme events and natural hazards such as floods, typhoons, tropical storms but also droughts. These events trigger serious impacts on physical water availability, both quantitatively and qualitatively.

This research analyzes the drought issues in Thailand in order to increase awareness about its management. Thailand is a country with well-established disaster management mechanisms. Nonetheless these mechanisms are a little imbalanced: indeed flood management has received much greater attention (not unjustified though) from the Government whilst drought management still struggles to find its way. This is not to say that drought is more important that flood (or vice versa), only that if one of the ultimate goals of water resources management is to overcome water crises, they both need to be considered by decision makers in an holistic manner. Therefore, this chapter explores what are the existing institutions and policy options to deal with this hazard and suggest a new framework, based on risk-reduction and proactive approach between national and local levels, which might be useful for its management.

The research has been mostly carried out under the SUMERNET's Regional Assessment on water scarcity and drought management in the Mekong Region,[1] led by the Stockholm Environment Institute-Asia Center in Bangkok (SEIA).

The work is structured as follows. Section 9.1 will provide a brief overview of Thailand's physical and socio-economic profile, with a special focus on its water

[1]SUMERNET-Sustainable Mekong Research Network is a long-term program of about 50 different institutions involved, focused on research and capacity building in the Mekong Region covering a wide range of research topics. For more information http://sumernet.org/.

resources status; regional climate change impacts and future projections for the region will be presented, including El Niño-Southern Oscillation (ENSO) influence of the regional climate. Section 9.2 describes briefly the methods and the materials used in this research. Section 9.3 describes an overview of the institutional framework that governs water resources and disaster management at the national level. Section 9.4 introduces laws, policies, strategies and plans of drought's management in Thailand. Section 9.5 analyzes a 4-step risk-based drought management framework, modeled from international guidelines (in particular guidelines for Eastern Europe and the Near East countries), to be tailored to "fit" in the Thai context. Ideal institutional and policy adjustments will be suggested for each of the four steps in particular analyzing in the first step the passage between the national to the local level. Then the challenges and opportunity for Thailand of the framework's application at national and local levels are analyzed in Sect. 9.6. Finally, Sect. 9.7 will look at potentialities and critical challenges of the proposed framework with critical lens, reflecting on the actual feasibility of its application and considering also a possibility for scaling the framework up to (but not only) the regional level.

9.1.2 Thailand and Increasing Threat of Drought

Thailand's climate can be classified as tropical where seasonal monsoon winds, namely the southwest (SW) monsoon and the northeast (NE) monsoon, influence rainfall patterns, in addition to the passage of the Inter Tropical Convergence Zone (ITCZ) and tropical cyclones (TMD 2007).

Meteorologically, Thailand's climate is divided into three main seasons:

- rainy season (or SW monsoon season): usually lasting from mid-May to mid-October. Abundant rain falls all over the country, with August-September being the wettest period of the year (with the exception of southeast region where the rainfall continues until December)
- winter season (or NE monsoon season): usually lasting from mid-October to mid-February. This is the mild period of the year, in the northern regions it can become quite cold
- summer season (or pre-monsoon season): this is considered that usually lasting from mid-February to mid-May. In this period it becomes very warm, with April being the hottest month of the year (especially in the northern part).

Rainfall is a significant, if not the most important source of water for Thailand. Usual rainfall patterns vary considerably across the country as well as Thailand has been historically considered a country with abundant water resources and relies heavily on monsoon rainfall patterns.

Thailand's fluvial system comprises 25 river basins, divided into 254 sub-basins. Of these river basins, 7 have been considered "hot spots" given the economic and social pressure combined with limited freshwater resources: Chao Phraya and Thajeen (central part), Chi and Moon rivers (northeast), Bang Pakong and easter seaboard (east) and Songkla lake (south) (The World Bank 2011).

Mainly used for domestic consumption, agriculture and industries, this water source is getting rapidly overexploited. As of 2014, according to information provided by the latest DWR's Water Resources Management Strategy of Thailand, water demand can be summarized as following:

- The agricultural sector remain the major water user with more that 80% of the country's total water use, through a mix of large- medium- and small-scale infrastructure system. 65,000 million m^3 are delivered to irrigated areas across the nation. Where there is no irrigation system, agriculture is rain-fed, this area corresponds to 19.2 million ha. In order to support the dry-season cultivation, groundwater extraction is some areas is deeply practiced.
- The 2014 water demand for the household consumption is estimated at 6490 million m^3 with projections for 2027 of an increase of 8260 million m^3.
- Concentration of manufacturing and industrial estates is around Bangkok and in the eastern part of the country. For the water sector is estimated around 4202 million m^3 (projected 7515 million m^3 by 2027).
- The minimum water flow requirement for the dry season should not be less than 27,090 million m^3.

Although the rapid economic growth is not at same rates of 80s and early 90s, rapid urbanization and industrialization are still happening. These processes, while beneficial for poverty reduction and income increase, are putting serious pressure over natural resources, especially water resources.

Significant water quality issues arising from an uncontrolled exploitation of both surface and ground water resources as well as pollution of water bodies from industrial discharges affect both surface and groundwater resources (The World Bank 2011). Some areas experience saline intrusion problems, not only in coastline areas but also in the Northeast, where salinity encroachment combines with soil acidity issues. Salinity intrusion is a concern affecting not only water supply but also crop cultivation fisheries and domestic uses (The World Bank 2011).

It is worth to bear in mind that there is a particular climatic phenomenon that has strong repercussions on the Thailand and South East Asia's climate and weather. This phenomenon is called El Niño Southern Oscillation (ENSO). It is noticed how, during the El Niño period, the South and Southeast Asia suffer rainfall deficits as well representend in Fig. 9.1. In the same time significant increases in intensity (but reduction in the number) of tropical cyclones will lead to more intense storms in the region. Extreme rainfall events will increase the level of flood risk.

Narrowing the focus on the Mekong Region (Cambodia, Lao PDR, Myanmar, Thailand and Vietnam), studies conducted at regional level have shown how a greater inter-seasonal and inter-annual variability in rainfall patterns and hydrology across the region are expected, with large uncertainties on the monsoon pattern's behavior (Adamson and Bird 2010). This fact is witnessed by Thailand which is

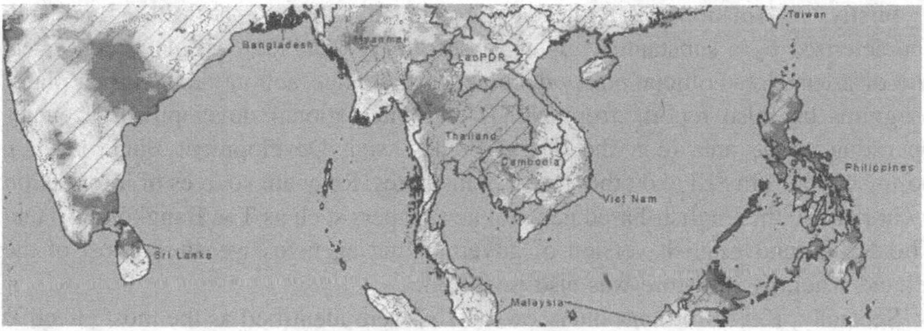

Fig. 9.1 Rainfall at end of June 2015: deficit (*brown*) and surplus (*blue*) (WFP 2015)

experiencing in recent years two extremes of water-related disasters, namely the 2011 floods and the ongoing drought started in 2014 that is causing severe water shortages issues across the country.

Although floods and storms are ranked as major hazards (Office of Environmental Policy and Planning 2000), water shortages arising from drought have always been part of Thailand's climatic conditions. Particularly, some parts of the country, such as the Northeastern region and the central plains, can be considered hot spots as concerns drought issues.

Uneven water resources distribution, climate change effects, increased water demand are progressively exacerbating trade-off among different water users and uses are all contributing to the worsening situation. Despite rapid industrialization, Thailand remains an agricultural-based country, relying mostly on rain-fed rice production. Regions which suffer the most are indeed the ones where irrigation canals and water distribution systems are still lacking.

According to the Department of Water Resources (DWR), villages lacking of water supply system are around 7000 (interview with DWR in 2015). Drought and water shortages are threatening directly Thailand's food and energy security. The country's rice production has been hit very hard; its vulnerability could affect global prices on the rice markets. Furthermore, the ongoing severe drought could also result in significant geopolitical consequences since the government is floating the idea to divert new water sources from near river basins (so a transboundary water context) such as the Salween and Yuam at the border with Myanmar or from the Mekong river to pump water to major dams in the Northeastern part (Lovelle 2016).

9.2 Method and Materials

The methodology applied in this research consisted mainly in a qualitative approach. Four main phases can be identified along the information process and most of it was gathered in Thailand, with the support of the Stockholm Environment Institute-Asia Centre.

Firstly, information was collected through an extensive desk-based research characterized by a substantial review of secondary data sources, including water- and disaster-related official policy documents, strategies, action plans, master plans, programs but also reports from NGOs and international development agencies operating in the area (e.g. the World Bank, Asian Development Bank, etc.…), policy briefs from SEI and other research institutes. Relevant sources of information included also the English-based national newspapers such as The Bangkok Post and The Nation and English version of governmental agencies websites. Most of the related scholarly literature was also reviewed.

Secondly, Key Informant Interviews (KIIs) were identified as the most effective tool to collect primary data. Eight KIIs took place in Bangkok from May 2015 to September 2015 and most of the interviewee at the time were high-ranking government officials from the Ministry of Natural Resources and Environment, Department of Water Resources, Department of Groundwater Resources, Department of Disaster Prevention and Mitigation but also researchers and academics from the Hydro and Agro Informatics Institute, Chulalongkorn University and the Mekong River Committee Drought Management Programme. All of these interviews were audio-recorded and transcribed. Seven of these interviews were conducted in English whilst one (with the Department of Disaster Prevention and Mitigation) was conducted in Thai and translated into English.

Some of the information has been also collected with the participation to international conferences and workshops such as the ASEAN Drought workshop and the SUMERNET Regional assessment workshop on water scarcity and drought management in the Mekong region and through personal conversations with colleagues and experts met in Bangkok.

Thirdly, a review of guidelines available, both at the international and regional level, to develop a potential drought management framework was performed. Particularly, three key documents were identified and utilized as a basis for the development of a tailored management framework for Thailand: the 2009 Drought Risk Reduction Framework and Practices: contributing to the implementation of the Hyogo Framework for Action produced by UNISDR; the 2014 Integrated National Drought Management Policy Guidelines: A Template for Action published by the IDMP, a joint program of the Global Water Partnership (GWP) and the UN's World Meteorological Organization (WMO); and the 2015 Guidelines for the Preparation of Drought Management Plans. Development and Implementation in the context of the EU Water Framework Directive elaborated by the GWP for Central and Eastern Europe. The last one was especially useful to see a precedent application of the framework adapted to a more localized context.

Lastly, as a result of this review, a 4-step process was elaborated (see Sects. 9.5.1, 9.5.2, 9.5.3, 9.5.4). For each one of these key steps, drawing on the indications of the conceptual framework, the current situation was highlighted and suggestions for improvement along with organizational arrangements and policy directions were discussed.

Once the framework was developed, a critical reflection on potential possibilities and challenges was presented employing the SWOT-analysis model.

9.3 Water Resources and Disaster Management Institutions and Mechanisms in Thailand

At this stage, a brief overview of the institutional framework that governs water resources and disaster management at the national level is provided. A quite complex institutional net emerges, thus without the pretention to be exhaustive, we will attempt to identify key responsibilities related to water resources management (and consequently drought) for each of the main[2] ministries and agencies involved to provide the reader with a better sense of existing institutional dynamics. As commonly observed in other developing countries, water resources management tasks in Thailand are widespread among several committees, ministries and specialized agencies reflecting every administrative level namely the national, the provincial (76 provinces) and the local (district, sub-district and village) level. In facts, there are about thirty agencies and bureaus involved in water resources governance, reflecting a high fragmentation of roles and responsibilities (ADB 2013).

A first, national committee related to water resources was the National Water Resources Committee (NWRC). Firstly established in 1989 (then revised in 2002 and 2007) and chaired by the Prime Minister, the NWRC was responsible for supervision and monitoring of water resources management and policy formulation. This committee was not based on any legal act that can guarantee a permanent status, therefore NWRC has been often subjected to various changes, without having a real power to implement policies (ADB 2013).

The catastrophic event which triggered a significant reform of high-level water related institutions was the 2011 flood, pushing the Prime Minister, at that time Yingluck Shinawatra, to take action. Hence, with a Prime Minister's Act on "Reconstruction and Future Development" she established two committees for flood prevention and control: a Strategic Committee for Reconstruction and Future Development (SCRFD) and a Strategic Committee for Water Resources Management (SCWRM). The last one was specifically tasked to draft a water and flood management master plan (Funatsu 2014). According to this plan, with another Prime Minister Act on "Water Resources and Flood Control Management Committee", known also as the "Single Command Authority Act" (Funatsu 2014), three other bodies were established:

- A National Water Resources and Flood Policy Committee (NWRFPC) with the mandate to formulate flood management policies and provide recommendations to the Cabinet, chaired by the Prime Minister;

[2]Other ministries consulted and involved in decision-making during a disaster management are the Ministry of Defense (MOD), Ministry of Social Development and Human Security (MOSDHS), Ministry of Industry (MOI), Ministry of Labor (MOL), Ministry of Public Health (MOPH), Ministry of Commerce (MOC), Ministry of Finance (MOF) and the Royal Thai Army (NDPMC 2009).

- A Water Resources and Flood Management Committee (CWRFM) chaired by the Deputy Prime Minister responsible for the execution of adopted policy, measures and approval of water management investment projects action plans, approval of water management investment projects, and monitoring and evaluation the implementation of investment projects. All relevant line agencies are represented in NWFPC and WFMC;
- An Office of National Water Resources and Flood Committee (ONWF) acting as secretariat body for the two committees and serves as single command authority (ADB 2013).

This landscape of committees had been in place until May 2014, when the Royal Thai Army overthrew the Prime Minister, taking control of the Government, and all the long-term flood rehabilitations plans were halted by the National Council of Peace and Order (Funatsu 2014).

A key success towards the introduction of Integrated Water Resources Management process in Thailand has been the establishment of River Basin Committees (RBCs). First steps were taken in mid-nineties when in 1998 the Chao-Phraya River Basin Committee was formed (Anukularmphai, n.d.). In 2002, after the establishment of the Department of Water Resources (Fig. 9.2), twenty-five RBCs were created, each major river basin was further divided into sub-basins to better cope with the necessities and diversity of the hydrological features (Anukularmphai, n.d.). Among the main roles, RBCs have mandate to formulate water resources management plans at basin level and coordinate those plans with relevant agencies, prioritize water allocation in equitable and efficient measures, monitoring and evaluation of other agencies' performance in the basin. Members of RBCs are local authorities, water user groups and community stakeholders (NGOs/academics). Despite their establishment, it cannot be said that River Basin Committees in Thailand do have a proper legal status (Anukularmphai, n.d.).

A special river committee is the Thailand National Mekong Committee (TNMC) in charge of representing the country within the Mekong River Commission (MRC), main intergovernmental agency with mandate of transboundary cooperation of the Mekong River Basin. Other national committees relevant to our analysis, both chaired by the Prime Minister, are the National Committee on Climate Change, established in 2006 as the highest policy-making body on climate issues and the National Disaster Prevention and Mitigation Committee, established in 2007, with the mandate to lay down policy for the formulation of the National Disaster Prevention and Mitigation Plan (NDPMP).

Each of these Ministries along with several departments, bureaus, offices and centers are reflected into lower administrative levels. It goes that, for instance, the DWR has a branch at the provincial level, at district and sub-district level and so has the DDMP etc. It is intuitive to understand the complexity of institutional arrangements as well as the number of government officials involved in the hierarchical Thai bureaucracy.

When a disaster strikes and is officially announced, the mechanism in accordance with the 2007 National Disaster Prevention and Mitigation Plan (NDPMP)

Ministry	Department	Responsibility/ Mandate
Natural Resources and Environment (MNRE)[a]	Water Resources (DWR)	Water resources policy, plans and management outside irrigated areas. it hosts the Water Crisis Prevention Centre, in charge proposing and coordinating action plans to solve water crises in disaster areas
	Groundwater Resources (DGR)	Groundwater resources policy, plans and management
	Office of Natural Resources and Environmental Policy Planning (ONEP)	Natural resources and environmental conservation policy, plans and management. National focal point for climate change
	Pollution Control Dept. (PCD)	Environmental quality standards
	Marine and Coastal Resources (DMCR)	Marine and coastal areas policy, plans and management
Agriculture and Cooperatives (MOAC)	Royal Irrigation (RID)	Water resources management and allocation for irrigation of agricultural areas, dams and storage reservoirs under its commanded areas
	Land Development (LDD)	Land use, management and research. National focal point of United Nations Convention to Combat Desertification
	Agriculture (DOA)	Crop research and development
	Agricultural Extension (DOAE)	Agricultural production and promotion
Interior (MOI)	Disaster Prevention and Mitigation (DDPM)	Policy-making body and state agency for disaster management.
	Metropolitan / Provincial Waterworks Authorities	State-owned companies in charge of water supply provision and distribution respectively in Bangkok and at provincial level
Information and Communication Technology (MICT)	Thai Meteorological (TMD)	Weather and climate forecasting
	National Disaster Warning Centre (NDWC)	Monitor and control over warning towers. Disseminates warnings to line agencies
Science and Technology (MOST)	Geo Informatics Space and Technology Development Agency (GISTDA)	Provides satellite remote sensing and GIS data to public and private sector
	Hydro and Agro Informatics Institute (HAII)	Research, develop and apply science and technologic for agricultural and water resources management
Energy (MOE)	Electricity Generation Authority of Thailand (EGAT)	state-owned utility responsible electric power generation and transmission. In charge of dams/reservoirs operations and maintenance

[a] Initially, the State management over the water resource belonged to MOAC. Between late 1990s and early 2000s the country went through a general institutional reform process that led to the reorganization of some ministries and included the establishment of new departments. Thus, in 2002 a new MNRE was established in order to provide an holistic control, management and exploitation of all natural resources, included water (interview with DWR).

Fig. 9.2 Main institutional actors involved in water resources and disaster management in Thailand

prescribes the activation of Emergency Operation Centers (EOCs) at all administrative levels (national, provincial, district, local plus Bangkok). Administratively, the provincial level plays a crucial role for disaster management. In particular, the provincial governor, who chairs the provincial Disaster Relief Committee is the person with the right to announce an area stricken by a disaster and has the mandate to approve/reject budget requests for action plans and projects coming from lower levels (e.g. district, sub-district) (interview with DDPM).

The provincial Disaster Relief Committee is responsible for organizing efforts among sectorial agencies (e.g. RID, DWR, LDD, DOAE etc.) coordinates lower levels (districts and sub-districts) while, on paper, the EOCs are in charge of command, control, support and coordination of response and relief measures (NDPMC 2009). However, this do not seem to always match the reality as "EOCs during a drought disaster mainly focus on monitoring" (interview with DDPM). Apart for these institutional arrangements, a very important role, at local scale, during a disaster emergency is played by stakeholders of the civil society such as the Thai Red Cross, non-governmental organizations, charitable organizations and volunteers and the private sector who actively help and cooperate with the local authorities to carry out relief measures for people affected by the disaster (NDPMC 2009).

In times of disaster, important (and often unpopular) decisions need to be taken as the ones concerning water allocation practices.

During normal times, water allocation in Thailand mainly takes place annually at river basin level. For the Chao Phraya river basin, this process usually begins slightly before the end of each rainy season (around October) when an overall assessment of total water volume available in the major multi-purpose reservoirs, Bhumipol and Sirikit dams, is carried out by the central authorities. Once the total amount of water available is known, a pre-seasonal allocation plan to match the irrigation area is carried out (Takeda et al. 2015; Divakar 2011). Consequently the responsible authorities announce farmers on what crops are not recommended for the next dry season (UNESCAP 2000).

Two key actors involved in this process are RID, under MOAC, and EGAT, under MOE which need to come up with a joint decision in order to release the water, since RID regulates water within the irrigation systems whilst EGAT is in charge of dams and reservoirs operations and maintenance (UNESCAP 2000).

Usually the amount of water planned does not correspond to the actual amount of water released (Divakar et al. 2011, personal conversations). Since there is no law defining water rights, what happens is that some users, along the river basin can divert water stealing in this way part of the resources destined to other downstream users (UNESCAP 2000). When a water crisis manifests itself, emergency inter-ministerial meetings are held and normally the highest water allocation priority is always assigned to domestic consumption (interviews with Chulalongkorn University and DWR).

9.4 Drought Management in Thailand: Laws, Policies, Strategies, Plans

Water resources in Thailand are governed by several water-related laws and numerous amendments, regulations and decrees. Nonetheless, at the time of writing, a proper Water Resources Law, which long and tormented approval process dates back to the early nineties, has never reached the consensus of all political parties, leading unavoidably the country to be characterized by an highly-fragmented legal framework. This is mainly due to an atmosphere of continuous political instability and lack of political will characterizing various governments as well as to the diversity of stakeholder interests in all water-related sectors, amongst other reasons. Hence, given the absence of a comprehensive legally-binding instrument for water resources, this sector is ruled by a number of sectoral laws governing different aspects of this precious resource.

In order to be promulgated by the Royal Thai Government, any policy, plan, program and strategy must be consistent with a broad socio-economic development framework. This overarching development framework assumes the shape of a 5-year National Economic and Social Development Plan, currently at its 11th edition (2012–2016). The plan promotes development strategies to be implemented according to the concept of "sufficiency economy",[3] the guiding principle of Thailand's development since 1974, presented by H.M. King Bhumibol Adulyadej (NESDB 2012).

When dealing with water resources, the milestone documents of Thailand's waterscape are represented by the National Water Vision and Policy both promulgated in 2000. Before being approved, these two documents went through a drafting process based on an extensive participatory approach which involved water-related stakeholders (government official, researchers, NGOs and private sector), in line with the Dublin Principles of Integrated Water Resources Management (Anukularmphai, n.d.). The policy, articulated in 9 pillars, calls for actions to be undertaken such as a suitable and equitable water allocation for all water use sectors, the need to meet water demand for agricultural and domestic uses, the approval and enforcement of the Water Law, insurance of sufficient budget allocation for water-related issues and encouragement of effective stakeholder engagement (Anukularmphai, n.d.).

The only document issued specifically after last year's drought is the Integrated Plan for Drought Management for 2015 which comprises strategies on prevention and mitigation of drought impact by focusing on predictions of drought-prone areas and development of a quick, reliable alert warning system; preparation for disaster by supplying water in drought-prone areas with highest priority to water for consumption; emergency management by establishing operation centers and post-disaster management by providing financial compensation, employment and livelihood for victims (Royal Thai Government 2015).

Climate change is a cross-sectoral issue, well acknowledged in Thailand's policy as demonstrated by the 2000 Initial Communication under the UNFCCC and by the

[3]The Sufficiency Economy is a fundamental concept for the Thai society, especially because it has been developed and promulgated by the king in person. It fosters a sustainable development dominated by a "happy society with a equity, fairness and resilience" vision (NESDB 2012).

Second Communication under the UNFCCC (ONEP 2010). In order to provide guidelines to face challenges posed by climate change the National Strategic Plan on Climate Change 2008–2012 first (Pipitsombat, n.d.; UNDP and Overseas Development Institute 2012) and the National Climate Change Master Plan 2015–2050 then, were also promulgated.

In terms of disaster management Thailand's Climate Change Master Plan 2015–2050 then, were also promulgated. Efforts seem to be based on a sound legal and policy framework, at least in principle. The 2007 Disaster Prevention and Mitigation Act provides the overarching legal mechanisms for disaster management in Thailand (ADPC 2013). According to this Act, a National Disaster Prevention and Mitigation Plan (2010–2014), promulgated in 2009, represents the national policy framework for disaster management within which all the other lower-level plans have to be formulated. The plan envisages all phases of the disaster management cycle, namely prevention and impact reduction, preparedness arrangements, disaster emergency management and post-disaster management (NDPMC 2009). The document also calls for formulation of specific drought-related plans such as Drought Prevention and Mitigation Integrated Action Plan or Dry Season Crops Cultivation Promotion Plan, in line with agricultural water management in the dry season plans. As this plan has expired, a new National Disaster Prevention and Mitigation Plan was approved in 2015 but unfortunately, at the time of writing an English version of this policy document is not available. Beside the national plan, a Strategic National Action Plan (SNAP) on Disaster Risk Reduction 2010–2019 was formulated to fulfill the requirements of the Hyogo Framework for Action (HFA) 2005–2015, adopted right after the tsunami in the Indian Ocean in 2004 (DDMP 2010).

9.5 An Institutional/Organizational Framework to Include Water-Related Disaster in Ordinary Water Resources Management

This section seeks to rethink water-related disaster management in Thailand by proposing a possible re-shaping of institutional and policy landscapes, envisaging more holistic coordination mechanisms and information flow which would engage all administrative levels (from national level to local level) and concerned stakeholders. In the following sections a 4-step risk-based drought management framework, modeled from international guidelines (in particular guidelines for Eastern Europe and the Near East countries) to be tailored to fit in the Thai context, is presented.

9.5.1 Establish a Drought Committee and Related Organizational Arrangements

International guidelines suggest that the first step towards an effective drought management system would be to identify or confirm an appropriate competent authority (GWP CEE 2015). After identifying this authority has been chosen, a formal government resolution or legislation should introduce the legal and institutional framework for the entire drought planning process, establishing roles and responsibilities at all administrative levels (GWP CEE 2015). Once the institutional frameworks along with a suited legislation have been approved, a Drought Committee should be formed. This Drought Committee, regulated by the national authority, should become a permanent body with a strong mandate, which tasks include establishing its organizational structure, setting up specific working groups, defining clear competencies and responsibilities of both the Committee and individual members, providing a communication strategy among all administrative levels, coordinating specific measures among government and stakeholders, assigning tasks during all drought stages (normal, pre-alert, art, emergency) and supervising the overall process evaluation procedures (GWP CEE 2015). Moreover, it is fundamental that the composition of the Drought Committee represents the multi-disciplinary nature of this issue: representatives ranging from national-level decision makers, local authorities, professional institutions providing technical expertise, key stakeholders such as farmers, local communities, NGOs but also energy, tourism, industry sectors and water providers and suppliers should all be able to sit on this national body. Besides, a specific drought task force consisting in Working groups/Committees addressing all the necessary technical assessments and procedures should also be established (GWP CEE 2015).

With the 2007 National Disaster Mitigation and Prevention Act, Thailand has already officially adopted an institutional framework for disaster risk reduction (DRR), which includes drought. According to the 2007 Act, disaster management is under the mandate of the Ministry of Interior, specifically, the Department of Disaster Mitigation and Prevention (DDMP) has the responsibility to coordinate, support and enhance all disaster related activities and formulate national disaster plans. Even though, in principle, Chap. 15 of the 2009 National Disaster Prevention and Mitigation Plan (NDPMP) prescribes a procedure which fully takes into account mitigation and preparedness, putting emphasis on all the disaster cycle management (i.e. pre-disaster, during the disaster and post-disaster), one might observe that, in reality, governance mechanisms concerning drought measures are put in place only when drought strikes. This means that, in practice, full attention is given only to the central phase of disaster management cycle, during the disaster itself. Being drought primarily a water resources management issue, we argue that drought management mandate of the Ministry of Interior should be equally divided —at ministerial level—with the Ministry of Natural Resource and Environment (MNRE), in charge of state management of water resources.

Regarding the set up of a National Drought Committee, there may be no need to establish a further new commission, given the high number of committees and sub-committees that proliferate within the Thai institutional landscape. Rather, it might be suggested a reform of the current[4] National Water Resources and Flood Policy Committee (NWRFPC), and the Committee for Water Resources and Flood Management (CWRFM) established by the Water Resources Master Plan after the 2011 devastating floods. This would lead to the creation of a National Water Resources Flood and Drought Policy Committee (NWRF&DPC), integrating in this way mandates on flood and drought management policy and, accordingly, the CWRFM could be readjusted in a Committee for Water Resources, Flood and Drought Management (CWRF&DM). As concerns the organizational arrangements, in order to make sure that this renewed NWRF&DPC would appropriately include all stakeholder representatives of water-related sectors, reflecting the multi-disciplinary nature of the issues, the current inter-ministerial composition[5] could be extended to representatives of local governments, research organizations and academia, technical service providers (e.g. assessments, modelling, remote sensing etc.), non-governmental organizations and the private sector.

The NWRF&DPC, chaired by the Prime Minister, would have the authority to formulate polices related to flood and drought management, supervising the overall preparedness and mitigation process for both hazards. Under its authority operates the CWRF&DM, in charge of formulating and implementing water management action plans following policy guidelines provided, approving water-related investment projects, endorsing fiscal budget and manpower mobilization and, lastly, preparing the National Drought Management Plan (DMP).

The CWRF&DM could be jointly co-chaired by Ministry of Natural Resources and Environment (MNRE) and Ministry of Science and Technology (MOST).[6] The justification is that MNRE holds the State Management of water resources and, as such, it should promote an overall sustainable integrated water resources management whilst MOST controls two of the most relevant agencies in terms of monitoring and forecasting capacity.

Lastly, the drought task force should be supported by three different working groups or committees that provide services and implement different tasks of the planning process:

[4]After the military junta took the leadership in 2014, the functions of the single-command authority established by the previous government after the 2011 floods have been halted and a new National Water Resources Committee established. For simplicity, here it is preferred to maintain the structure previously utilized, acknowledging that, on reality, the future of these bodies is highly uncertain. It is very complex to deal with non-permanent commissions that keep changing with every government, so in this framework it is assumed that NWRFPC and CWRFM have a permanent nature.

[5]National Water-related Committees usually include representatives from water-resources related ministries such as MNRE (DWR, DGWR), MOAC (RID, DOAE), MOI (DDPM, Public Work), MIST, MOPH etc.

[6]In reality this Committee is chaired only by the Minister of Science and Technology (MOST).

- A Monitoring and Early Warning Committee (M&EWC), responsible for collecting all data related to water resources scattered around different agencies and establishing a Drought and Flood information system and database;
- A Risk Assessment Committee (RAC), responsible of performing risk, impact and vulnerability assessments along with evaluations of past drought events;
- A Mitigation and Relief Committee (MiReC), in charge of coordinating assistance and mitigation measures when critical thresholds are crossed and water shortages become drought. This Committee would act only when activated (during the disaster and post-disaster) and would be chaired by DDPM, under Ministry of Interior.

All these committees mentioned above have primarily an inter-governmental nature. Nonetheless, it would be advantageous to expand their meetings to external consultants from academia, NGOs and other research institutes that perform studies on the same topics, improving their mutual understanding and ensuring a more transparent information sharing.

What has been presented lies at the inter-ministerial and national level. A lower, suitable level to prepare and implement a Local Drought Management Plan would be the river basin level, with the 25 River Basin Committees (RBCs) to be the designated authorities. Following the national-level organization, a drought (and flood) task force (here named River Basin Flood and Drought Management Committee-RBF&DMC) should be created within the RBC as well as the Drought management plan should be adopted as integral part of the River Basin Management Plan, essential requirement as every river basin has different hydro-morphological features, water users and water needs. Accordingly, three similar working committees (RB-M&EWC, RB-RAC and RB-MiReC) should be established. Policy guidelines, methodologies for assessments, drought indicators and thresholds will be performed as requested by the national level, in order to apply an harmonic a holistic approach throughout the country. Furthermore, given that different provinces lie within a river basin boundaries, the RB-MiReC could be chaired by a rotation of provincial governors, maintaining the current disaster management structure (Fig. 9.3).

9.5.2 *Ingredients to Develop a National Drought Risk-Based Management Policy*

Once the National Drought Committee has been appointed, a second step has to be developed. A risk-based drought policy should be established with a strategy to implement it, then the resulting policy document shall be endorsed by the government (GWP CEE 2015). This policy document may contain a first part regarding the overall framework and principles, pointing out the proactive and risk-based

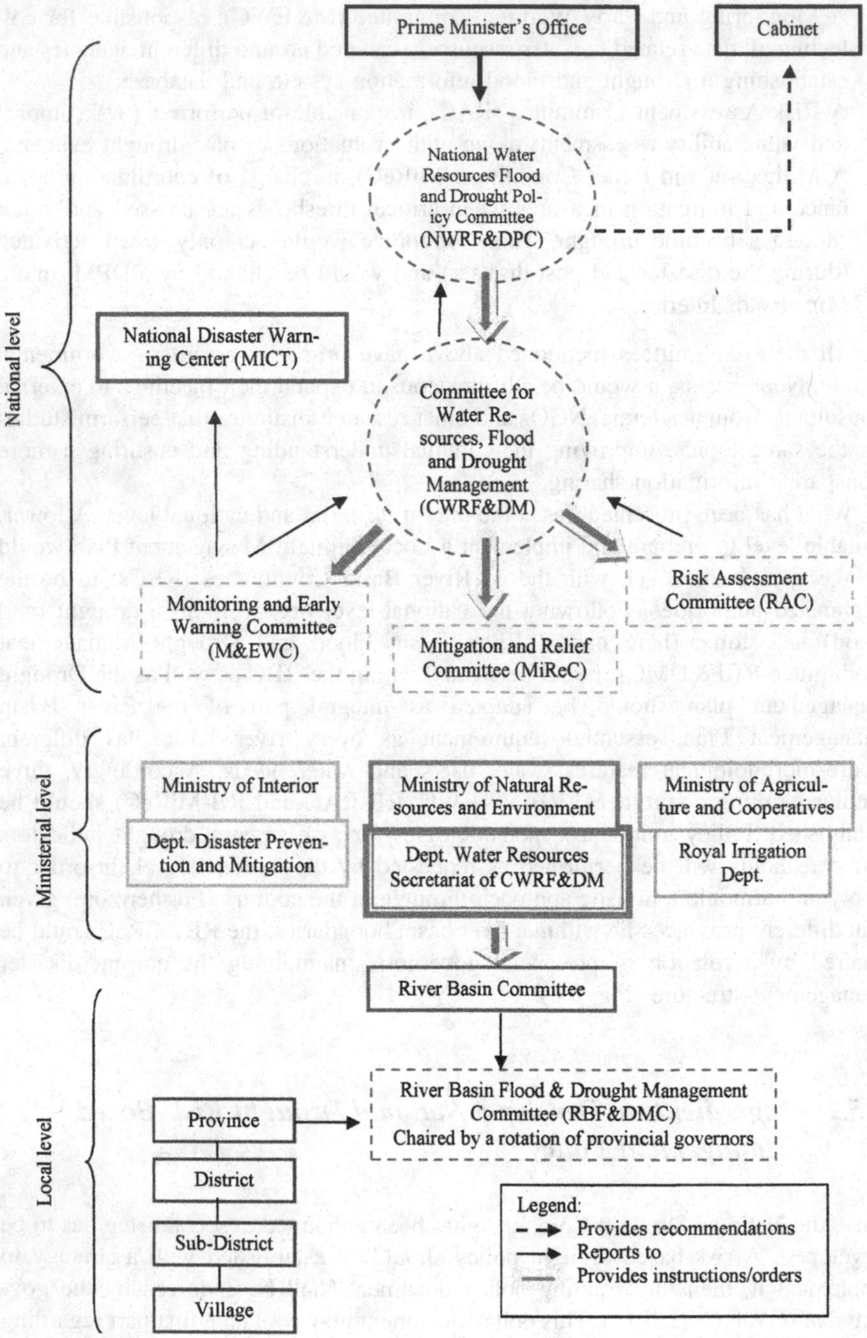

Fig. 9.3 Suggested organizational arrangements (Franzetti 2016)

approach, together with a generic roadmap for its implementation in addition to key phases to develop a Drought Management Plan (DMP) (e.g. administrative, financial, organizational, institutional, etc.) (GWP CEE 2015). Defining objectives along with specific and achievable goals of a drought-risk reduction policy is a decisive step as it represents the basis for the implementation of the DMP (WMO and GWP 2014). A critical passage is that the policy can only be effective if there is a prior identification that grasps the most vulnerable groups, activities and drought-prone areas nationwide (GWP CEE 2015).

As emerged from the policy review conducted, policies that can be related directly or indirectly to drought management have to be researched in sectoral policies and plans; in fact to date the country does not present any single, intergraded and comprehensive drought management policy and strategy. Nevertheless, in the context of disaster management, Chap. 15 of the 2009 NDPMP contains some relevant policy objectives including "(i) to prevent and mitigate adverse impact of drought on well-being and property of the people; (ii) to facilitate and ensure effective and timely provision of assistance to affected people as well as rehabilitation to normalcy as soon as possible and (iii) to identify the clear and integrative tasks and responsibilities to be taken by all government agencies as well as non-government network organizations for dealing with drought" (NDPMC 2009). Surely, the objectives presented already reflect a disaster-risk based approach and this gives us important indications about the direction which Thailand would like to move towards.

A first recommendation is that it would be necessary to officially consider and recognize drought as a normal and recurrent feature of Thailand's climate, providing in this way political acknowledgement of the water shortage issue permanently, and not just when and where the problem arises. Usually as soon as the drought situation withdraws, drought relief measures are dismissed and water level conditions become acceptable again, all media and political attention on this problem suddenly vanishes until another water shortage happens, leading to another water crisis.

In terms of drought policy formulation, there are plenty of experiences, more or less successful, in drought-prone countries that could be mentioned.

A significant example worth to investigate may be represented by the Australian Drought Policy, which has been constantly revised and improved over time.

A significant point to outline is that this National Drought Policy places great emphasis of farmers responsibilities to manage climate risks (Stone 2014). Thailand could take some inspiration from the way Australia has adopted, among its policy goals, the "self-reliance" concept to encourage farmers when it comes to climate risk management. Besides, given the great importance of the agrarian basis of the country's economy, surely a number of considerations on other sectors, especially agricultural productivity, must be included when drafting the policy.

9.5.3 *Drought-Related Data Inventory and Gaps Identification*

Regarding data inventory, a third step is necessary. For this reason the GWP CEE guidelines present the information to collect and analyze an effective DMP grouped into six macro categories, namely meteorological, hydrological, agricultural, socio-economic, environmental impact and drinking water supply data. In order to establish a drought information system, it would be necessary to collect and update constantly this information. Nevertheless, it is well known that gathering all these data together is not as easy as it might seem: this type of information is usually scattered around different departments and agencies that not very often share and integrate data, especially when it comes to different ministries (GWP CEE 2015). For the abovementioned reasons, a careful analysis of the constraints (such as weaknesses and obstacles) in terms of data availability should be carried out in the early stage of a drought management process, before developing a DMP. Lastly, it is worth to mention that relying on complete and integrated data also allows to create an appropriate national drought indicators system during the planning phases, which is a crucial condition to ensure the success of the plan.

Some indications on responsibilities to conduct drought risk assessments analyzing risk factors and possible drought areas as well as to verify, update and prepare the drought database and hazard mapping are included in the Thai National Disaster Prevention and Mitigation Plan (Praneat 2014).

Availability and transparent sharing of this type of information determines the conditions for success of a Drought Management Plan. To date, Thailand does not present any comprehensive integrated[7] national drought database: in fact, drought-related data are scattered around various agencies. Every government agency monitors and collects data according to its mandate of competence so, for instance, the Thai Meteorological Dept. is in charge of monitoring weather stations; surface water resources status is be monitored by Dept. of Water Resources; Royal Irrigation Dept. monitors only water status within irrigated areas while EGAT monitors water levels in reservoirs; groundwater resources are monitored by the Dept. of Groundwater Resources; disaster information is monitored by Dept. of Disaster Prevention and Mitigation; Land Development Dept. monitors soil conditions and the Royal Forest Dept. is in charge of forest monitoring and so forth. It is evident, in order to be effective, a strong and harmonic data management capacity is needed and given the low level of communication among all the agencies concerned, current data flowing among agencies cannot be considered sufficient.

The National Disaster Prevention and Mitigation Plan (NDPMP) 2010–2014 prescribed three important actions that fit into this point, namely to review and update existing data on drought-prone areas and create drought hazard maps along with a water resources database; to prepare a database on relevant personnel and

[7]By "integrated" here we are referring to the presence of a single database where all the drought-related information converges.

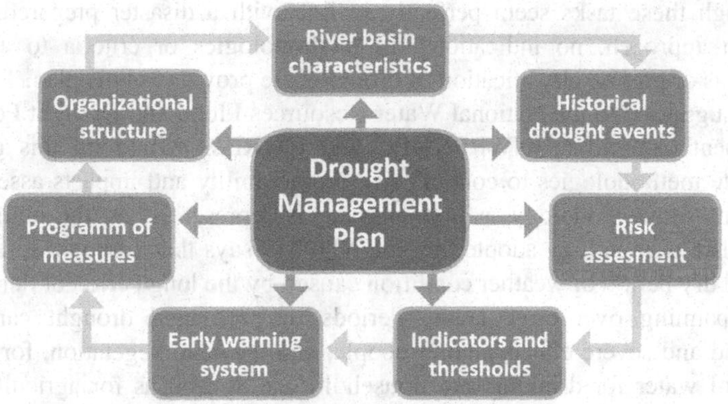

Fig. 9.4 Elements of a drought management plan (GWP CEE 2015)

mechanical equipments in order to ensure their immediate availability and serviceability during a drought event and, lastly, to develop a drought information system. Thus, according to the NDPMP, one of the points required would be the development of a drought database and GIS as well as a drought information sharing system along with the identification of main agencies that should be involved (NDPMC 2009).

9.5.4 Developement and Update a Drought Management Plan

This fourth step might be considered as the heart of a drought planning process. After making sure to possess the right information, the development of a Drought Management Plan (DMP) can begin. According to the GWP CEE Guidelines, an ideal DMP should follow five sub-steps including a definition of the content, a characterization and evaluation of historical drought events, the establishment of appropriate drought indicators and thresholds, the creation of an Early Warning System (EWS) followed by the development of a program of measures (WMO and GWP 2014) (Fig. 9.4). Analyzing Thailand's efforts written in its main policy document for disaster management, the National Disaster Prevention and Mitigation Plan (NDPMP) 2010–2014,[8] one may observe that interestingly most of the key elements highlighted in the conceptual guidelines are already included in this official document along with competent agencies in charge of implementing these tasks.

[8]The NDPMP has been recently updated for the timeframe 2015–2019. However, in this work the previous version will be considered as no official translation of the plan in English is yet available.

Although these tasks seem perfectly in line with a disaster preparedness and mitigation approach, no indications on methodologies or criteria to adopt for assessments or precise classification of droughts are provided in the plan. Therefore, it can be suggested to the National Water Resources Flood and Drought Policy and Management Committee (NWRF&DPC and CWRF&DM) to fill this gap with appropriate methodologies to conduct risk, vulnerability and impacts assessments.

In terms of methodology, a first obstacle is represented by the definition of drought. The terminology adopted by the NDPMP says that "drought refers to the prolonged dry period of weather condition caused by the long period of deficit or no rainfall spanning over large areas. Periods of prolonged drought can trigger widespread and severe effects among people, animals, and vegetation, for instance shortage of water for drinking and household use as well as for agricultural and industrial purposes, substandard or highly limited crop or yield productions, death of livestock etc." (NDPMC 2009).

Now, it can be argued that this definition is quite vague in the sense that no specific thresholds to identify drought stages are provided. This is confirmed by the policy review: most of the documents mention the problem of water shortages and measures that concerned agencies intend to apply, but no thresholds. It follows that it is not clear how different types of drought are identified and treated. In our opinion, this is a significant gap to acknowledge and address: drought affects groups and activities in different ways, requiring ad hoc measures and definitions for every area. Common and shared definitions between government and stakeholders should be discussed and adopted, making possible to begin working collectively on appropriate drought indicators and thresholds.

Looking at drought monitoring system, a recent study of NASA Goddar Space Flight Center highlighted how "the available drought monitoring system in Thailand looks only at the agricultural drought (…) and this was insufficient for analyzing accurate risk management and decision-making" (McCartney et al. 2015). Once again, it is likely that several agencies' technical divisions perform their own analyses but results are not shared among other agencies. It is necessary to come up with unified and acceptable indicators utilized throughout the country, maybe improving also the international cooperation channels through technical staff exchange with countries highly advanced in this field.

Being a country that often experiences extreme climate events, it can be said that Thailand has a good capacity with early warning systems (EWS). This is true for some types of natural hazards including floods, storms, tsunami, tropical cyclones and earthquakes, but it does not apply to drought. The 2004 tsunami that caused tremendous impacts on South and Southeast Asia is considered the trigger point for the adoption and implementation of a disaster EWS, followed by the 2011 Great flood. After the establishment of a National Disaster Warning Center in 2004, in 2012 an agreement was reached in order to develop a Decision Support System for operational flood risk management for the Chao Phraya river basin (central Thailand) (DHI 2012). Recent past drought events have strongly outlined the necessity to establish an effective drought forecasting system and, consequently, an EWS for this hazard. To date, Thailand "does not have a drought forecasting system but this year (it) will start to implement a project to develop it" (interview with HAII 2015). Hopefully, learning from the dramatic drought experiences, Thailand would invest more resources in the near future.

Regarding the program of measures, it has been already said that the NDMP embeds several measures covering, in principle, all phases of the disaster management cycle. However, preparedness measures found in the plan, are suitable but certainly not enough to cope with drought in a pro-active way. It would be necessary to identify and improve in all water-related policies specific strategies and measures to cope with drought, so every agency would better know in its field what options are effective and which do not work. Every agency could report its best strategies to the CWRF&DM which will incorporate them in a single program of measures, to be reviewed and constantly updated.

Thailand has a fairly good experience in dealing with public hearing and consultations, even though it has not always been effectively implemented. For instance, the country went through a long process of stakeholders consultations to introduce the Integrated Water Resources Management (IWMI) that led to the National Water Policy and Vision not to mention the consultations for drafting the Water Resources Bill (Anukularmphai, n.d.). Consultations could take place at national level, within the NWRF&DPC, where all stakeholders are represented and at the River Basin level, hosted by the River Basin Committees (RBCs), that will report to the Dept. of Water Resources, secretariat of the Flood and Drought Policy and Management Committees. Comments and suggestions will then be analyzed by the NWRF&DPC that will make sure they will be incorporated in the final draft.

As for educational awareness, the NDPMP already prescribes, amongst other activities, to organize "public education training and campaign to raise awareness among all members of community at risk and to provide important information to help them understand potential drought as well as an instruction of drought mitigation strategies" that, for instance, include water restriction measures such as to regulate water use and planting less water—dependent crops in dry season (NDPMC 2009). Thailand is doing a relatively fair job in terms of awareness increase in some fields. For example, the Land Development Department has created programs like the "Volunteer Soil Doctors" with the purpose to train people on a voluntary basis to take care of and deal with soil problems. This idea could be replicated also in specific training programs for drought.

To be fully effective, evaluation procedures should be accompanied by the gap analysis presented in Sect. 9.5.3. Monitoring each of the steps is vital to identify strengths and weaknesses of the whole process. A suggestion here might be to appoint an official responsible for this task within CWRF&DM (the Management Committee) to keep track of the ongoing evaluation whilst it might be useful to produce two separate post-drought evaluations: one conducted within the institutional environment and another one assigned to an external auditor (e.g. a non-profit research institute, or a private consultancy agency). In this way the two evaluations can be crossed, allowing new elements to emerge. During the research period no trace of evaluation procedures on drought-related in English was found.

9.6 Challenges and Opportunities for Thailand

Before reflecting on potential opportunities and challenges of an hypothetical application of the framework, a mention to the assumptions lying behind the framework is necessary. The first one is the approval of the Water Law, providing the country with a proper legal support for water resources management. Secondly, it was assumed that the suggested NWRF&DPC and CWRF&DM have a permanent nature, assuring the continuity of the organizational framework. Thirdly, the Drought Management Plan would be issued as a complementary document of the River Basin Management Plan as the river basin is considered the ideal boundary to deal with water resources management. Lastly, we propose a paradigm shift as drought is not to be considered as a disaster (implying that measures are undertaken when it is already too late) but its recurrent nature is acknowledged and integrated into water resources policies and plans. Hence, keeping these assumptions in mind, in order to perform the following analysis, factors were grouped into five categories representing both challenges and windows of opportunities on institutional, technical, political, economic and financial and socio-cultural aspects (Fig. 9.5).

A first group to be considered is the one of institutional factors that could certainly benefit from a concrete application of the framework. First of all, River Basin Committees would be in the position to confidently implement their mandate and would manage water resources in a more holistic and appropriate manner. Furthermore, a permanent nature of the NWRMF&DPC would involve all stakeholders on a regular basis, providing great room for discussion and consistency, since it is suggested to expand meetings that would allow to academia, NGOs, research institutes and civil society organizations to participate in decision-making

Factor	Opportunity	Challenge
Institutional	• Permanent nature of Committees • Improved communication among agencies • Paradigm shift on drought • Increased stakeholder engagement	• Highly fragmented institutional and legal framework
Technical	• Single and comprehensive database on water resources • Shared drought definition and classification • Gaps identification • Increased assessment capacity	• Lack of qualified personnel
Political	• Ongoing drought represents a (policy) window of opportunity as it was for 2011 floods.	• Unstable political context • Water as object of bargaining
Economic/Financial	• Shift towards pro-active risk management is more cost-effective than the costs of inaction	• Lack of water markets and water-saving incentives
Socio-cultural	• Increased awareness and better preparedness at all levels • Self-reliance concept	• Water as a gift of God • Very hard to change farmers' behavior

Fig. 9.5 Opportunities versus challenges (Franzetti 2016)

processes while formalizing a knowledge exchange full of mutual benefits. This would also lead to improved collaboration and coordination among concerned agencies, creating a constant and more transparent information flow.

To undermine these positive changes there is still a highly-fragmented institutional and legal framework.

A second group examines some technical factors such as, for instance, the concentration of several agencies' efforts under a single Monitoring and Early Warning Committee (M&EWC) in charge of establishing a single and comprehensive database on water-related issues and disasters. More effective early warnings could be provided, not to mention the adoption of common drought definitions, classification and indicators to be implemented throughout the country. Enacting the harmonization of data collection and analysis would be advantageous as it is also linked with technical capacity to perform risk, impacts and vulnerability assessments. Engagement in process evaluation procedures would allow the authorities to support the entire process and identify strengths and weaknesses where an intervention is needed.

And again, the ability to conduct assessments on current and past drought events would shed light on past experiences, serving as a basis for a more tailored program of measures to be implemented.

All these opportunities, however, could be hindered by the lack of highly-qualified and trained personnel, especially at local level. This point has been highlighted by several studies and sectoral capacity assessments (e.g. ADB 2013) as one of the main bottlenecks when it comes to maintaining and implementing measures on the ground.

Implications of political factors become evident if one observes the institutional arrangements that were created after the 2011 floods. It can be hypothesized that the ongoing water crisis in Thailand could potentially act as a trigger to take proactive actions against drought, providing a window of opportunity for the country. A policy statement recognizing that drought is not to be considered only a disaster but needs to be addressed under the ordinary water resources management would reflect an increased awareness among high-level government officials. Moreover, in order to effectively perform a drought preparedness and mitigation process, political stability is an indispensable precondition. Unfortunately, this is definitely a critical, if not daunting, challenge in Thailand: history of several water-related committees established under different governments has shown how the lifetime of these bodies is short. Hence, it would be important that the nature of the suggested committees can become permanent, in order to ensure continuity of the organizational framework.

Naturally, in a very unstable political context, it is hard for government agencies to plan and respond to these changes effectively. This is why expectations from policies, strategies and plans' performance do not always turn out as desired.

Political will is another crucial factor as it influences the outcome of any type of action related (but not only) to water resources. Investments in water infrastructure development to solve water shortages and drought issues, not to mention the constant objective to expand irrigated areas, have always dominated political discourses but have never definitely solved the problem (Molle 2001). Probably, one of the many complex implications behind this, is that sometimes water, especially in agriculture, can become an object of political bargaining. For example, regarding slow expansion of irrigated areas that does not seem to match the continuous claims from the authorities (RID) to increase irrigation, Achara Deboomne in an opinion on The Nation wrote that "the root of this long-standing problem lies in the fact that policymakers treat irrigation as a political tool, channeling budgets to temporary projects that only address short-term hurdles. No government has ever embarked on a grand-scale project to tackle long-term problems" (Deboonme 2015).

Economic and financial aspects related to this framework's application include some significant constraints that Thailand may face such as the lack of budget availability, even though a shift towards pro-active risk management can be more cost-effective if compared to the costs of inaction. The budgeting system in Thailand is very centralized and it takes a long time before projects can be approved, with the results that very few of them see the light (interview with Chula 2015). For these reason, it is likely that funds for drought related projects would encounter several difficulties along their way. Another recurrent challenge, often mentioned in the literature, is the lack of a real water economics and market-based mechanisms to incentivize water savings, in addition to water restriction measures usually applied in times of crisis. In Thailand, water itself does not have a price as it is considered an "open access" resource (interviews with DWR and Chula 2015). In the agricultural sector, by far the most water-intensive sector, the only fees collected are those for the water service and distribution system, not even enough to cover operation and maintenance costs (Molle 2001).

If the framework were to be fully implemented could also have a significant impact on civil society. Socio-cultural opportunities represented by committing investments in human and financial resources for educational programs, awareness campaigns and drought preparedness meetings at local level could help enormously in preparing people to cope with drought.

Public involvement in the approval process of a Drought Management Plan would give voice to affected local communities and would allow the government to be more sensible to specific local needs, usually ignored at national level, whilst in a long-term perspective it would perhaps increase people's trust in governmental mechanisms. Besides, programs targeting different groups can sensitize people to the importance of water saving practices, where the government does not arrive with market-based interventions. It would mean to cultivate that "self reliance" concept which would provoke a change in farmers' perspectives and behavior, as what usually happens today is that "even if they are advised that there will not be water to grow a second crop because there is no water (…) they will do it anyway because they rely on compensation and assistance schemes from the government" (interview with Chula 2015).

9.7 Conclusion

Nowadays Thailand, and more generally Southeast Asia, is experiencing one of the worst water shortage crises of the past 15 years and drought is representing a serious threat for the country's food and energy security. Main drivers include population growth, rapid urbanization, industrialization and economic integration, unsustainable agricultural practices and depletion of natural resources (especially water) which, combined with increasing water demands, are exacerbating trade offs amongst all water-related sectors (e.g. household consumption, agriculture, industry, hydropower production, tourism, fisheries, ecological flows maintenance). To complicate the situation, global climate change and the growing threat of extreme natural events are presenting serious challenges for the whole humankind. As a consequence, water resources are increasingly becoming more scarce and finite. Ensuring water availability (quantity and quality) for all sectors of an economy is a key concern of water resources management.

Since, until recently, most of the Thai government's attention has been focusing on flood mitigation and control, this chapter had the objective to suggest a possible management framework as well as to increase general awareness on this insidious, slow-onset hazard.

Results from the reviews have highlighted how fragmented water-related duties and responsibilities are among ministries and line agencies, with overlapping responsibilities, amplified by a low level of communication and information sharing. Furthermore, integrated flood and drought management is acknowledged in most documents, but unfortunately it does not seem to reflect the practice: institutional arrangements exist for flood management but not for drought. It has also been stressed that, even though no single and comprehensive policy to address drought is explicitly presented, some indications can be found scattered around sectorial policies. We have seen that this phenomenon is considered as a disaster and, although in national documents measures to be adopted during all phases of a disaster cycle are prescribed, again this is not reflected on reality as only disaster relief and emergency measures are effectively carried out, confirming the traditional crisis approach in disaster management.

To overcome this, a shift towards a more proactive, risk-based approach would be advisable and in order to succeed, a tailored 4-step drought management framework has been suggested, looking at international guidelines applied elsewhere as a basis. The final outcome of the proposed framework would be the preparation of a Drought Management Plan based on important pillars such as forecasting, monitoring and evaluation system, comprehensive database, risk assessments and early warning system. Organizational arrangements along with a re-shaped scope of existing national committees on water resources have also been discussed presenting a possible reorganization, especially at national level. Opportunities deriving from an actual application of this drought management framework have been identified (e.g. increased awareness, better cooperation among all stakeholders, improved public participation etc.). Nevertheless, it is predictable that these opportunities could be hindered by structural constraints and barriers such as highly fragmented institutional framework, political unrest and the lack of an appropriate comprehensive water legislation.

References

Adamson, P., and J. Bird. 2010. The Mekong: A drought-prone tropical environment? *International Journal of Water Resources Development* 26 (4): 579–594. doi:10.1080/07900627.2010.519632.

ADB. 2013. *Thailand sector assessment (Summary): Water resources. Country partnership strategy: Thailand, 2013–2016*. Manila.

ADPC-Asian Disaster Preparedness Center. 2013. *Assessment for disaster management planning, policies and responses in Thailand*. ADPC website. http://www.adpc.net/igo/contents/adpcpage.asp?pid=2. Accessed Sept 2015.

Anukularmphai, A. n.a. *Implementing integrated water resources management (IWRM): Based on Thailand's experience*. IUCN.

DDMP-Department of Disaster Mitigation and Prevention. 2010. *Strategic national action plan (SNAP) on disaster risk reduction 2010–2019*.

Deboonme, A. 2015. Thailand's rice basket shrivels amid drought of long-term policy. *The Nation*, 3 Mar. http://www.nationmultimedia.com/opinion/Thailands-rice-basket-shrivels-amid-drought-of-lon-30255170.html. Accessed Aug 2015.

DHI. 2012. *Protecting Thailand from floods. Using DSS to forecast floods in the Chao Phraya River Basin*. DHI case story. Danish Hydraulic Institute.

Divakar, L., M.S. Babel, S.R. Perret, and A.D. Gupta. 2011. Optimal allocation of bulk water supplies to competing use sectors based on economic criterion—An application to the Chao Phraya River Basin, Thailand. *Journal of Hydrology* 401 (1–2): 22–35. doi:10.1016/j.jhydrol.2011.02.003.

Franzetti, F. 2016. *Towards the development of a drought management framework: Challenges and opportunities for Thailand*. Master's thesis, University of Turin.

Funatsu, T. 2014. Organizational reformation on water resources management after the 2011 Thailand Great Floods. In *Politics of the Environment - the formation of "late-comer" Public Policy*, ed. Tadayoshi Terao. Chiba: Institute of Developing Economies.

Ganter, C. 2015. Water crises are a top global risk. *World Economic Forum*. http://www.weforum.org/agenda/2015/01/why-world-water-crises-are-a-top-global-risk. Accessed 11 Aug 2016.

GWP CEE. 2015. *Guidelines for the preparation of drought management plans. Development and implementation on the context of the EU Water framework directive*. Global Water Partnership Central and Eastern Europe.

Lovelle, M. 2016. Thailand: Drought, water sources and the potential implications. *Future Directions International*. http://www.futuredirections.org.au/publication/thailand-drought-water-sourcing-and-the-potential-implications/.

McCartney, S., et al. 2015. The Trifecta of drought: monitoring three types in Thailand. *Earthzine*, July 30. https://earthzine.org/2015/07/30/the-trifecta-of-drought-monitoring-three-types-in-Thailand. Accessed 24 Feb 2017.

Molle, F. 2001. *Water pricing in Thailand: Theory and practice*. DORAS DELTA, 7. Bangkok: Kasetsart University.

MNRE. http://webeng.mnre.go.th/ewt_news.php?nid=5. Accessed Mar 2016.

NDPMC-National Disaster Prevention and Mitigation Committee. 2009. *National disaster prevention and mitigation plan 2010–2014*.

NESDB-National Economic and Social Development Board. 2012. *The eleventh national economic and social development plan 2012–2016*.

Office of Environmental Policy and Planning. 2000. *Thailand's initial communication under the United Nation Framework Convention on Climate Change*. Bangkok: Ministry of Science, Technology and Environment.

ONEP-Office of Natural Resources and Environmental Policy and Planning. 2010. *Thailand's Second Communication under the United Nations Framework Convention on Climate Change*. Bangkok: Ministry of Natural Resources and Environment.

Pipitsombat, N. n.a. *Policies related to climate change in Thailand*. http://www.ostc.thaiembdc.org/atpac2011/Presentation_Ms_Nirawan.pdf.

Praneat, V. 2014. Drought conditions and management strategies in Thailand. http://www.droughtmanagement.info/literature/UNW-DPC_NDMP Country_Report_Thailand_2014.pdf. Accessed May 2015.

Royal Thai Government. 2015. *Integrated plan for 2015 drought management*. http://www.thaigov.go.th/en/cabinet-synopsis-/item/90223-id90223.html.

Stone, R. 2014. Constructing a framework for National drought policy: The way forward—The way Australia developed and implemented the national drought policy. *Weather and Climate Extremes* 3: 117–125.

Takeda, M., A. Laphimsing, and A. Putthividhya. 2015. Dry season water allocation in the Chao Phraya River basin, Thailand. *International Journal of Water Resources Development*. doi:10.1080/07900627.2015.1055856 .

TDM-Thai Meteorological Department. 2007. *The climate of Thailand*. http://www.tmd.go.th/en/archive/thailand_climate.pdf.

The World Bank. 2011 *Thailand environment monitor. Integrated water resources management: A way forward*. Washington D.C.: World Bank. http://documents.worldbank.org/curated/en/367151468303847751/Thailand-environment-monitor integrated-water-resources-management-a-way-forward.

UNDP and Overseas Development Institute. 2012. *Thailand climate public expenditure and institutional review*. https://www.cbd.int/financial/climatechange/g-cpeirmethodology-undp.pdf.

UNESCAP. 2000. *Principles and practices of water allocation among water-use sectors*. Water Resource Series, 80.

WFP. 2015. *El Nino: Implication and scenarios for 2015*. http://documents.wfp.org/stellent/groups/public/documents/ena/wfp276236.pdf.

WMO and GWP. 2014. *National drought management policy guidelines: A template for action (D.A. Wilhite)*. Integrated Drought Management Program-IDPM Tools and Guidelines Series 1. Geneva and Stockholm: WMO and GWP.

[illegible] 2011 [illegible]

[illegible] 2015 [illegible]
[illegible] 2013 [illegible] Journal of [illegible]
TMD [illegible] 2015 [illegible]

UNDP [illegible] 2011 [illegible]
UNFCCC [illegible]
WRI. 2014 [illegible]
[illegible]

Part II
Decision Making Tools for Climate Planning

Chapter 10
Relevance and Quality of Climate Planning for Large and Medium-Sized Cities of the Tropics

Maurizio Tiepolo

Abstract In the last seven years, the number of plans with climate measures for tropical cities has increased 2.3 times compared to the previous seven years as a result of the initiatives of central and local governments, multi-bilateral development aid and development banks. The plans matter in achieving the 11th United Nations' Sustainable development goal. Therefore, the objective of this chapter is to ascertain the relevance and quality of climate planning in large and medium-sized cities in the Tropics. The chapter proposes and applies the QCPI-Quality of Climate Plans Index, consisting of 10 indicators (characterization of climate, number, quantification, relevance, potential impact, cost, funding sources, timetable and responsibility of measures, implementation monitoring and reporting). It is revealed that 338 tropical cities currently have a local development, emergency, master, mitigation, adaptation, risk reduction plan or a resilience or smart city strategy. These tools were unquestionably more common in large cities, especially in OCDE and BRICS countries, while they were rare in Developing Countries. Local development plans (Municipal development, general, comprehensive) were the most common in medium-sized cities, along with those with the lowest quality, while stand-alone strategies and plans (resilience, mitigation, sustainable, adaptation), applied mostly in big cities, present much higher quality.

Keywords Climate change · Municipal development plan · Master plan · Emergency plan · Mitigation plan · Adaptation plan · Risk reduction plan · Resilience strategy · Smart city · Plan quality · Tropics

---

M. Tiepolo (✉)
DIST, Politecnico and University of Turin, Viale Mattioli 39, 10125 Turin, Italy
e-mail: maurizio.tiepolo@polito.it

M. Tiepolo et al. (eds.), *Renewing Local Planning to Face Climate Change in the Tropics*, Green Energy and Technology,
DOI 10.1007/978-3-319-59096-7_10

10.1 Introduction

Climate plans are used by a growing number of cities to reduce the emission of Green House Gases (GHG) into the atmosphere (mitigation, sustainable action plans) and the impacts of climate change (emergency, adaptation, risk reduction, resilience plans). Municipal development and master plans should also be considered among climate plans when they contain measures aimed at altering the urban spatial configuration, density, land cover and physical building characteristics, all factors that can modify the urban micro climate (Alcoforado and Matzarakis 2010).

In the Tropics, climate plans began to be developed in around 2003, based on the initiatives of several multilateral bodies (development banks, United Nations), bilateral aid, associations and movements of local governments (ICLEI-International Council for Local Environmental Initiatives, C40), certain foundations (Rockefeller, Bloomberg), and commitments made by individual countries within the framework of international agreements (UN Framework Conference for Climate Chance, Hyogo Framework for Action). These commitments are often converted into national laws that enforce local climate planning.

One of the recurring topics of the debate on climate planning is which tools to use to plan the reduction of GHG emissions and the impacts of CC on cities: stand-alone plans or mainstreaming of climate measures in existent plans (UN-Habitat 2009, 2011, 2015). According to Aylett (2014), climate planning is not achieved with isolated plans. This said, the mainstreaming of climate measures requires the existence of plans which, according to Fraser and Lima (2012), are absent in small towns.

A second topic is the quality of plans. Until a few years ago, according to the results of the surveys carried out by Wheeler (2008), Tang et al. (2010) and Preston et al. (2011), the quality was rather low. The first quality factor is the solidity of the preliminary analysis that precedes planning. According to Hunt and Watkiss (2011), it was very imbalanced in relation to flood risk. Another quality factor concerns measures. According to Buckley (2010) and Anguelowski and Carmin (2011), those for mitigation of emissions prevail over those for adaptation to impacts. Among the measures envisaged, those effectively applied are only the short-medium term ones, those that can be accomplished within the space of a mandate (Lethoko 2016). Wheeler (2008) then found that, in the United States, the measures envisaged by climate action plans were insufficient to significantly reduce GHG emissions.

The above information is taken from surveys with a rather varied reach: from a few case studies (Vergara 2005; UN Habitat 2015) to a few hundred cities (Carmin et al. 2012; CAI 2012; CDP 2014) without distinguishing between towns and mega cities, tropical, sub-tropical or boreal settlements, least-developed countries and wealthiest economies. Carried out in this way, these surveys do little to help identify the points on which to strengthen climate planning. To find a remedy for this absence, in 2015 we investigated two climatic zones (Sub-tropics and Tropics), a specific class of city (large), concentrating attention only on stand-alone plans (Tiepolo and Cristofori 2016). It turned out that climate planning was still rather scarce (24% of cities) despite being on the increase (+1.5 times in the last five years). Emergency, mitigation, adaptation and resilience plans were, in that order, the most widespread in big cities but quality was poor in 70% of cases. The results of that survey looked promising. This is why, in 2016, we extended the survey to medium-sized cities and to all the existing plans, but narrowed observation to the Tropics only, where there was a higher concentration of Developing Countries (DCs) and Least-Developed Countries (LDCs).

This chapter aims to ascertain (i) the relevance of climate planning and (ii) its quality. The methodology followed starts with the identification of the cities that are populated by over 0.1 million inhabitants (1388) in tropical zone and then, among these, those that have enforced climate a planning tool (338). This is followed by the download of the plans and the construction of the database. Lastly, the analysis: types of plans, their relevance (by class of city, by country and economy), and quality, through the specially conceived QCPI-Quality of Climate Plans Index. The following paragraphs are going to look at the methodology, the results (rise of climate planning in the Tropics, planning categories, quality), discussion, conclusions (focusing on possible improvements in climate planning and on recommendations to put them into practice).

10.2 Materials and Methods

This chapter is based on the results of a survey on climate plans in Chinese, English, French, Portuguese and Spanish, for cities with over 100,000 inhabitants, in 95 tropical countries. The survey covers 63% of tropical cities, while the remaining 37%, which belongs to countries from other linguistic areas (especially Philippines, Indonesia, Malaysia, Bangladesh and Viet-Nam), is covered on a limited basis by plans accessible in English. The term big cities refers to jurisdictions with over one million inhabitants. Medium-sized cities are administrative jurisdictions with a population of between 0.1 and 1 million people. Some big cities are split into municipalities. If these jurisdictions have over 100,000 inhabitants and have a plan, like those of Lima, Miami, Niamey, Phoenix, Port-au-Prince, Rio de Janeiro, Santo Domingo and Yaoundé, then they were considered.

The plans were identified on the websites of the municipalities. Unavailable open access plans were excluded from the survey.

The tropical zone was defined with Köeppen-Geiger's classification based on temperatures and rainfall observed over the period 1971–2000 (Rubel and Kottok 2010) on a 0.5° latitude/longitude regular grid, as presented on the website http://koeppen-geiger.wu-wien.ac.at/shifts.htm. Categories and subcategories were used according to Trewartha's classification (Belda et al. 2014), which is adopted by the FAO, the Joint Research Centre of the European Commission and by IPCC (www.fao.org/docrep006/ad6528/ad652e07.htm). The tropical zone includes the categories wet-tropical rain forest and tropical wet in dry called savanna. A city is considered tropical if at least part of it is included in the tropical climatic grid. For cities on the edges of the climatic zone of interest, the built-up area was recognized on Google Earth. In addition to the demographic class, cities with a climate plan were divided by economy: OECD (Australia, overseas French jurisdictions, Mexico and USA) and Singapore, BRICS (Brazil, India, China and South Africa), DC-Developing Countries (23 countries), LDC-Least Developed Countries (9 countries). All the measures for the reduction of emissions and adaptation to CC were registered (123 for medium-sized cities and 147 for big ones). When we had finished, we had created a database with 53,000 data items containing 355 types of information.

We created a specific QCPI-Quality of Climate Plans Index to assess the action driven nature and the potential impact of climatic plans. The index gathers 10 indicators (climate characterization, number, quantification, relevance, potential impact, cost, funding sources, timetable and responsibility for each measure, implementation monitoring and reporting) (Table 10.1). Every indicator is assigned an equal weight (1 point).

The value of the QCPI can, therefore, vary between 0 and 10. In identifying the indicators, we referred to previous works which had examined plan quality (Baer 1997; Norton 2008) and reduced the number of indicators to effectively succeed in measuring them in a high number of plans often very different from one another. The indicator of potential impact refers the existence in every plan of at least one of the mitigation measures susceptible for significantly reducing emissions of CO_2 or for reducing the hydro-climatic risk according the mitigations plans of Belo Horizonte, Fortaleza, Miami and Phoenix (Table 10.2). It is not possible to appreciate the degree of implementation of the plans through this survey, as the municipalities rarely publish annual monitoring reports on this matter.

Table 10.1 Description of the indicators used for the QCPI

Indicators	Concept
1. Climate characterization	Local climate trends and changes expected over the next 20 years
2. Number of measures	Plans with over 10 climatic measures
3. Quantification of measures	Specification of the quantity of each measure
4. Relevance of measures	At least one measure that would significantly reduce CO_2 emissions or risk
5. Potential impact	Estimated impact of any measure on emissions or risk
6. Cost of measures	Estimated cost of each measure
7. Funding	Specification of the funding sources for each measure
8. Responsible	Specification of the structure responsible for implementing each measure
9. Monitoring & Reporting	Description of the monitoring and reporting system
10. Timetable	Distribution of measures over time

Table 10.2 Soundness of measures in climate plans for big and medium-sized tropical cities

Sector	Measure	Potential reduction of CO_2 emissions %
Mitigation	Less polluting fuels	0.9–7.5
	Bike/car sharing	1
	LED street lighting	2
	Green roof	3
	LEED Certification	2–3
	RRR waste	1–2.1
	Solar in buildings	5.3
	Subway	4
	BRT	5.7–8
	Methane capture (landfill)	20
	Renewable energy	33
	Waste water reuse	33
Adaptation	Resettlement	100

10.3 Results

10.3.1 Relevance of Climate Planning in the Tropics

Reducing the impact of CC entails (i) reducing the causes, (ii) protecting the population and assets exposed both before and (iii) during a hydro-meteorological or climate-related disaster.

Usually, the local governments plan these activities as the application of specific national or regional laws, as in the case of Colombia, Mexico, Niger, Philippines, etc. In the remaining cases, they do so after signing the US Conference of Mayors' Climate Protection Agreement (2005), the C40 or other unilateral initiatives to reduce the risk.

In the Tropics, today, at least 338 cities in 41 countries have a climate plan (Table 10.3; Figs. 10.1 and 10.2). This is almost a quarter of the cities. The highest number (79%) is in the OCDE countries, followed by BRICS (49%), LCDs (35%) and DCs (just 22%).

Since 2010, the number of climate plans has increased 2.3 times the number produced during the previous seven years (Fig. 10.3). In 2012, climate plans

Table 10.3 Tropical cities provided with climate plan

Tropical cities (class)	a	b	a/b * 100
	with climate plan	all	
Big, > 1MP	65	166	39
Medium, 0,1–1 MP	273	1222	22
Big and medium	338	1388	24

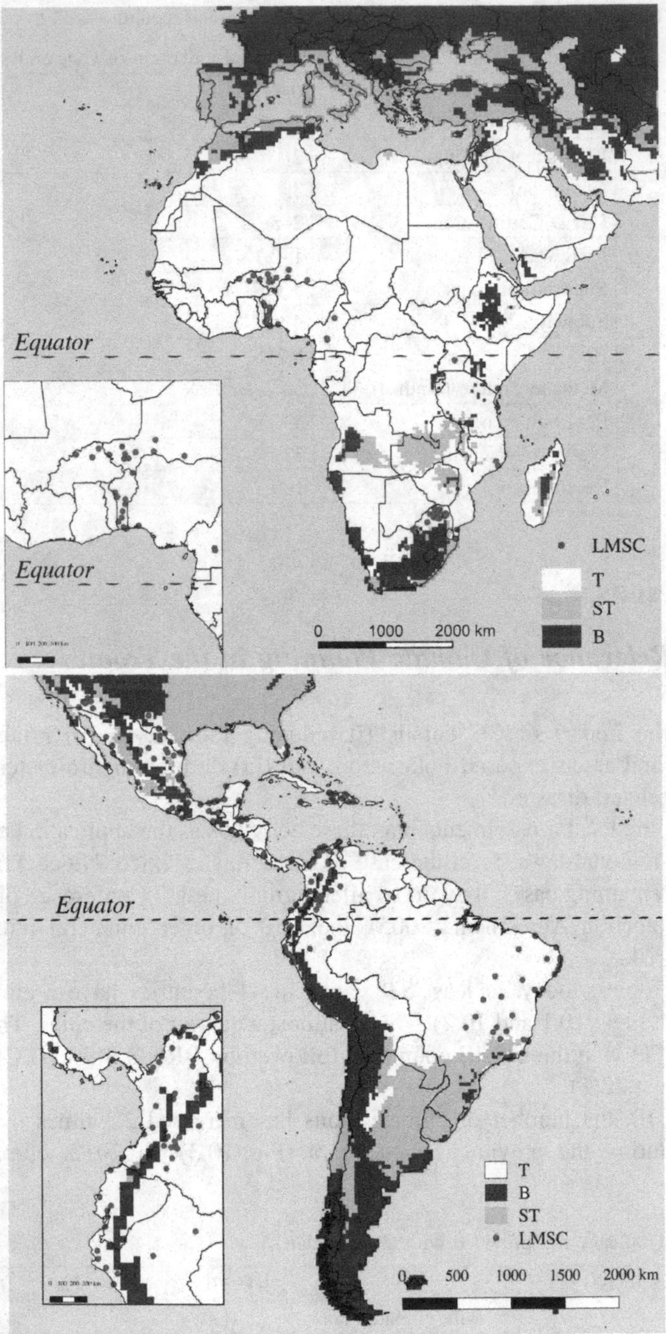

Fig. 10.1 Big and medium-sized cities of Tropical (*T*) Africa an Latin America provided with climate plan

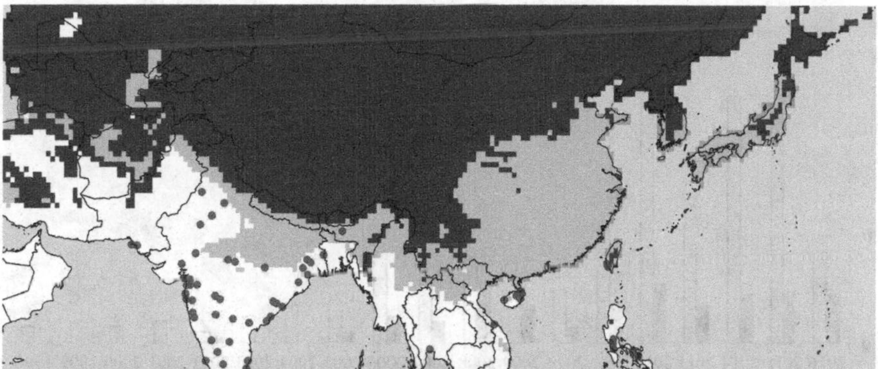

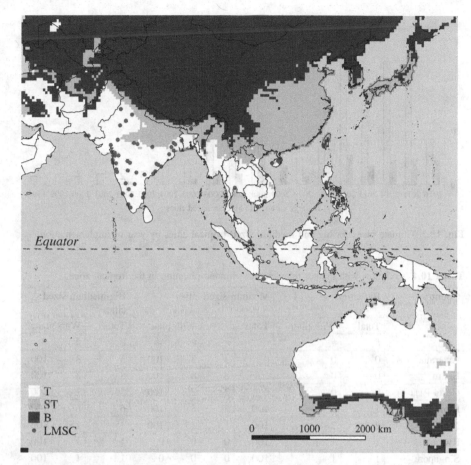

Fig. 10.2 Big and medium-sized cities of Tropical (*T*) Asia provided with climate plan

suddenly increased, following the entry into force of national legislations (Colombia and Mexico) which required the consideration of risk in municipal development plans (MDPs). Nevertheless, cities with climate plans in the Tropics are half those in the Sub-Tropics (Tiepolo and Cristofori 2016).

In 13 countries, climate planning is practiced by over half the cities (Table 10.4).

10.3.2 Plan Categories

Climate plans unite numerous tools, which sometimes have no equivalents in countries belonging to different linguistic areas (Table 10.5).

Municipal development plans are the most common tools (44%), followed by master plans (13%), emergency plans (12%), sustainable action and mitigation

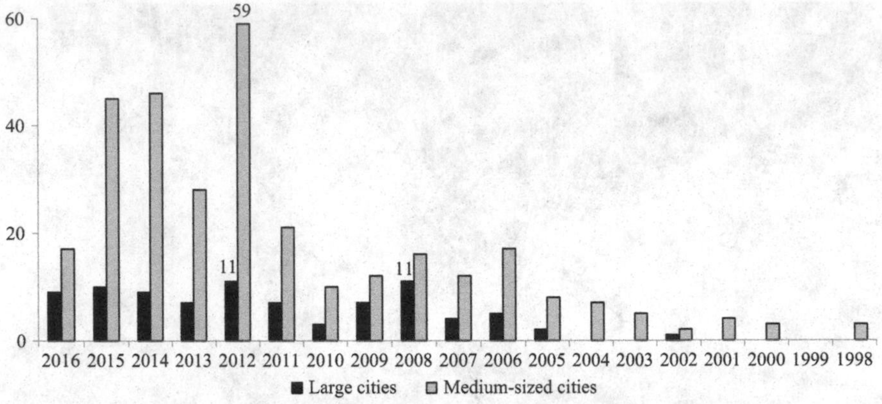

Fig. 10.3 Climate plans for big and medium-sized tropical cities by year of implementation

Table 10.4 Countries with greater relevance of climatic planning in the tropical zone

Country	Big cities		Medium-sized cities				Big-medium-sized cities		
	Total	With plan	Total		With plan		Total	With plan	
	#	#	%	#	#	%	#	#	%
Australia	0	0	0	3	3	100	3	3	100
Benin	0	0	–	8	8	100	8	8	100
Costa Rica	0	0	0	2	2	100	2	2	100
Honduras	1	1	100	5	5	100	6	6	100
Peru	2	2	100	13	13	100	15	15	100
Rwanda	1	1	100	0	0	0	1	1	100
Singapore	1	1	100	0	0	0	1	1	100
USA	2	2	100	20	20	100	22	22	100
South Africa	0	0	0	29	28	97	29	27	97
Colombia	5	5	100	45	38	84	50	43	86
Brazil	14	11	79	42	31	74	56	42	75
Mexico	1	0	0	60	45	73	61	45	74
Nicaragua	1	1	100	6	3	50	7	4	57
Taiwan	2	2	100	2	0	0	4	2	50

plans (12%), risk reduction plans (8%), smart city proposals (4%), adaptation plans (3%), resilience strategies (2%) and other types of plans (2%) (Table 10.6).

Sixteen cities have both a mitigation/adaptation or resilience plan, as well as an emergency or risk reduction plan.

Big cities use a broad range of tools: municipal development plans (21%), mitigation plans (17%), emergency plans (13%) and smart city proposals (13%) and

Table 10.5 How to say climate plan in English, Spanish, French and Portuguese speaking countries. Asteriscs show long term plans

English	Spanish	French	Portuguese
Comprehensive development* Integrated development* General plan*	P. desarollo communal/municipal/local concertado, P. territorial desarollo integral	P. développement communal, P. investissement communal, P. développement strategique	
Contingency p. Emergency p. Crisis management p.	P. contingencia P. emergencia, PLEC	P. contingence	P. contingência
Climate action p.	Plan accion climatica municipal	Plan climat energie territoire	Ação e metas para a redução de Gases de Efeito Estufa
Smart city			
Recovery & rehabilitation			
Sustainable action p. S. development strategy			P. ação sustentável
Resilience action p.			
Disaster management p.	P. municipal gestión riesgo desastre		P. municipal de redução do risco
Adaptation p.			
Green print	Estrategico		
Environment resources management			
Master p. City p.	P. directeur d'urbanisme, P. local d'urbanisme		P. diretor
Land use p.			P. de ordenamiento territorial

* Long term plans

emergency plans (13%). Medium-sized cities, on the other hand, use municipal development plans (50%), master plans (15%) or risk reduction plans (9%).

Municipal Development, General and Compehensive Plans
These are the most popular plans (44%). They organize the measures for every area of jurisdiction of local administration: transportation, infrastructure, housing, land use, conservation, economic development, education, healthcare. Municipal Development Plans (MDP) are medium term tools, that define actions within the space of a mandate (4–5 years). General and comprehensive plans, common in the

Table 10.6 Types of climate plan for big and medium-sized tropical cities

Type of plan	Climate plans for cities					
	Big/medium-sized		Big		Medium-sized	
	n.	%	n.	%	n.	%
Municipal development	164	44	13	21	151	50
Master plan	48	13	4	7	44	15
Emergency	45	12	10	13	35	12
Risk reduction	29	8	2	3	27	9
Sustainable	25	7	6	5	19	6
Mitigation	23	6	13	17	10	3
Adaptation	11	3	7	9	4	1
Smart city	15	4	10	13	5	2
Resilience	6	2	4	5	2	1
Other	10	3	5	7	5	2
Total	376	100	74	100	302	100

USA, are long term tools that define policies for the next 20 years. A diagnosis (participated) of the shortcomings precedes planning of measures. The most accurate plans spread measures over time, defining the cost and the sources of funding, the body responsible and the system of monitoring and evaluation. The MDP was born to promote local development, not to respond to CC, so it does not characterize the local climate, unless obliged by law to do so (Niger). However, different measures fall between those that are typical for mitigation (tree planting, pedestrian and cycle mobility, waste reducing/reusing/recycling) or adaptation (storm water drainage, resettlement from hazard prone areas). In the Tropics, MDPs are the first tools to contain measures for adaptation to CC: Djougou and Malanville, Benin (2003 and 2004), La Ceiba, Honduras (2005), Estelí, Nicaragua (2005), Thane (2005) and Mysore (2006), India. MDPs with climate measures have been generalized in Colombia, India, Mexico and other tropical countries since 2012. Odessa and Laredo are the first medium-sized cities to adopt a comprehensive plan (1988 and 1991). Glendele and Honolulu are the first to have a general plan (2002).

Master Plans

Master plans (13%) are the second most widespread tools. They first started to be implement in big cities (Belo Horizonte 1996, Kigali 2004, São Luís 2005) and then in medium-sized cities (Petrolina, São João Meriti, Uberlandia 2006). These plans are medium-long term and don't just define land use, but also transportation, water and sanitation, waste, economic and social development (education, health, access to housing, tourism) and environment, without, however, specifying the costs of the measures or their distribution over time, and without foreseeing the impacts or a device for monitoring and reporting (M&R).

Emergency Plans

This group (12%) is especially common in South America but not in tropical Africa. 61% of plans traced have been drawn up over the past three years.

There are two types of plan: those that specify the arrangements a local government must make to become operative in the event of a disaster (La Esperanza, Sincelejo), and those that are operational (Santiago de Cali), describing the sequence of established operations that the various actors are called upon to implement in order to respond to the emergency. These plans are, at times, limited to one hazard only (heat in Ahmedabad, drought in Campinas, floods in Machala). Though they do not solely refer to CC-related disasters, they overlap mitigation/adaptation plans in relation to the definition of zones that are either exposed or at risk. In cities that have both an emergency plan and an adaptation plan, the two tools refer to different bodies, specifically civil defence or fire brigade in the first case, city council/environmental sector in the second.

The plan is implemented after an early warning. The warning should be strategically communicated to areas that are exposed to the hazard, especially in the event of floods. This only occurs in 30% of cases. However, only 18% of plans define the early warning threshold. Barely 20% define hazard prone areas and only in half the cases do they make use of maps. Emergency or contingency plans "analyse specific potential events or emerging situations that might threaten society or the environment, and establish arrangements in advance to enable timely, effective and appropriate responses to such events and situations" (UNISDR 2009: 7). Once again, a national law envisages the disaster prevention device (Colombia, India, Philippines, etc.) or civil defence (Brazil), and subsequently the creation of municipal emergency committees to draw up dedicated plans. In other cases (Texas), the contingency plan is drawn up in compliance with rules defined by state commissions. The first medium-sized cities to implement an emergency plan were El Progreso, Honduras (1998), Granada, Nicaragua (2005), San Andres de Tumaco, Colombia (2006) and Tacna, Peru (2007), followed by two big cities: Tegucigalpa, Honduras (2007) and Hyderabad, Pakistan (2008).

Risk Reduction Plans

These scarcely used plans (8%), contain structural and non-structural measures described in detail (costs and methods of financing), which respond to the hazards to which the city is exposed. Sometimes these plans lack climate characterization. Risk reduction plans are in place in Brazil, Cameroun, Colombia, Nicaragua and Philippines. One of the first medium-sized cities to adopt these plans is Granada, Nicaragua (2005), followed, among the big cities, by Maçeio and Vitoria, Brazil (2007).

Mitigation Plans

This group collects 6% of the planning tools traced. These plans focus on the need "to reduce the sources or enhance the sinks of greenhouse gases" (IPCC 2014: 19). Mitigation plans comprise the emission inventory, GHG reduction goals and measures. The most detailed plans estimate the expected reduction of emissions, the risks resulting from CC, the cost, funding sources and the timing of every measure. The capability assessment, which considers the technical and political ability to implement the measures, is rarely carried out. The first mitigation plans for big tropical cities were prepared for Bangkok, Thailand (2007) and Miami, USA

(2008), and authentic climate plans, with numerous measures, for medium-sized cities are those of Cairns and Darwin, Australia (2009 and 2010).

Sustainable Action Plans
These plans (6%) were introduced by the Inter-American Development Bank starting in 2012. They contain mitigation measures (tree planting, pedestrian and cycle mobility) and adaptation measures (early warning and stormwater drainage) effective in the midterm. Their greatest limit is that they are stand-alone plans, not envisaged by law, which the cities are not required to set up.

Smart City Proposals
Smart city proposals are the ultimate planning tool. Employed in India (in about twenty cities) and in Indonesia (Bogor 2015), these tools are still little used in the Tropics (4%). They are based on broad consultation and, in general, concern one or just a few parts of the city. The accent is on technological innovation (for example, sensors which tell you when trash bins are full, etc.) more than tackling big urban problems (slum upgrading and informal settlement regularization, poverty and risk reduction). However, they also contain measures for mitigation (reduction of vehicle traffic) and adaptation (storm water drainage), which are not deriving from climate characterization or from risk assessment.

Adaptation Plans
Adaptation planning focuses on "the adjustment in natural or human systems in response to actual or expected climatic stimuli or their effects, which moderates harm or exploits beneficial opportunities" (UNISDR 2009). Plans include non-structural measures (e.g. early warning, flood drills) and structural measures (flood barriers, stormwater drainage, and resettlement of inhabitants from flood-prone areas). In the latter case, if local adaptation plans are not expressly envisaged by the general environment protection law or by a specific national law (Philippines 2007), they are implicit in the Local Government Act of various countries (Australia 1999) when council functions are specified and often extended to "protect its area from natural hazards and to mitigate the effects of such hazards." In other cases, the individual cities implement the specific national commitments made at the United Nations Framework Convention for Climate Change (UNFCCC). Adaptation planning is funded by dedicated programmes, but is still scarcely practised (3%) as it has not been able to be implemented on the same global mobilisation levels as GHG mitigation. The first examples of adaptation plans are those of Darwin (2010) for medium-sized cities and of Semarang (2010) for big cities.

Resilience Strategies
Resilience strategies (2%) are common in Asia only. The target of these strategies is to strengthen planning, organisational and management skills in view of a disaster, rather than implementing structural measures. Aside from structural measures, they define many actions based on the accumulation of information (databases on hazards, hazard prone area identification), management (establishing a CC coordination office), training and awareness. The most accurate resilience strategies identify

vulnerabilities, list the actions required to reduce them, define implementation phases and the relevant financial mechanism. Since 2010, UNISDR has supported local resilience strategies, especially partnered in Asia by the Rockefeller foundation. Can Tho (Viet Nam) was one of the first big cities to implement a Resilience plan (2010), with Sorsogon (Philippines) being one of the first medium-sized cities to do so (2010).

10.3.3 *Quality of Climate Planning*

Several parameters, such as internal consistency between objectives, priorities and measures (Baer 1997; Norton 2008; Baker et al. 2012), or between climate characterization and measures, can be considered to assess the quality of plans. In our case, we are interested in assessing whether the currently formulated plans are able to guide the implementation of measures that reduce the impact of CC, and whether the measures they propose are really appropriate to reduce GHG emissions and impacts of CC. The focus will then be on 10 indicators: climate characterization, number, quantification, relevance, expected impact, cost, funding sources, timetable and responsibility for each measure, monitoring plan implementation and reporting, brought together in the QCPI, which can theoretically reach the maximum value of 10.

Tropical cities have a QCPI of 2.5. Such a low score is due to the fact that just 2% of the plans specify the impact that the measures are expected to have, 6% characterize local climate, 18% quantify each measure, 20% of the plans indicate the cost of the measures and include a timetable for implementation, 22% have more than 10 climate measures and 24% specify the source of funding (Table 10.7).

Table 10.7 Frequency of indicators used in the QCPI according to tropical city size

Indicator	Climate plans according city' size					
	Big		Medium		Big and medium-sized	
	#	%	#	%	#	%
1. Climate characterization	9	14	11	4	20	6
2. Number of measures	24	36	42	19	76	22
3. Quantification of measures	11	17	50	18	61	18
4. Relevance of measures	49	74	109	39	158	46
5. Potential impact	3	5	4	1	7	2
6. Measure cost	25	38	118	42	143	42
7. Measure funds	15	23	69	25	84	24
8. Measure responsible	14	21	56	30	70	20
9. Monitoring and reporting	5	8	86	31	91	26
10. Time table	17	26	122	44	70	20
Plan considered #	66	100	280	100	346	100

Table 10.8 QCPI in tropical cities according to city size and plan category

Climate plan categories	City size		
	Big QCPI	Medium QCPI	Big and medium QCPI
Resilience	4.8	3.3	4.0
Sustainable action	1.0	3.7	3.4
Mitigation	3.2	4.0	3.5
Sustainable action	1.0	3.7	3.4
Adaptation	2.8	5.0	3.1
Smart city	2.7	3.2	2.9
General	–	2.9	2.9
Municipal development	2.9	2.7	2.7
Risk reduction	1.0	1.8	1.8
Comprehensive	–	1.6	1.6
Master	1.1	1.0	1.0
All	2.7	2.2	2.5

However, the QCPI varies depending on the size of the cities. Big cities have higher quality plans (QCPI 2.7) than those of medium-sized cities (QCPI 2.2) (Table 10.8). This is due to the greater frequency in the former of resilience strategies, mitigation, adaptation plans, tools which present the highest QCPI (4.8, 3.2 and 2.8 and 2.7 respectively). Medium-sized cities on the other hand are characterized by municipal developments and master plans, which have a low QCPI (2.7 and 1.0). Although the average QCPI values are fairly low, certain categories of plans have some high quality tools such as Cairns' mitigation plan (QCPI = 8), the MDPs of Concepcion de la Vega and Tanout, the resilience plan of Semarang, the risk reduction plans of Palmira and Puerto Plata (QCPI = 7).

In short, stand-alone plans (resilience, mitigation, sustainable action, adaptation, smart city) have higher QCPI than the general plans (municipal development, general, comprehensive, master plans), regardless of the size of the city.

The relevance indicator can, however, be misleading. When climate planning is carried out, not all cities start from the same baseline. For example, separate glass, metal, paper and food waste and recycling programs or the use of LED bulbs for street lighting can be innovative measures in one city, while they are so consolidated in another that they need not even be mentioned among the measures established by the plan. Hence, the absence of certain measures does not always indicate lack of detail or visionary planning.

The 346 climate plans of tropical cities traced (the remaining are emergency plans with no measures and other plans with little detailed measures), concern the main jurisdiction only, with the sole exception of Miami, where the whole metropolitan area was considered. Planning on a metropolitan scale stems from the need to harmonise the measures of many jurisdictions and authorities (water, etc.), whose consent is required (Revi et al. 2014: 44).

Plans for metropolitan areas include building awareness, studies and assessments to ensure that mitigation/adaptation measures become rooted in each jurisdiction,

Table 10.9 Aim of measures in 271 climate plans for tropical cities

Measure goal	Measures for Big and medium sized tropical cities	
	N°	%
Adaptation	96	62
Mitigation	47	31
Adaptation & mitigation	11	7
All	154	100

Table 10.10 Structural and non-structural measures in 271 climate plans for tropical cities

Measure nature	Big and medium-sized cities		Big cities		Medium-sized cities	
	N°	%	N°	%	N°	%
Structural	886	61	245	66	650	61
Non-structural	256	39	125	34	420	39
Total	1142	100	370	100	1070	100

and fundraising initiatives. Municipal plans focus instead on direct impact, especially on municipal facilities (offices, transportation, employees), and on sectors in which the Municipality has regulatory authority (private construction works, road systems, waste, education, etc.).

10.3.4 Climate Measures

As regards measures, there are two aspects of interest: knowing the main aims of the measures (mitigation or adaptation) and checking their nature (structural or non-structural). The survey of 346 climatic plans ascertained that adaptation prevails (62%) over mitigation (Table 10.9).

Secondly, structural measures (e.g.: tree planting, storm water drainage, cycle lanes, resettlement) prevail (61%) over those of a non-structural nature (e.g.: early warning system, emergency plan, risk maps, air quality monitoring, etc.) (39%) and this prevalence is valid for both big and medium-sized cities (Table 10.10).

10.4 Discussion

The survey on climate plans in the Tropics has allowed in-depth understanding of the intentions of cities to tackle climate change. Its results belie all previous knowledge. Firstly, climate planning in the Tropics has risen considerably in the past seven years, and now concerns a fourth of all cities.

The option between stand-alone and existing plans (UN-Habitat 2015) is soon unravelled. Three quarters of climate-action is carried out using existing tools: MDPs and master plans, especially in medium-sized cities. Stand-alone plans are used only by big cities. The mainstreaming of climate measures in the existing tools has several advantages in theory (UN-Habitat 2015; Revi et al. 2014; Basset and Shandas 2010): decrease of hazard exposure (prohibition of building on hazard prone areas), increased chances of implementation (MDP), mitigation burden shared with private sector (mitigation imposed upon developers), cost reduction (measures already funded in other sectors), reduced contribution to CC (regulations on building materials). In practice, we have found that mainstreaming presents several limitations compared to stand-alone plans. Specifically, in the case of municipal development plans, we notice very few climatic measures (an average of 4). In the case of master plans other limitations include lack of characterisation of the hazard, prevalence of measures that only concern land use (setback to be complied with during construction works, and land use allowed), lack of priority and scheduling, no reference to the potential impact of works, rare monitoring and reporting of the plan implementation.

Secondly, previous knowledge of climate measures is belied. In the Tropics, the focus isn't on flood (Hunt and Watkiss 2011). It is on air quality (47 measures), drought, heat, fire, landslide and wind (28 measures), with flood coming last (16 measures). The equivalent measures aren't related to mitigation (Buckley 2010; Anguelowsky and Carmin 2011) but to adaptation. Lastly, contrary to the claims made by Wamsler et al. (2013), the measures are not always the same in the different contexts with a prevalence of non-structural measures. They depend on the economy to which the city belongs (OECD, BRICS, DC, LDCs) and on the size of the city, and are mainly structural.

Thirdly, the QCPI confirms the observations of Wheeler (2008), Tang et al. (2010) and Preston et al. (2011) on smaller contexts: plan-quality is still low. Climate plans are not particularly action driven and their measures are inadequate to significantly reduce GHG emissions and the impacts of CC.

The existing plans, in which to carry out the mainstreaming of climate measures, present the lowest quality, regardless of the size of the city and the economy to which it belongs. Mainstreaming requires a considerable amount of work to raise the quality of MDPs, MPs and RMPs.

Our survey has two main limits. The first is that it considers no plans other than those accessible in Chinese, English, French, Spanish and Portuguese. Secondly, the QCPI can be misleading in relation to the relevance of measures: the absence of measures in some sectors (LED, drainage) can mean that the city in question is already sufficiently well-equipped.

10.5 Conclusion

This chapter aimed to characterize climate planning (dissemination and trend) and to ascertain its quality in a homogeneous context: the big and medium-sized cities of the Tropics. The survey of 338 tropical cities (Table 10.11) identified two important trends. First, strong growth of climate plans, especially in big cities. Second, dissemination of climatic measures mainly in MDPs.

The assessment of the quality of plans using the QCPI made it possible to identify the weaknesses that could be eliminated in the second-generation plans.

Local climate characterization (temperature, precipitations, sea level rise) is absent in all classes of city, as are indications on its future trend: it is paradoxical that 84% of plans define climate measures without knowing the local impacts of the climate and the trends expected for the next 20 years.

The potential impacts of the measures are estimated by just 2% of plans.

A measure which could significantly reduce emissions and impacts is present in just 46% of plans. For example, in relation to buildings, we have found that measures rarely make the construction of carbon-neutral structures compulsory.

Only 26% of climate plans describe the monitoring and reporting system and 22% of plans for medium-sized cities envisage more than 10 climate measures.

These considerations should sound as a recommendation for the multi-bilateral bodies that finance local climate plans, for the NGOs that accompany their formation, for the local governments that approve them and for the central governments that draw up the guidelines for their preparation.

This survey can be furthered in three ways: (i) passing from the occasional survey to tracking, (ii) checking whether the second-generation plans have overcome the weaknesses highlighted by the ten indicators of the QCPI, (iii) passing from the survey on the plan quality to that on plan implementation. Sometimes, the absence of details in planning is a choice made by local governments, to ensure long life to the plans for instance.

Table 10.11 Big and medium-sized cities of the Tropics provided with A-Adaptation, CD-City Development, C-Comprehensive, Ci-City, DM-Disaster management, E-Emergency, G-General, ID-Integrated development, M-Mitigation, MI-Municipal Investment, M-Master, R-Resilience, RR-Risk reduction, SAP-Sustainable Action, S-Strategic, SC-Smart city plans

City	Country ISO 3166-1	Population million	Plan type	Year
Big				
Agra	IND	1.6	MDP	2006
Ahmedabad	IND	6.3	E, SC	2015, 16
Aurangabad	IND	1.1	MDP	2012
Bangalore	IND	8.4	MDP, O	2011, 2015
Bangkok	THA	8.2	M	2007
Barranquilla	COL	1.2	A	2012
Belem	BRA	1.4	MP	2008
Belo Horizonte	BRA	2.5	MP, M	1996, 2013
Bhopal	IND	2.8	MDP	2006
Bogor	IDN	1.0	SC	2015
Brasilia	BRA	2.6	MP	2009
Bucaramanga	COL	1.0	MDP	2016
Can Tho	VNM	1.2	R	2010
Cartagena das Indias	COL	1.0	A, MDP	2012, 13
Cebu City	PHL	2.5	M	2013
Chennai	IND	4.2	SC	2015
Coimbatore	IND	1.6	MDP	2002
Dakar	SEN	2.5	MP	2010
Delhi	IND	16.8	E, SC	2015
Dubai	UAE	1.8	M	2014
Fortaleza	BRA	2.6	MP, M	2009, 2012
Goiania	BRA	1.3	SAP, RR	2012
Guayaquil	ECU	2.3	A	2012
Haikou	CHN	2.2	E, M	2014, 2015
Ho Chi Minh City	VNM	8.2	R	2013
Hyderabad	IND	6.8	E	2013
Hyderabad	PAK	1.1	E	2003
Indore	IND	2.0	R	2015
Jabalapur	IND	1.1	SC	2015
Jaipur	IND	3.3	SC	2012
Kampala	UGA	1.7	Other	2014
Kaoshiung	TWN	3.0	M	2011, 2014
Karachi	PAK	23.0	A	2013
Kigali	RWA	1.1	MP	2004
Lagos	NGA	9.0	A	2013
Lima	PER	8.5	E	2011

(continued)

Table 10.11 (continued)

City	Country ISO 3166-1	Population million	Plan type	Year
Ludhiana	IND	1.4	SC	2015
Maçeio	BRA	1.0	RR	2007
Managua	NIC	1.3	MDP	2013
Manaus	BRA	2.0	MP	2014
Maracaibo	VEN	2.0	MDP	2005
Miami	USA	2.6	M	2008
Mombasa	TZA	1.1	MDP	2013
Montería	COL	1.0	M	2011
Nagpur	IND	2		
Phoenix City	USA	1.5	M	2009
Pimpri Chinchwad	IND	1.7	MDP	2006
Puducherry	IND	1.2	MDP	2013
Pune	IND	5.1	SC	2015
Raipur	IND	1.4	MDP	2014
Recife	BRA	1.6	MDP, SAP	2008, 2014
Rio de Janeiro	BRA	6.5	MP, M	2011, 2014
Salvador de Bahia	BRA	2.9	E, MP	2015, 2016
San Juan Lurigancho	PER	1.1	MDP	2011
Santa Cruz de la Sierra	BOL	2.5	MDP	2016
Santiago de Cali	COL	2.1	E	2009
São Luís	BRA	1.0	MP	2006
Semarang	IDN	1.6	R	2010
Singapore	SGP	3.8	M	2012
Surat	IND	4.8	R	2011
Tainan	CHN	1.9	A, E	2010
Tegucigalpa	HND	1.2	E	2007
Thane	IND	1.3	MDP	2005
Vasai Virar	IND	1.2	MDP	2009
Visakapatnam	IND	2.0	SC	2016
Medium-sized				
Abomey Calavi	BEN	0.7	MDP	2005
Acapulco	MEX	0.7	MDP	2012
Acuña	MEX	0.2	MDP	2014
Aganang	ZAF	0.1	ID, DM	2009, 2012
Agartala	IND	0.4	DP	2011
Agglomération centrale Martinique	MTQ	0.2	M	2012
Altamira	MEX	0.2	MDP	2011
Ananindeua	BRA	0.5	MP	2006

(continued)

Table 10.11 (continued)

City	Country ISO 3166-1	Population million	Plan type	Year
Anapolis	BRA	0.4	MP	2016
Apartado	COL	0.2	MDP	2012
Aracaju	BRA	0.6	MP	2010
Armenia	COL	0.3	MP	2009
Asansol	IND	0.6	MDP	2006
Babahoyo	ECU	0.1	E	2009
Bacolod	PHL	0.5	A	2013
Ba-Phalaborwa	ZAF	0.2	DM, IDP	2012, 2015
Barrancabermeja	COL	0.3	MDP	2012
Belagavi	IND	0.5	SC	2016
Bello	COL	0.4	MDP	2012
Berhampur	IND	0.4	CDP	2011
Betim	BRA	0.4	MP	2007
Blouberg	ZAF	0.2	MDP	2015
Boa Vista (Roirama)	BRA	0.3	MP	2006
Bohicon	BEN	0.1	MDP	2008
Buenaventura	COL	0.4	MP	2001
Cairns	AUS	0.2	M	2010
Campeche	MEX	0.9	SAP	2015
Campina Grande	BRA	0.4	MP	2006
Campo Goyatacazes	BRA	0.5	MP	2007
Campos	BRA	0.5	MP	2007
Cancun	MEX	0.7	MDP	2014
Cariacica	BRA	0.4	MP	2007
Carmen	MEX	0.1	MDP	2012
Carrefour	HTI	0.5	MIP	2011
Cartago	COL	0.2	MP	2000
Chandler, AZ	USA	0.3	E, GP	2006, 2008
Chetumal see Othon P. Blanco	MEX	0.2	MDP	2013
Chiclayo	PER	0.5	MDP	2015
Chilón	MEX	0.1	MDP	2012
Choloma	HND	0.2	MDP	2003
Cienaga	COL	0,1	MDP	2012
Ciudad Madero	MEX	0.2	MDP	2013
Ciudad Valles	MEX	0.2	MDP	2004
Concepción de la Vega	DOM	0.2	MDP	2016
Contagem	BRA	0.6	RR	2007
Coral Springs	USA	0.2	CP	2008

(continued)

Table 10.11 (continued)

City	Country ISO 3166-1	Population million	Plan type	Year
Cordoba	MEX	0.1	MDP	2014
Cotonou	BEN	0.8	MDP	2008
Cuiabá	BRA	0.6	MP	2008
Cumana	VEN	0.8	PAS	2015
Dargol	NER	0.1	MDP	2013
Darwin	AUS	0.1	M, S	2011, 2012
Davanagere	IND	0.4	SC	2016
Ding'an	CHN	0.3	M	2016
Dioundiou	NER	0.1	MDP	2009
Ditsobotla	ZAF	0.2	IDP	2015
Djougou	BEN	0.3	MDP	2003
Dongfang	CHN	0.4	E	2015
Dori	BFA	0.1	MDP	2008
Dos Quebradas	COL	0.2	E	2011
Duque de Caxias	BRA	0.9	RR	–
El Porvenir	PER	0.1	MDP	2014
El Progreso	HND	0.2	E	1998
Envigado	COL	0.2	MP	2011
Ephraim Mogale	ZAF	0.1	IDP	2015
Esmeraldas	ECU	0.5	E	2012
Estelí	NIC	0.1	MDP, A	2005, 12
Feira de Santana	BRA	0.6	MP	2013
Florencia	COL	0.2	E	2013
Floridablanca	COL	0.2	MDP	2012
Fort Lauerdale	USA	0.2	E, CP	2008, 2015
Frances Baard	ZAF	0.4	DM, IDP	2006, 2015
Fusagasugà	COL	0.1	MP	2001
Gilbert	USA	0.2	GP	2012
Glazoué	BEN	0.1	MDP	2014
Glendale, AZ	USA	0.2	GP	2002
Gomez Palacio	MEX	0.3	MDP	2010
Granada	NIC	0.1	E	2005, 09
General Escobedo	MEX	0.4	MDP	2010
Greater Giyani	ZAF	0.2	IDP, DM	2012, 2013
Greater Letaba	ZAF	0.2	E, IDP	2012, 2015
Greater Tubatse	ZAF	0.3	MDP	2016
Greater Tzaneen	ZAF	0.4	DMP, IDP	2012, 2013
Guadalajara de Buga	COL	0.1	MP	2000

(continued)

Table 10.11 (continued)

City	Country ISO 3166-1	Population million	Plan type	Year
Guadalupe	MEX	0.7	MDP, MP	2005, 2016
Guntur	IND	0.5	CDP	2006
Guwahati	IND	0.8	SC	2016
Hat Yai	THA	0.2	A	2016
Heredia	CRI	0.1	MDP	2014
Hermosillo	MEX	0.8	MP	2015
Hialeah	USA	0.2	CP	2007
Honolulu	USA	0.3	GP, E, M	2002, 2008, 2014
Hoshanghad	IND	0.1	MDP	2011
Hué	VNM	0.3	A	2014
Ibagué	COL	0.5	E, MDP	2011, 2016
Iguala	MEX	0.1	MDP	2015
Iloilo city	PHL	0.4	A	2014
Indipendencia	PER	0.2	MDP	2011
Iquitos	PER	0.4	MP	2011
Itagui	COL	0.2	E, MDP	2012, 2016
Jaboatão de Guararapes	BRA	0.7	MP	2008
Jamshedpur	IND	0.7	MDP	2006
Jean Rabel	HTI	0.1	MDP	2013
Jiutepec	MEX	0.2	MDP	2016
João Pessoa	BRA	0.8	MP, SAP	2008, -
Jozini	ZAF	0.2	IDP	2013
Kagisano-Molopo	ZAF	0.1	IDP	2005
Kakinada	IND	0.3	MP	2016
Kalfou	NER	0.1	MDP	2014
Klouékanmè	BEN	0.1	MDP	2010
Kochi	IND	0.6	SC	2016
Kollam	IND	0.4	CDP	2014
La Ceiba	HND	0.2	MDP	2005
La Esperanza	PER	0.1	E	2015
La Paz	MEX	0.2	M	2012
Lepelle-Nkumpi	ZAF	0.2	IDP	2016
Lephalale	ZAF	0.1	DMP, IDP	2012, 2015
Limassol	CYP	0.1	M	2013
Los Mochis	MEX	0.3	MP	2014
Machala	ECU	0.2	E	2009
Magangué	COL	0.2	MDP	2012

(continued)

Table 10.11 (continued)

City	Country ISO 3166-1	Population million	Plan type	Year
Makhado	ZAR	0.5	IDP, RR	2012, 2014
Makhuduthamaga	ZAR	0.3	IDP, DMP	2010, 2013
Malambo	COL	0.1	MDP	2012
Malanville	BEN	0.1	MDP	2004
Managua districts	NIC	0.3	MP	2013
Manzanillo	MEX	0.2	MDP	2015
Matamoros	MEX	0.5	MP	2005
Matola	MOZ	0.9	E	2015
Mazatlan	MEX	0.4	MDP	2014
Merida	MEX	0.8	MDP	2015
Mesa	USA	0.4	E, GP	2009, 2014
Mexicali	MEX	0.7	MDP	2014
Miami City	USA	0.4	CP	2015
Miami Gardens	USA	0.1	CP	2006
Miniatitlan	MEX	0.4	MDP	2014
Moca	DOM	0.2	MDP	2012
Mogalakwena	ZAF	0.3	DMP, IDP	2012, 2014
Molemole	ZAF	0.1	DMP, IDP	2010, 2014
Molina (La)	PER	0.1	MDP	2012
Monclova	MEX	0.2	MP	2012
St. James-Montego Bay	JAM	0.1	PAS	2015
Moretele	ZAF	0.2	IDP	2015
Moses Kotane	ZAF	0.2	IDP	2014
Muntinlupa	PHL	0.5	R	2014
Muriaé	BRA	0.1	RR	2010
Mysore	IND	0.9	CDP	2006
Naga	PHI	0.2	E	2013
Nampula	MOZ	0.5	M	2015
Nanded	IND	0.4	MP	2006
Natal	BRZ	0.8	MP, RR	2007, 2008
Natitingou	BEN	0.1	MDP	2004
Navojoa	MEX	0.2	MDP	2016
Neiva	COL	0.3	MP	2009
Ngaoundere	CMR	0.2	MDP	2014
Niamey1	NER	0.2	MDP	2012
Niamey4	NER	0.1	MDP	2014
Niamey5	NER	0.1	MDP	2013

(continued)

Table 10.11 (continued)

City	Country ISO 3166-1	Population million	Plan type	Year
Nicosia	CYP	0.1	M	2014
Niteroi	BRA	0.5	MP	2004
Noumea	NCL	0.1	MP	2014
Nova Iguaçu	BRA	0.8	MP	2008
Nueva Laredo	MEX	0.4	MDP	2014
Ocaña	COL	0.1	RR	2012
Odessa	USA	0.1	CP	2008, 11, 15
Olinda	BRA	0.4	RRP	2004
Ouahigouya	BFA	0.1	MDP	2009
Ouessé	BEN	0.1	MDP	2011
Palmas	BRA	0.3	PAS	2015
Palmira	COL	0.3	E, RR	2012
Panama	PAN	0.9	PAS	2015
Parakou	BEN	0.2	MDP	2007
Pembroke Pines, FL	USA	0.2	CP	2013
Peoria	USA	0.2	E, SAP, GP	2003, 2009, 2013
Pereira	COL	0.5	MDP, PAS	2012, 15
Petrolina	BRA	0.3	MP	2006
Petropolis	BRA	0.3	RR	2007
Piedecusta	COL	0.1	MDP	2016
Piedras Negras	MEX	0.2	MDP	2014
Pissila	BFA	0.1	MDP	2008
Piura	PER	0.4	MDP	2014
Polokwane	ZAF	0.6	IDP, DMP	2012, 2013
Popayan	COL	0.3	MDP	2016
Porto Novo	BEN	0.3	MDP	2005
Porto Viejo	ECU	0.3	E	2008
Poza Rica	MEX	0.2	MDP	2014
Puerto Cortés	HND	0.1	MDP	2013
Quibdó	COL	0.1	RR	2012
Qiunghai	CHN	0.5	E	2015
Rajpur Sonarpur	IND	0.3	DP	2007
Ranchi	IND	0.8	MDP	2016
Ratlam	IND	0.3	CDP	2010
Ratlou	ZAF	0.1	IDP	2016
Resende	BRA	0.1	E	2014
Reynosa Tamaulipas	MEX	0.6	MDP	2012

(continued)

Table 10.11 (continued)

City	Country ISO 3166-1	Population million	Plan type	Year
Ribeirão das Neves	BRA	0.3	RR	2009
Rio Branco	BRA	0.4	E	2016
Rio Hacha	COL	0.2	MDP	2011
Rourkela	IND	0.3	MDP	2015
Rustenburg	ZAR	0.5	IDP	2012
Saint Louis du Nord	HTI	0.1	MDP	2012
Saint Marc	HTI	0.2	MIP	2011
Saltillo	MEX	0.7	MDP	2014
San Andrés de Tumaco	COL	0.2	E	2004
San Diego	VEN	0.1	MDP	2014
San Felipe Puerto Plata	DOM	0.2	RRP	2013
San José	CRI	0.3	MDP	2012
San José de Cucuta	COL	0.6	MP	2001
San Jose del Monte	PHL	0.4	E	2014
San Juan Maguana	DOM	0.2	MDP	2012
San Luis Rio Colorado	MEX	0.2	MP	2013
San Pedro Garza Garcia	MEX	0.1	MDP	2012
San Pedro Macorís	DOM	0.2	MDP	2013
San Pedro Sula	HND	0.5	MDP	2015
Santa Ana	SLV	0.2	PAS	2012
Santa Marta	COL	0,4	PAS	2012
Santiago de los Caballeros	DOM	0.7	PAS	2015
Santiago de Surco	PER	0.3	MDP	2009
Santo Domingo Este	DOM	0.9	MDP	2015
Sanya	CHN	0.7	E	2014
São João Meriti	BRA	0.5	MP	2006
Scottsdale	USA	0.2	GP	2014
Sincelejo	COL	0.2	MDP	2012
Solarpur	IND	0.9	MDP	2015
Soledad	COL	0.5	MDP	2012
Sol Plaatje	ZAF	0.2	IDP	2012
Sorsogon	PHL	0.2	R	2010
Sousse	TUN	0.2	MDP	2014
Tacloban	PHL	0.2	RR	2014
Tacna	PER	0.2	E	2007
Tanout	NER	0.1	MDP	2005
Tapachula	MEX	0.2	MDP	2005
Tarapoto	PER	0.1	MDP	2012
Taytay	PHL	0.3	R	2015

(continued)

Table 10.11 (continued)

City	Country ISO 3166-1	Population million	Plan type	Year
Tempe	USA	0.2	GP	2013
Tepic	MEX	03	MP	2010
Teresina	BRA	0.8	MP	2015
Triruvananthapuram	IND	0.8	MP	2012
Thulamela	ZAF	0.6	IDP, DMP	2012, 2013
Tierralta	COL	0.1	RR	2012
Torreon	MEX	0.6	MDP	2014
Townsville	AUS	0.2	A, E, GP	2013, 2015
Trinidad	COL		MDP	2012
Trujillo	PER	0.8	MP	2013
Tswaing	ZAF	0.1	IDP	2014
Tucson	USA	0.5	E	2014
Tuluá	COL	0.2	MP	2000
Tuxtepec	MEX	0.1	MDP	2011
Tuxtla Gutierrez	MEX	0.4	MDP	2012
Uberlandia	BRA	0.7	MP	2006
Ujjain	IND	0.5	MDP	2010
uMhlabuyalingana	ZAF	0.2	IDP	2016
uPhongolo	ZAF	0.1	IDP	2008
Urapan	MEX	0.3	MDP	2015
Uribia	COL	0.1	MDP	2012
Valledupar	COL	0.4	PAS	2015
Veracruz	MEX	0.6	MP	2008
Victoria	MEX	0.3	MDP	2013
Villa Alvarez	MEX	0.1	MDP	2014
Villa el Salvador	PER	0.4	MDP	2001
Villa Hermosa	MEX	0.6	MDP	2012
Villa Maria Triunfo	PER	0.4	MDP	2007
Villavicencio	COL	0.4	MP	2015
Vitoria (Espirito Santo)	BRA	0.4	PAS	2015
Vitoria da Conquista (Bahia)	BRA	0.4	MP	2006
Warangal	IND	0.8	MDP	2012
Wengcheng Town	CHN	0.2	M	2015
West Palm Beach	USA	0.1	CP	2008
Xalapa	MEX	01	SAP	2015
Yaoundé 1	CMR	0.3	MDP	2012
Yaoundé 6	CMR	0.3	RR	2014
Yopal	COL	0.1	MP	2013
Zipaquira	COL	0.1	MP	2003

References

Alcoforado, M.-J., and A. Matzarakis. 2010. Planning with urban climate in different climatic zones. *Geographicalia* 57: 5–39.

Anguelowsky, I., and J.A. Carmin. 2011. Something borrowed, everything new: Innovation and institutionalization in urban climate governance. *Climate Opinions in Environmental Sustainability* 3: 1–7. doi:10.1016/j.cosust.2010.12.017.

Aylett, A. 2014. *Progress and challenges in the urban governance of climate change: Results of a global survey*. Cambridge, MA: MIT.

Baer, W.C. 1997. General plan evaluation criteria: An approach to making better plans. *JAPA* 63 (3): 329–344. doi:10.1080/01944369708975926.

Baker, I., A. Peterson, G. Brown, and C. McAlpine. 2012. Local government response to the impacts of climate change: An evaluation of local climate adaptation plans. *Landscape and Urban Planning* 107: 127–136. doi:10.1016/j.landurbplan.2012.05.009.

Basset, E., and V. Shandas. 2010. Innovation and climate action planning. Perspectives from municipal plans. *Journal of the American Planning Association* 76 (4): 435–450. doi:10.1080/01944363.2010.509703.

Belda, M., E. Holtanová, T. Holenka, and J. Kalvová. 2014. Climate classification revisited: From Köeppen to Trewartha. *Climate Research* 59 (1): 1–13. doi:10.3354/cr01204.

Buckley, H. 2010. Cities and the governing of climate change. *Annual Review of Environment and Resources* 35: 229–259. doi:10.1146/annurev-environ-072809-101747.

CAI Clean air initiative and Cities development initiative for Asia. 2012. *Climate Change Plans and Infrastructure in Asian Citis*. Pasig city: CAI and CDIA.

CDP-Carbon Disclosure Project. 2014. *CDP Cities 2013. Summary Report on 110 Global Cities*. London: CDP.

Carmin, J.A., N. Nadkami, and C. Rhie. 2012. *Progress and Challenges in Urban Climate Adaptation Planning. Results of a Global Survey*. Cambridge, MA: MIT.

Fraser, A., and D.V. Lima. 2012. Regional technical assistance initiative on climate adaptation planning in LAC cities. Survey results report.

Hunt, A., and P. Watkiss. 2011. Climate change impacts and adaptation in cities: A review of the literature. *Climatic Change* 104 (1): 13–49. doi:10.1007/s10584-010-9975-6.

IPCC-International Panel on Climate Change. 2014. Annex II: Glossary. In *Climate change 2014: Synthesis report. Contribution of working groups I, II and III to the fifth assessmeny report of the Intergovernmental panel on climate change*, IPCC. Geneva: IPCC.

Lethoko. 2016. Inclusion of climate change strategies in municipal integrated development plans: A case from seven municipalities in Limpopo province, South Africa. *Jàmbá-Journal od Disaster Risk Studies* 8 (3). doi:10.4102/jamba.v8i3.245.

Norton, R.K. 2008. Using content analysis to evaluate local master plans and zonng codes. *Land Use Policy* 25: 432–454. doi:10.1016/j.landusepol.2007.10.006.

Preston, B.L., R.M. Westaway, and E.J. Yuen. 2011. Climate adaptation planning in practice: An evaluation of adaptation plans for three developed nations. *Mitig Adapt Strateg Glob Change* 16: 407–438. doi:10.1007/s11027-010-9270-x.

Revi, A. et al. 2014. Urban areas. In *Assessment report 5*, IPCC-WGII, 1–113.

Rubel, F., and M. Kottok. 2010. Observed and projected climate shifts 1901–2100 depicted by world maps of the Köppen-Geiger climate classification. *Meteorol. Zeitschrift* 19 (2): 135–141. doi:10.1127/0941-2948/2010/0430.

Tang, Z., S.D. Brody, C. Quinn, L. Chang, and T. Wei. 2010. Moving from agenda to action: evaluating local climate change actions plans. *Journal of Environmental Planning and Management* 53 (1): 41–62. doi:10.1080/0964056090339977.

Tiepolo, M., and E. Cristofori. 2016. Climate change characterisation and planning in large tropical and subtropical cities. In *Planning to cope with tropical and subtropical climate change*, ed. M. Tiepolo, E. Ponte, and E. Cristofori, 6–41. Berlin: De Gruyter Open. doi:10.1515/9783110480795-003.

UN-Habitat. 2009. *Climate change. The role of cities. Involvement, influence, implementation*. UN-Habitat and UNEP-United Nations Environmental Program.

UN-Habitat. 2011. *Cities and climate change. Global report on human settlements 2011*. London-Washington, DC: Earthscan.

UN-Habitat. 2015. *Integrating climate change into city development strategies. Climate change and strategic planning*. Nairobi: UN-Habitat.

UNISDR. 2009. *2009 UNISDR terminology on disaster risk reduction*. Geneva: UNISDR- United Nations.

Vergara, W. 2005. *Adapting to climate change. Lessons learned, work in progress, and proposed next steps for the World Bank in Latin America*. The World Bank.

Wamsler, C., E. Brink, and C. Rivera. 2013. Planning for climate change in urban areas: From theory to practice. *Journal of Cleaner Production* 50 (1): 68–81. doi: 10.1016/j.jclepro.2012.12.008 .

Wheeler, S.M. 2008. State and municipal climate change plans. *Journal of the American Planning Association* 74 (4): 482–496. doi:10.1080/01944360802377973.

Chapter 11
Local and Scientific Knowledge Integration for Multi-risk Assessment in Rural Niger

Maurizio Tiepolo and Sarah Braccio

Abstract In the rural Tropics, the participatory risk assessment, based on local knowledge only, is very widespread. This practice is appropriate for hazard identification and for raising the awareness of local communities in relation to the importance of risk reduction, but it is still imprecise in determining risk level, ranking and treatment in a context of climate change, activities in which technical knowledge is unavoidable. Integration of local and technical-scientific knowledge within the framework of an encoded risk assessment method (ISO 31010), could favour more effective decision making with regard to risk reduction. The aim of this chapter is to verify the applicability of a multi-risk local assessment-MLA which combines local knowledge (participatory workshop, transect walk, hazard and resource mapping, disaster historical profile) and scientific knowledge (climate downscaling modelling, hazard probability and scenarios, potential damages, residual risk). The test is carried out in two villages of the Western Niger, particularly exposed to flooding and agricultural drought. The risk (hazard probability * potential damages) is identified, analysed (level of risk) and evaluated (residual risk, adaptation measures compared with potential damage costs). The MLA is feasible. The two villages, while bordering on one another, have a different risk ranking. Depending on the village, the risk treatment could reduce the risk level to 17 and to 41% of the current risk, with costs equating to 34 and 28% of the respective potential damages.

Keywords Climate change · Local and scientific knowledge · Multirisk · Risk analysis · Risk evaluation · Flood · Drought · Gothèye · Niger

---

Maurizio Tiepolo is author of all the sections with the only exception of 11.3.3, written by Sarah Braccio.

M. Tiepolo (✉) · S. Braccio
DIST, Politecnico and University of Turin, Viale Mattioli 39, 10125 Turin, Italy
e-mail: maurizio.tiepolo@polito.it

S. Braccio
e-mail: sarah.braccio@polito.it

M. Tiepolo et al. (eds.), *Renewing Local Planning to Face Climate Change in the Tropics*, Green Energy and Technology,
DOI 10.1007/978-3-319-59096-7_11

11.1 Introduction

The risk assessment is the overall process of risk identification, analysis and evaluation. Identification consists in finding, recognising and describing risks. The analysis allows us to comprehend the nature and determine the level of risk, expressed in terms of a combination of the consequences of an event and the associated likelihood of occurrence. Evaluation determines whether the risk is acceptable or tolerable (ISO 2009).

In the Tropics, local risk assessment took hold after the World conference on natural disaster reduction (Yokohama 1994), which indicates it as a required step in disaster reduction (UNDHA 1994). NGOs and multi-bilateral donors conformed the risk assessment to the participatory rural appraisal-PRA of Chambers (1992), which used different tools, including community maps, gradually evolved into participated GIS-pGIS and 3D participated modelling (Rambaldi 2010).

The second Conference on disaster risk reduction (Hyogo 2004) recommended the integration of disaster risk reduction into plans. The first risk assessment experiences were consolidated in the Community Risk Assessment-CRA, introduced by the Provention consortium, a global coalition of governments, international organizations, academic institutions, private sector and civil societies organizations (Abarquez and Murshed 2004), and in the Vulnerability and Capacity Assessment-VCA of the International Federation of Red Cross and Red Crescent Societies (IFRCRC 2006; Van Aalst et al. 2008), recently adapted to consider climate change (IFRCRC 2014). Today, in the Tropics, a growing number of countries integrates risk analysis into local development plans: Madagascar and South Africa (since 2002), Philippines (since 2008), Mexico and Nepal (since 2011), Colombia and Honduras (since 2013), Nicaragua (Faling et al. 2012; Florano 2015; GoN 2011; Guardiola et al. 2013; Ruiz Rivera et al. 2015; Vermaak et al. 2004).

The recommendations of the Fourth African Regional Platform for Disaster Risk Reduction to quantify risk, standardize risk assessments and create and strengthen risk databases (UNISDR 2013: 55) are incorporated into certain pilot examples of risk analysis at meso-scale (Pezzoli and Ponte 2016; Tiepolo and Braccio 2016b) and at micro-scale (Ponte 2014, Tiepolo and Braccio 2016a).

The 3rd World Conference on DRR (Sendai 2015) eliminated every reserve regarding the integration of scientific and local knowledge (UN 2015: 11) on which different experiences had already been accumulated, especially in meteo/climate and early warning (Victoria 2008; Ziervogel and Opere 2010; Fabiyi et al. 2012; Masinde et al. 2012). To follow-up on this recommendation, it is necessary to recognize the fields that are best investigated using participatory methods (mobilization of communities in hazard identification and in the organization of bottom-up responses) and those better suited to scientific-technical investigation (risk level, hazard probability, scenarios, potential damages, measures identification residual risk) to allow not only greater efficiency but also the up-scaling of local information for the purposes of policy monitoring (e.g.: sustainable development goals 11 and 13) and design.

The aim of this chapter is to check whether or not it is possible to improve the community risk assessment according to an encoded definition of risk assessment

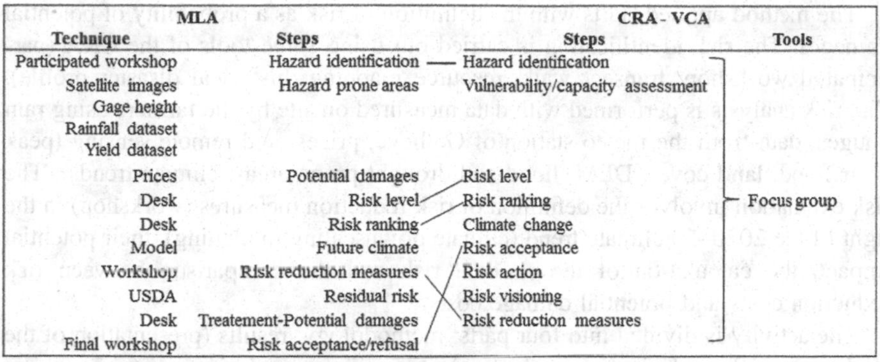
MLA
CRA-VCA
Technique
Steps
Steps
Tools
Participated workshop
Satellite images
Gage height
Rainfall dataset
Yield dataset
Prices
Desk
Desk
Model
Workshop
USDA
Desk
Final workshop
Hazard identification
Hazard prone areas
Potential damages
Risk level
Risk ranking
Future climate
Risk reduction measures
Residual risk
Treatement-Potential damages
Risk acceptance/refusal
Hazard identification
Vulnerability/capacity assessment
Risk level
Risk ranking
Climate change
Risk acceptance
Risk action
Risk visioning
Risk reduction measures
Focus group

Fig. 11.1 Steps and tools of the MLA and the CRA compared

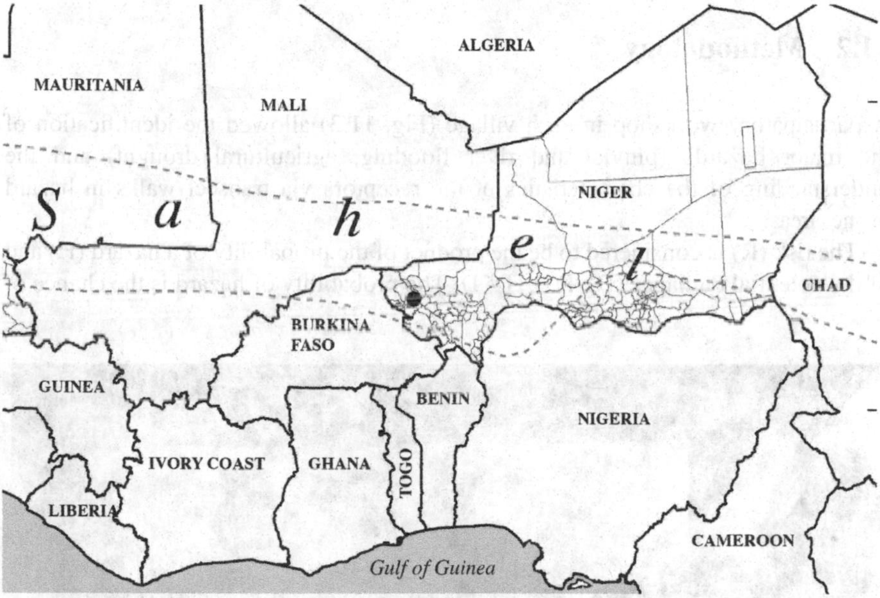

Fig. 11.2 The study area (*dot*) in Wester Africa context

(ISO 31010), using an integrated information system (local knowledge, remote sensing, information collected on the ground) (Fig. 11.1).

The multi-risk local assessment is carried out in the villages of Garbey Kourou (population 4800, 27 km^2) and Tallé (population 2700 and 15 km^2) in the Gothèye municipality, in Niger (Fig. 11.2). The two villages are among those worst hit by river and pluvial flooding in the municipality and are also exposed to agricultural drought. They are two neighbouring administrative villages, situated on the left bank of the Sirba river, just before the confluence with the Niger river, 12 km from the municipal capital and 60 km from Niamey, near the Niamey-Burkina Faso national road.

The method applied starts with the definition of risk as a probability of potential damages. The risk identification is carried out using some tools of the CRA (participated workshop, transect walk, resource mapping, historical disaster profile). The risk analysis is performed with data measured on site by the farmers using rain gauges, data from the meteo station of Gothèye, prices, and remote sensing (peak river flood, land cover, DEM, flood and drought prone areas, climate trends). The risk evaluation involves the definition of risk reduction measures (workshop) in the light of the 2020–50 climate trend (climate downscaling modelling), their potential impact, the calculation of the residual risk and the comparison between risk reduction costs and potential damage costs.

The activity is divided into four parts: methodology, results (presentation of the study area, risk analysis, risk evaluation), discussion, conclusions.

11.2 Methodology

A participatory workshop in each village (Fig. 11.3) allowed the identification of the major hazards (pluvial and river flooding, agricultural drought) and the understanding of the characteristics of the receptors via transect walks in hazard prone areas.

The risk (R) is considered to be the product of the probability of a hazard (H) and of the potential damages (D): $R = H * D$. The probability of hazard is the chance of

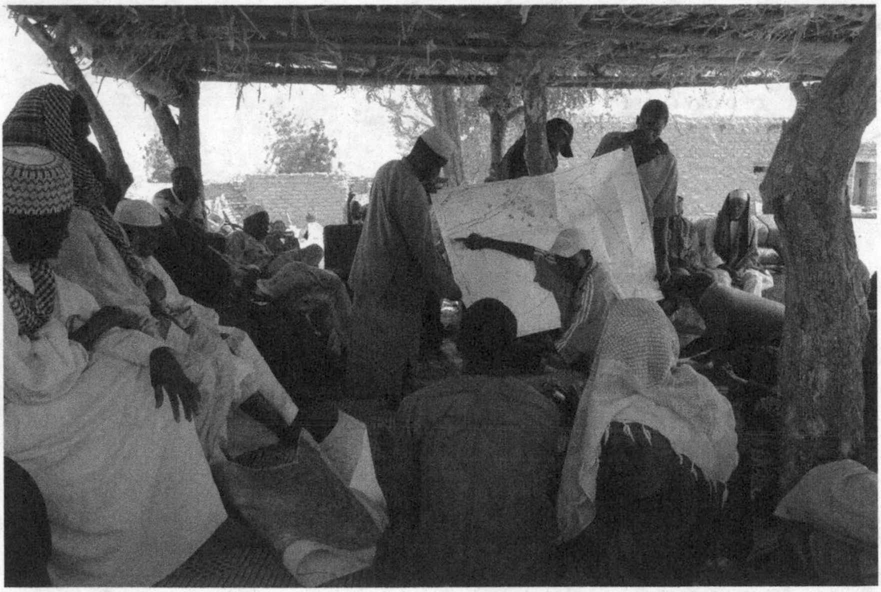

Fig. 11.3 Tallé, November 2014. Flood prone area during the participatory workshop

occurrence (inverse of the return period) expressed as a number between 0 (impossibility) and 1 (absolute certainty) that critical rain will cause pluvial flooding, that there will be a repeat of the record flood peak of the river Sirba (531 cm), that there will be a sequence of more than 10 dry days or a wet season of less than 65 days (fatal to millet). On this matter, we must remember that millet has two critical phases: germination and heading-flowering. The lack of water during these phases causes panicle abortion. For cultivars with a cycle that is shorter than 90 days, the critical phase lasts from the 40th to the 10th day before harvesting. For cultivars with a cycle that is longer than 90 days, the phase lasts from the 50th to the 10th day (Edlin 1994: 19). In the daily rainfall dataset of Gothèye, we identified daily total precipitation of up to 2.5 mm, below the daily minimum required by millet (Edlin 1994: 19) and considering the effect of the last rain before a sequence of dry days. The last millimetres of rainfall are divided by the mean value of the daily potential evapotranspiration in the wet season (6 mm/day), as calculated in Tillabéri, 54 km north of Garbey Kourou (Sivaakumar et al. 1993). For example, a sequence of 16 consecutive dry days is reduced to 13 due to the 16 mm that fell the day before the dry spell began. Obviously this is a simplification which does not consider the water that has actually seeped into the ground due to runoff. The receptors have been identified and measured in the area:

- flooded on 11 August 2014, after 40 mm of rain
- flooded by the river Sirba on 22 August 2012, when the water reached a height of 531 mm at Garbey Kourou (record flood peak of the last 69 years)
- farmed on 28 August 2014, at the end of one of the worst wet seasons in recent years

The receptors exposed to flooding were identified (Satellite image Stereo WorldView 2014) only within the sub-watersheds to which the two villages belong, because they contain the receptors with the highest value. Those exposed to drought were identified in the watersheds (interpretation of a satellite image at high geometrical resolution) and in the rest of the village lands (estimation).

The flooded zone was obtained by projecting the height corresponding to the 531 cm recorded at the gauge height in Garbey Kourou (Fig. 11.4) on the DEM obtained from the satellite image ortho-rectified with control points on the ground (satellite 2014).

The cropped area was determined considering that Garbey Kourou and Tallé can still satisfy their yearly cereal needs (YCN) with the village lands alone when the wet season is favourable, like in 2012, when they exceeded 9%. Consequently, the cropped area (CA) in the village lands becomes: $CA_{GK+T} = (P_{2015} * YCN)$/millet yield.

The potential damages in the event of flooding are the "maximum damages that could occur should a flood eventuate. In assessing potential damages, it is assumed that no actions are taken by the flood-affected population to reduce damage, such as lifting or shifting items to flood-free locations and shifting motor vehicles" (NSWG 2005: 5). The potential damages are calculated as the value of the receptors (houses, garden fences, millet, gombo, oseille, and sesame crops). Depth-damage and

Fig. 11.4 The Sirba hydro station at Garbey Kourou in June 2014

duration-damage curves were not used as all the buildings have walls made of crude earth which collapse as soon as the water crosses the threshold. The indirect effects of flooding are not considered. The value of the receptors is obtained from the cost of every construction element and labour recorded using the price schedule (Zaneidou 2013) and then checked locally.

Drought on the other hand does not cause the complete loss of the harvest. It has been assumed that the maximum cereal deficit of the Department of Gothèye in the last six years (–53% in 2011) can be used as the value of unproductive fields in the two villages, assuming that the farms only sow the area necessary to meet their cereal needs. The value of the crop potentially damaged is obtained considering the price of the crop on the market of Gothèye in September 2014 (RN, MCPSP 2014a, b) multiplied by the yield/ha (RN, MF, INS 2013a) and by the rate of unproductive fields.

The risk evaluation requires knowledge of the risk level after the risk treatment (residual risk) and comparison of the cost of the risk treatment with that of potential damages. The residual risk should allow acceptance or refusal of the risk. The risk treatment required the identification of the risk reduction measures in workshops, in the presence of the mayor, the municipal technical services and experts of the ministry of agriculture, CNEDD, DIST-Politecnico di Torino, Ibimet-CNR and the National Directorate for Meteorology. The choice of the measures prioritises the reduction of the receptors, prevention of the installation in hazard prone areas, integration of risk reduction measures with development measures (irrigated agriculture) and consistency with the climate (improved seeds, WSC). The National Meteorological Directorate of Niger has performed a specific analysis of the climate change expected to take place by 2025 and by 2050 in the region to which the villages belong (Bacci et al. 2016).

The sizing of the measures was carried out in consideration of the demographic dynamics of the villages revealed in the 3rd and 4th population census (RN, MF, INS 2006, 2014) and climate change.

The expected effect of the WSC measures is calculated using the USDA method (1986) adapted to the local context by Fiorillo and Tarchiani (Chap. 12), assuming that the reduction of the runoff generates an equivalent reduction of damages (e.g.: 30% of run-off reduction = 30% of damage reduction).

The sources of information used are a satellite image (Satellite image Stereo WorldView 2014) taken towards the end of the wet season (28 August 2014), the relative DEM with a contour interval of 1 m, the population census 2001 and 2012, the river daily discharge and gauge height between 1956 and 2014 (Ministère de l'hydraulique et de l'assainissement, hydrological forecasting system for the Niger river basin), the daily rainfall recorded in Gothèye between 1954 and 2014 and in the two villages in 2012 and in 2014 (Direction de la Météorologie Nationale), the prices of construction materials and of the works, as shown in the price schedules (RN, MDA unknown; Zaneidou 2013), those of crops (RN, MCPSP, SIMA 2014a, b) and relative yields (RN, MF, INS 2014).

11.3 Results

11.3.1 Study Area

Gothèye municipality administers an area of 3600 km^2 delimited by the rivers Niger, in the east, and Sirba to the south. Just 7500 people live in the municipal capital, 85,000 are distributed in 25 administrative villages and 121 hamlets.

Garbey Kourou (population 4800) and Tallé (population 2700) are among the biggest and oldest villages in the Municipality, having been founded in 1301 and 1845 respectively. The two villages lie in a complex of sub-watersheds on a slight slope (about 1%) towards the river Sirba, 4 km from the confluence with the river Niger (Figs. 11.5 and 11.6).

The administrative rank (it is a place where taxes are paid and justice is served), the presence of a health centre, rural radio, the black smith, a welder, a gas station and the weekly market make it the reference point for a vast area. The village lands also stretch along the right bank of the river, in the Namaro municipality, and include some hamlets, used during the wet season to cultivate the fields furthest away. In recent years, the two villages have reached a good level of infrastructure thanks to the construction of numerous boreholes and to the arrival of electricity (Commune de Gothèye 2009).

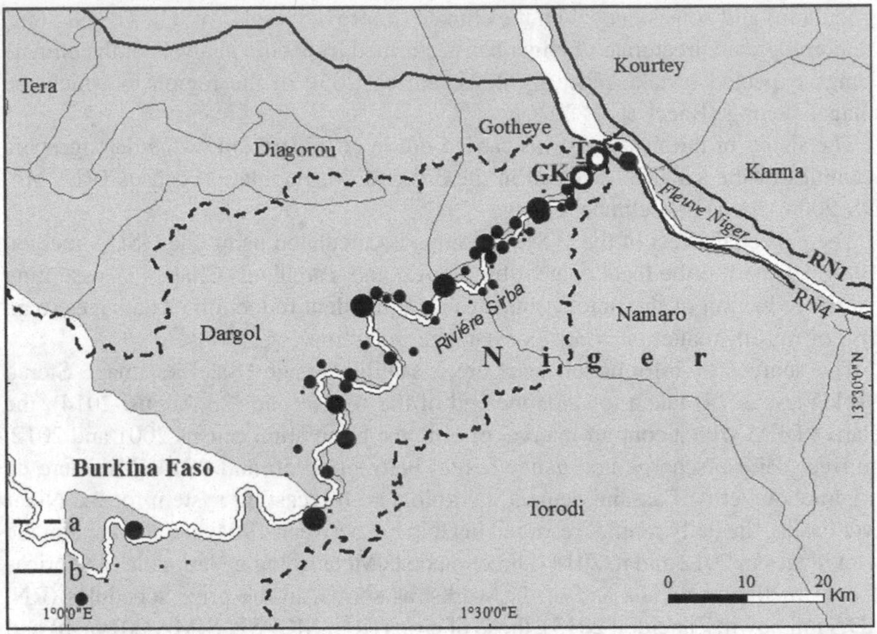

Fig. 11.5 Garbey Kourou (GK) and Tallé (T): Sirba watershed (a), National road n. 4 (b)

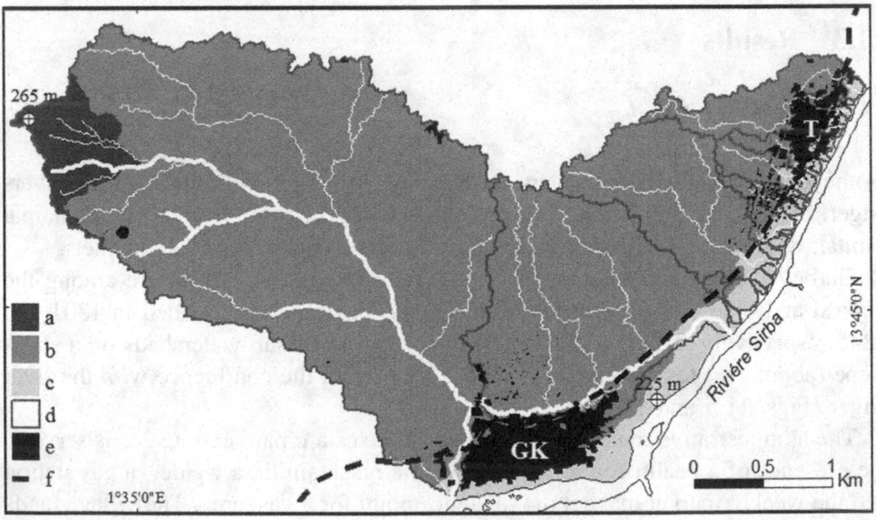

Fig. 11.6 The morphology of the Garbey Kourou (GK) and Tallé (T) sub-watersheds, 2015: glaze (a), lower glaze (b), valley bottom (c), sub watersheds boundaries (d), built-up (e), laterite track (f)

Table 11.1 Demographic trends for Garbey Kourou and Tallé, 2001–2022

Village	Gender	2001	$r_{2001-12}$	2012	$r_{2012-15}$	2015	$r_{2015-22}$	2022
Garbey Kourou	MF	3990	1.36	4634	1.36	4827	1.36	5382
	H	488		520		542		597
	MF/H	8.2		8.9		8.9		8.9
Tallé	MF	2235	1.39	2603	1.39	2713	1.39	2990
	H	273		345		360		396
	MF/H	8.2		7.5		7.5		7.5

F Females, *H* Households, *M* Males, *r* Annual variation rate
Sources RN, MF, ISN 2006, 2014

In Niger, a wet season of 90 days allows millet and sorghum rain-fed crops to grow. Sometimes the wet season is shorter and, more and more often, there are long gaps between rains, reducing the crop yield. To compensate for losses, the farmers practice irrigated crops during the dry season (November–May). Garbey Kourou is a pioneer centre of female associations for the irrigated production of gombo, oseille, paprika and sesame (Hadizatou 2012), which spread from this village to its neighbours about twenty years ago.

This is why the rural radio, which played an important role in spreading this practice, broadcasts from Garbey Kourou.

The demographic growth (1.37% annually between 2001 and 2012) and the slight reduction in the dimension of households (Table 11.1) have led to increased demand for land and for wood to burn, a reduction in tree coverage to less than 12 trees/ha (*Acacias, Balanites aegyptiaca* and *Ziziphus mauritiana*), the abandon of fallow lands with higher agricultural potential and the cultivation of lands used exclusively for grazing. Households keep goats and sheep in the village, sending cattle in transhumance during the wet season and bringing them back in after threshing to feed on crop residues.

11.3.2 Risk Identification

A participated workshop in each of the two villages has allowed the identification of three main hazards: pluvial flooding, river flooding, agricultural drought (Fig. 11.3). The causes of flooding are numerous: (i) changed rainfall pattern (climate change), (ii) opening of the dams in Burkina Faso during heavy rainfall, (iii) deterioration of the river banks due to the extraction of clay from the sides to produce bricks, (iv) sanding up of the riverbeds due to erosion triggered by deforestation. In the case of river flooding, the damages are due to the settlement in flood prone areas of households that cannot find safer sites at low cost. The cause of pluvial flooding is the deterioration of works for water and soil conservation-WSC upstream, which can no longer encourage the infiltration of rain water and reduce the runoff.

11.3.3 Risk Analysis

The drought prone areas are fields planted with crops at 28 August 2014: 353 ha (satellite image interpretation) in the sub-watersheds and 2500 ha (estimated) in the village lands of Garbey Kourou and Tallé. 40 mm of rain is 79% likely to fall in 2015. A river flood of 531 cm has a 1.6% likelihood of occurrence in 2015.

An agricultural drought is 13% likely to last eleven consecutive days (as in 2006) in 2015, and there is a 6% probability of having a wet season lasting less than 65 days and a 1.6% probability of having 16 consecutive dry days in 2015.

In the sub-watersheds considered, there are houses made fully from crude earth, roofed with corrugated metal sheets, with fenced gardens, millet, sorghum, gombo, oseille, and sesame fields. Despite being small and made of crude earth, the buildings make up the highest exposed value, considering the same surface area (Table 11.2; Fig. 11.7). In the event of drought, on the other hand, the crops have

Table 11.2 Potential damages at Garbey Kourou and at Tallé in 2015

Village hazard	Hazard prone quantity	Unit value	Total
Receptor	n.	FCFA	M FCFA
Garbey Kourou			
River flood			79.3
Crude earth house 33 m^2	79	953,375	75.3
Improved crude earth house 33 m^2	2	1,365,175	2.7
Garden fence	2	640,000	1.3
Pluvial flood			29.8
Crude earth house 41 m^2	22	1,201,890	26.4
Improved crude earth house 41 m^2	2	1,696,126	3.4
Drought, watershed			26.2
Millet, ha	(347 * 547) * 53/100	250,000	25.1
Gombo, oseille, sesame, ha	(6*)53/100	349,000	1.1
Drought, all village lands			147.7
Millet, ha	(4827 * 231) * 53/100	250,000	147.7
Gombo, oseille, sésame, ha	(6) * 53/100	349,000	1.1
Total, sub-watershed			135.4
Total, all village lands			257.9
Tallé			
River flood			59.7
Crude earth house 27 m^2	38 * ...		56.4
Improved crude earth house 27 m^2	2 * ...		3.5
Garden fence	164 * ...		0.3
Drought, sub-watershed			7.8

(continued)

Table 11.2 (continued)

Village hazard	Hazard prone quantity	Unit value	Total
Millet, ha	(75 * 547) * 53/100	250,000	5.4
Gombo, oseille, sesame, ha	(13) * 53/100	349,000	2.4
Drought, all village lands			83.0
Millet, ha	(2713 * 231 * 250) * 53/100		83.0
Gombo, oseille, sesame, ha	(13 * 349,000) * 53/100		2.4
Total sub watersheds			67.5
Total all village lands			145.1

1 FCFA = 0.565 €

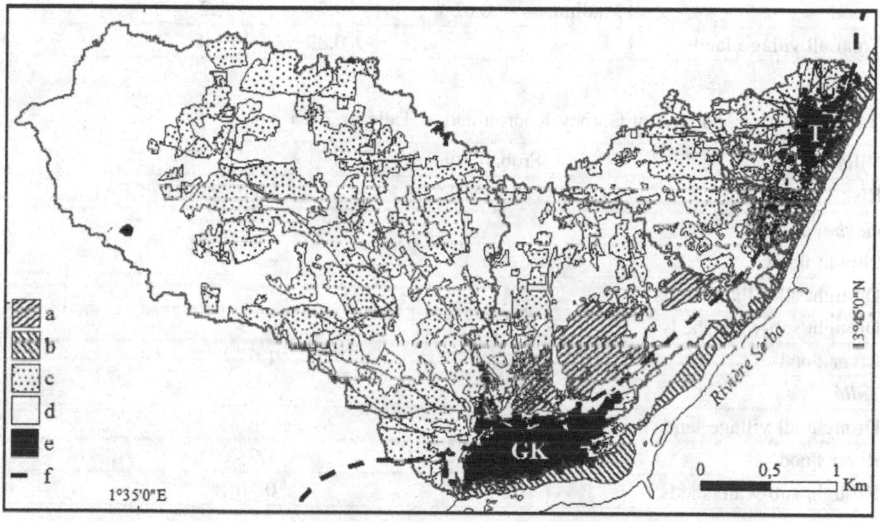

Fig. 11.7 Garbey Kourou and Tallé sub watersheds, 28 August 2014. Pluvial flood prone area (**a**), river flood prone area (**b**), agricultural drought prone area (**c**), sub-watershed limits (**d**), built up area (**e**), laterite track (**f**)

the highest exposed value. The 4827 residents of Garbey Kourou annually consume 231 kg of cereals per head (millet, sorghum, maize, fonio and rice) and, consequently, require 1000 tons (RN, MA, DS 2013). The worst drought (2011) in the last five years reduced production in the Department to 53%. If the same level of drought was to occur in Garbey Kourou too, the damages would amount to 500 tons which, at a price of 250,000 FCFA/ton, would have a value of 148 million FCFA, while in the sub-watersheds alone it would amount to 25 million FCFA. Similarly, the 2713 residents of Tallé require 627 tons, and drought could reduce the yield to 332 tons with damages amounting to 83 million FCFA (Table 11.2).

In Garbey Kourou (all village lands), pluvial flood risk is easily in first place, followed by drought risk and river flood risk. This result is obtained from a much

Table 11.3 Risk level at Garbey Kourou and at Tallé in 2015

Village hazard	Probability	H (%)	D (FCFA)	R = H * D (M FCFA)
Garbey Kourou				
River flood	Medium	0.02	79.3	1.6
Pluvial flood	High	0.79	29.8	23.5
Drought sub-watershed	High	0.13–0.06	26.2	3.4
Drought all village lands	High	0.13–0.06	147.7	19.2
Total all village lands				44.3
Tallé				
Drought sub-watershed	High	0.13	7.8	1.0
Drought all village lands	High	0.13	83.0	10.8
Flood	Medium	0.02	59.7	1.2
Total all village lands			0.89	12.0

Table 11.4 Risk ranking at Garbey Kourou and at Tallé in 2015

Village	Probability	
Risk	High (M FCFA)	Medium (M FCFA)
Garbey Kourou		
Pluvial flood	33.5	–
Drought all village lands	3.4	
Drought sub-watersheds	19.2	0.79
River flood	–	1.6
Tallé		
Drought all village lands	10.8	
River flood	–	1.2
Drought sub-watersheds	1.0	0.10

higher probability of occurrence of 40 mm rain than 11 consecutive days or compared to a flood of 531 cm (Table 11.3). In Tallé, on the other hand, drought risk is in first place, followed by river flood, if all the village lands are considered (Table 11.4).

11.3.4 Risk Evaluation

The evaluation requires knowledge of the cost and expected impact of the risk treatment on the current risk level, and comparison of the cost of potential damages before risk treatment with those of the treatment itself.

Twenty measures have been identified for Garbey Kourou and 14 for Tallé. The measures for pluvial flooding are the management of gullies, new WSC, existing WSC maintenance, installation of a gauge in the pond, pond drainage, a ridge

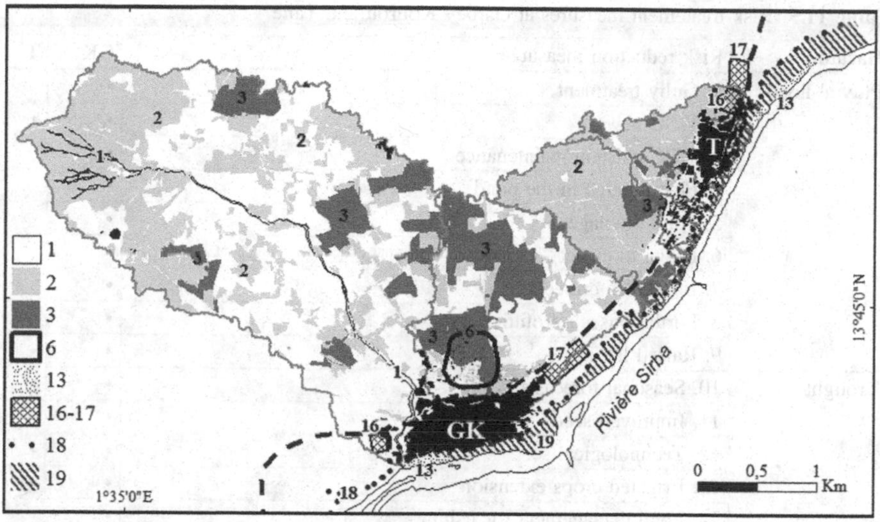

Fig. 11.8 Localisation of the measures of risk treatement at Garbey Kourou and at Tallé (numbers according Table 11.5)

(levee) around scattered houses in the pond, protected wells, retrofitting of crude earth construction, raised latrines (EAA 2012). Drought risk reduction measures are: seasonal forecasts, short cycle millet cultivar (Miko 2013), further development of irrigated crops (RN, CRAT 2012), pond management for fish breeding (Fermont 2013), livestock food banks. The river flood risk reduction measures consist in resettlement of the households currently settled in flood prone areas, prevention of further installations in the flood prone area by making building plots available in relation to the demographic dynamics of the two villages, flood prone area delimitation, land use regulation for the flood prone area, early flood warning system (Fig. 11.8; Table 11.5). The measures of WSC (trapezoidal bunds, zaï, contour stone bunding, semi-circular hoops) (BMZ 2012) are defined in relation to the type of soil (rocky, laterite, sandy, etc.), and irrigated gardens are sized to satisfy 10% of households. The unitary cost of these measures is determined on the basis of that sustained by the SEMAFO Foundation (2014) to create similar measures in the two villages, or consulting the price schedules available (RN, MDA unknown; Zaneidou 2013). This has made it possible to estimate a risk treatment cost of 101 million FCFA for Garbey Kourou (all village lands) and of 40 million FCFA for Tallé (sub-watershed only). These measures are consistent with the results of the climate change analysis in the Tillabéri region, which envisages a considerable inter-annual variability in total rainfall by 2025, an increase in minimum and maximum temperatures (+0.8 and +1 °C), a tendency to extend the farming season until the end of October-beginning of November, and drier conditions, with poor distribution of rainfall in phases when crops are vulnerable, suggesting the use of more drought-resistant cultivars, despite the fact that these are less productive (Bacci and Gaptia Lawan 2016: 115, Bacci et al. 2017).

Table 11.5 Risk treatement measures at Garbey Kourou and Tallé

Hazard	Risk reduction measure	GK	T
Pluvial flood	1. Gully treatment	•	
	2. WSC new	•	•
	3. WSC existent/maintenance	•	
	4. Flood gauge in the pond	•	
	5. Pond drainage	•	
	6. Ridge to protect settlement	•	
	7. Boreholes protection	•	
	8. Earth houses retrofitting	•	
	9. Raised latrines	•	•
Drought	10. Seasonal forecasts	•	•
	11. Improved seeds	•	•
	12. Technological set	•	•
	13. Irrigated crops extension	•	•
	14. Pond management for fishing	•	
	15. Livestock food banks	•	•
River flood	16. Resettlement of the population living in flood prone areas	•	•
	17. Avoiding reinstallation in flood prone areas	•	•
	18. Trees planting along the river flood prone area	•	•
	19. Regulation for the flood prone area	•	•
	20. Early warning river flood	•	•

To estimate the residual risk, we assumed that (i) the adoption of short cycle cultivars would reduce losses in the event of drought by 50%, also considering the lower yield compared to long cycle cultivars; (ii) the households that practice irrigated agriculture (10%) fully compensate for any losses of rain-fed crops with sales of the products of irrigated agriculture.

This said, in Garbey Kourou, the residual risk would fall to 6% (sub-watersheds) and to 17% (all village lands) of the 2015 risk level. In Tallé, on the other hand, it would fall to 23% (sub-watersheds) and to 41% (all village lands) (Table 11.6).

The cost of the treatment equates to 34% of the potential damages in Garbey Kourou and 28% in Tallé (Table 11.7).

11.4 Discussion

The aim of this chapter was to check the feasibility of the Multirisk Local Assessment-MLA carried out with the integration of local and scientific-technical knowledge in a context of climate change to improve current Community Risk

Table 11.6 Residual risk at Garbey Kourou and at Tallé, 2018 (* improved seeds (50%) + irrigated crops (10%)

Village	Risk	Damages reduction		Residual risk	
Hazard	MFCFA	%	MFCFA	MFCFA	%
Garbey Kourou					
Pluvial flood	23.5	100	23.5	0	
Drought sub-watershed	3.4	50	1.7	1.7	
Drought all village lands	19.2	60	11.5	7.7	
River flood	1.6	98	1.6	0.0	
Total sub-watershed	28.5	–	26.8	1.7	6
Total all village lands	44.3	–	36.6	7.7	17
Tallé					
Drought sub-watershed	1.0	50	0.5	0,5	
Drought all village lands	10.8	60	6.5	4.2	
River flood	1.2	98	1.2	0.0	
Total sub-watersheds	2.2		1.7	0.5	23
Total all village lands	13.0		7.7	5.3	41

Table 11.7 Garbey Kourou and Tallé. Risk reduction costs and potential damages

	a	b	a – b
Village Hazard	Treatement M FCFA	Potential damages M FCFA	Δ M FCFA
Garbey Kourou			
Pluvial flood	67	30	−37
Drought sub-watershed	16	26*	10
Drought all village lands	–	148	
River flood	18	79	61
Total sub-watersheds	101	135	34
Total sub-watershed/inhabitant	0.021934	–	–
Tallé			
Drought sub-watershed	15	8	8
Drought all village lands	–	83	–
River flood	25	60	35
Total sub-watershed	40	68	28
Total all village land	–	143	–
Total sub-watershed/inhabitant	0.015275	–	–

Assessment practices. The aim was triggered by the weaknesses found in the CRA in large communities, as Nigerien villages often are, and by the recommendations that emerged from the African platform for DRR (2013), WMO (2014) and Sendai conference on DRR (2015).

We flanked some tools of the CRA based on local knowledge (participated workshop, hazard and resource mapping, historical disaster profile, transect walk) with modern techniques (climate statistical downscaling modelling, hazard probability, potential damage, residual risk). This approach enabled:

- risk ranking based on hazard probability (pluvial and river flood, agricultural drought) to generate potential damages
- comparison of the risk level and ranking between different villages
- identification of risk reduction measures consistent with the local climate change trends at 2025 and 2050
- the decision to accept or refuse the risk, based on the results of residual risk and the comparison between risk reduction costs and potential damage costs.

The analysis produced unexpected results in the risk ranking and the evaluation reached unexpected results with regard to the convenience of risk reduction.

As far as suggestions for future research are concerned, the feasibility check on the MLA proposed in this chapter needs to be extended to a whole municipality (usually made up of dozens and dozens of villages).

11.5 Conclusions

This chapter has proven the feasibility and advantages of the MLA. We have verified the method in two adjacent villages, which are among those with the highest populations (7500 residents) along the river Sirba. Despite the two villages being next to one another, the risk ranking is very different: with pluvial flooding in first place in one and drought in the other.

The creation of known measures in the region would allow the reduction of the residual risk to 17% in Garbey Kourou and to 41% of the current risk in Tallé. Such a high reduction of the risk would require expenditure equating to just 34% of the potential damages in Garbey Kourou and 28% in Tallé: useful information which the VCA would not have supplied. This information also enables an understanding that the cost cannot be sustained with local resources only. It amounts to 22,000 FCFA per head in Garbey Kourou and 15,000 FCFA per head in Tallé. Every household would need to contribute 196,000 FCFA (65,000 FCFA/year) in Garbey Kourou, and 112,000 (37,000 FCFA/year) in Tallé, over a period of three years (the duration of the treatment). While this accounts for just 7 and 13% of the annual household income (504,000 FCFA), this would be enough to take the households too close to the 384,000 FCFA considered to be the food insecurity threshold in Niger (RN, INS, SAP 2012: 24).

In the absence of measures to introduce micro finance, the support of the government and so-called development partners is required.

At this point, there are four issues to look at.

- Perennial implementation of the risk treatment. The risk treatment is not permanent. The works require annual maintenance and renovation every 10 or 20 years. In the long term, the villages could do without outside help if they were to continue investing in commercial crops.
- Reproducibility of the MLA. The methodology can be re-proposed in municipalities at greatest risk (Tiepolo and Braccio 2016b) and, within these, in the bigger villages, as it is there that we find the receptors of higher value, which justify the costs of risk treatment. In the Tillabéri region, the second most exposed in Niger, there are 24 such villages, and about 140 in the whole country.
- Convenience of the LMA. The assessment allows the formulation of detailed risk treatment plans which would increase the impact of the measures.
- Who should develop the MLA. The NGOs which drew up the first generation local development plans ten years ago and are now busy drawing up the second generation could benefit from specific training and be supported by regional services (determination of the village lands boundaries, hazard prone areas, calculation of the hazard probability).

References

Abarquez, I., and Z. Murshed. 2004. *Community-based disaster risk management. Field practitioners' handbook*. Bangkok: Asia disaster preparedness center.

Bacci, M., and M. Mouhaimouni. 2017. Hazard events characterization in Tillaberi region: present and future projection. In *Renewing climate planning to face climate changing in the Tropics*, ed. M. Tiepolo, E. Pezzoli, and V. Tarchiani. Springer.

Bacci, M., K. Gaptia Lawan, and M. Moussa. 2016. Le climat de la région Tillabéri. In *Risque et adaptation climatique dans la région Tillabéri, Niger*, ed. V. Tarchiani, and M. Tiepolo, 79–98. Paris: L'Harmattan.

Bacci, M., and K. Gaptia Lawan. 2016. Variabilités et changements climatiques et leurs impacts sur les cultures pluviales dans la région Tillabéri, Niger. In *Risque et adaptation climatique dans la région Tillabéri, Niger*, ed. V. Tarchiani, and M. Tiepolo, 99–116. Paris: L'Harmattan.

BMZ. 2012. *Bonnes pratiques de CES/DRS. Contribution à l'adaptation au changement climatique et à la résilience des producteurs. Les expériences de quelques projets au Sahel*. GIZ.

Chambers. R. 1992. Rural appraisal: rapid, relaxed and participatory. *IDS Discussion Paper* 311.

Commune de Gothèye. 2009. *Plan de développement communal 2009–2012*.

EAA. 2012. *Étude sur l'assainissement productif Ecosan au Niger. Bilan et perspectives. Rapport définitif*, avril.

Edlin, M. 1994. Analyse des risques de déficit hydrique au cours des différentes phases phénologiques du mil précoce au Niger. Conséquences agronomiques. In *Bilan hydrique agricole et sécheresse en Afrique Tropicale*, eds. F.-N. Reyniers, and L. Netoyo, 17–30. Paris: John Libbey Eurotexts.

Fabiyi, O.O. et al. 2012. *Integrative approach of indigenous knowledge and scientific methods for flood risk anayses, responses and adaptation in rural coastal communities in Nigeria*. Final report for 2011 start grants for global change research in Africa.

Faling, W., J.W.N. Tempelhoff, and D. van Niekerk. 2012. Retoric or action: are South African municipalities planning for climate change? *Development Southern Africa* 29 (2): 241–257.

Fermont, Y. 2013. *La pisciculture de subsistance en étangs en Afrique: Manuel technique.* ACF-International Network.

Florano, E.R. 2015. Mainstreaming integrated climate change adaptation and disaster risk reduction in local development plans in the Philippines. In *Handbook of climate change adaptation*, ed. W. Leal Filho, 433–456. Springer. doi:10.1007/978-3-642-38670-1_36.

Fondation SEMAFO. 2014. *Rapport annuel 2013–2014.*

GoN-Government of Nepal. 2011. *National framework on local adaptation plans for action.* Ministry of science technology and environment.

Guardiola, L., J. Quiñónez, M. Domínguez, N. Jover, and R. Bernol. 2013. *Guía metodológica para incorporar la adaptacion al cambio climático en la planificación del dresarollo.* Tegucigalpa.

Hadizatou Alhassoumi. 2012. Dynamiques associatives et processus d'affirmation des femmes de Garbey Kourou (Ouest du Niger) atour des activités agricoles. In *Genre & agriculture familiale & paysanne. Regards Nord-Sud*, colloque international Toulouse-Le Mirail, 22–24 Mai 2012.

IFRCRC-International Federation of Red Cross and Red Crescent Societies. 2014. *Integrating climate change and urban risks into the VCA.* Geneva, IFRCRC: Ensure effective participatory analysis and enhanced community action.

IFRCRC-International Federation of Red Cross and Red Crescent Societies. 2006. *What is VCA? An introduction to vulnerability and capacity assessment.* Geneva: IFRCRC.

ISO-International Organization for Standardization. 2009. *Guide 73:2009. Risk Management – Vocabulary.*

Masinde, M., and A. Bagula. 2012. ITIKI: bridge between African indigenous knowledge and modern science of drought prediction. *Knowledge Management for Development Journal* 7 (3): 274–290. doi:10.1080/19474199.2012.683444.

Miko, I. 2013. *Multiplication et diffusion de semences de qualité des variétés améliorées et adaptées au changement climatique. Fiche de bonne pratique.* Niamey: FAO.

NSWG—New South Wales Government. 2005. *Floodplain development manual. Appendix M – Flood damage.*

Pezzoli, A., and E. Ponte. 2016. Vulnerability and resilience to drought in the Chaco, Paraguay. In *Planning to cope with tropical and subtropical climate change*, eds. M. Tiepolo, E. Ponte, and E. Cristofori, 63–88. Berlin: De Gruyter Open. doi:10.1515/9783110480795-005.

Ponte, E. 2014. Flood risk due to heavy rains and rising sea levels in the municipality of Maputo. In *Climate change vulnerability in Southern African Cities*, eds. S. Macchi, and M. Tiepolo, 187–204. Springer.

Rambaldi, G. 2010. *Modélisation participative en 3D-Principes directrices et applications.* Wageningen: Centre technique de coopération agricole et rurale ACP-UE.

RN, CRAT. 2012. *La production du sésame dans la zone de Gothèye: de la culture de consommation à celle commerciale.*

RN, MA, DS. 2013–2015. *Évaluation des récoltes de la campagne agricole d'hivernage 2012–2014 et résultats définitifs 2012–2014. Rapport national de synthèse.*

RN, MCPSP-SIMA. 2014a. *Bulletin Mensuel Volet Céréales*, 9/14, septembre.

RN, MCPSP-SIMA. 2014b. *Bulletin Mensuel Volet Produits de Rente*, 60, octobre.

RN, MDA. unknown. *Recueil des fiches techniques en gestion des ressources naturelles et de production agro-sylvo-pastorales.* PAC-Banque Mondiale.

RN, MF, INS. 2014. *Niger. Répertoire national des localités (ReNaLoc)*, juillet.

RN, MF, INS. 2013. *Annuaire statistique 2008–2012. Édition 2013.*

RN, MF, INS. 2006. *ReNaCom- Répertoire national des communes.*

Ruiz Rivera, N., J.M. Casado Izquierdo, and M.T. Sánchez Salazar. 2015. Los atlas de riesgo municipales en México como instrumento de ordenamiento territorial. *Investigaciones Geográficas. Boletín del Instituto de Geografía, UNAM* 88: 146–162.

Sivaakumar, M.V.K., A. Maidoukia, and R.D. Stern. 1993. *Agroclimatologie de l'Afrique de l'Ouest.* ICRISAT.

Tiepolo, M., and S. Braccio. 2016a. Flood risk preliminary mapping in Niamey, Niger. In *Planning to cope with tropical and subtropical climate change*, eds. M. Tiepolo, E. Ponte, and E. Cristofori, 201–220. Berlin: De Gruyter Open. doi:10.1515/9783110480795-013.

Tiepolo, M., and S. Braccio. 2016b. Flood risk assessment at municipal level in Tillabéri region, Niger. In *Planning to cope with tropical and subtropical climate change*, eds. M. Tiepolo, E. Ponte, E. Cristofori, 221–242. Warsaw/Berlin: De Gruyter Open. doi:10.1515/9783110480795-014.

UN. 2015. *Third World Conference on DRR. Sendai framework for disaster risk reduction 2015–2030.*

UNDHA-United Nations Department of Humanitarian Affairs. 1994. Yokohama strategy and plan of action for a safer world. Guidelines for natural disaster prevention, preparedness and mitigation. In *World conference on natural disaster reduction*. Yokohama, Japan 23–27 Mai 1994.

UNISDR-The United Nations Office for Disaster Risk Reduction. 2013. *Report of the 4th Africa regional platform on disaster risk reduction 13-15 February 2013*. Tanzania: Arusha.

Van Aalst, M.K., T. Cannon, and J. Burton. 2008. Community level adaptation to climate change: The potential role of participatory community risk assessment. *Global Environmental Change* 18: 165–179. doi:10.1016/j.gloenvcha.2007.06.002 .

Vermaak, J., and D. van Niekerk. 2004. Disaster risk reduction initiatives in South Africa. *Development Southern Africa* 21 (3): 555–574. doi:10.1080/0376835042000265487.

Victoria, L.P. 2008. Combining indigenous and scientific knowledge in the Dagupan city flood warning system. In *Indigenous knowledge for disaster risk reduction. Good practices and lessons learned from experiences in the Asia-Pacific Region*, 52–54. UNISDR.

WMO-World Meteorological Organization, GFSC. 2014. *Plan de mise en œuvre du cadre mondial pour les services climatiques*. Geneve.

Zaneidou, D. 2013. *Bordereaux des prix unitaires des principaux investissements réalisés par les projets FIDA au Niger – Capitalisation. Volume 1*, FIDA.

Ziervogel, G., and A. Opere (eds.). 2010. *Integrating meteorological and indigenous knowledge-based seasonal climate forecasts in the agricultural sector*. Ottawa: IDRC.

Satellite image Stereo WorldView-2ID:14AUG28105608-M2AS-053799140010 28 August 2014.

Chapter 12
A Simplified Hydrological Method for Flood Risk Assessment at Sub-basin Level in Niger

Edoardo Fiorillo and Vieri Tarchiani

Abstract Flood events are increasing year by year in the Sahel, mainly caused by climate and land use changes. New strategies and tools are necessary to optimize flooding risk reduction plans. This paper presents a new hydrological method (FREM, Flooding Risk Evaluation Method), based on the curve number runoff estimation. The method can be adopted for small-medium basins and is based on the integration of remote sensing techniques with field surveys and participatory mapping. It consists of preliminary identification of the areas and sub-basins that most contribute to the flood risk; scenarios can then be developed in order to: (i) optimize the placement of traditional water retention structures in the elementary sub-basins that contribute most to the overall risk, (ii) assess the contribution of each hydraulic structure to reduce the total risk, (iii) give a priority ranking to these structures identifying those most urgent. The main advantages of this method are that it is easy to use and can be implemented using free available land cover, soil and morphology data and open-source GIS (Geographic Information System) software. A case study for the Ouro Gueladjo basin (Tillabery Region, Niger) is presented.

Keywords Sahel · Remote sensing · Flood risk assessment · Water retention measures · Adaptation

---

E. Fiorillo (✉) · V. Tarchiani
National Research Council, Institute of Biometeorology (IBIMET), Via Caproni 8, 50145 Florence, Italy
e-mail: e.fiorillo@ibimet.cnr.it

V. Tarchiani
e-mail: v.tarchiani@ibimet.cnr.it

M. Tiepolo et al. (eds.), *Renewing Local Planning to Face Climate Change in the Tropics*, Green Energy and Technology,
DOI 10.1007/978-3-319-59096-7_12

12.1 Introduction

Western Africa Sahel is a region traditionally known for its droughts rather than its floods. After the droughts of the 1970s and 1980s, Sahelian countries are today also suffering from the effects of heavy rains and devastating floods. The authors investigated the Niger official database of flood damage (ANADIA NIGER DB) that has been compiled yearly since 1998 by the Niger government coordination unit "Système d'Alerte Précoce et Prévention des Catastrophes" (SAP). During the 1998–2014 period, more than 3600 localities and about 1.6 million people were affected by floods. The consequences of flooding don't concern only damage to goods, agricultural systems and economic infrastructure (roads, bridges, dams destroyed), but also affect population resilience against food insecurity, particularly in the most vulnerable groups of people. In addition, recent years have seen an alarming increase in flooding, particularly in 2012. The problem is concentrated in the southwest of the country, but affects almost the entire country with 216 municipalities involved out of a national total of 263, among them 81 with more than 5000 people affected over the years.

The increased number of floods per year is mainly due to two factors: a higher frequency of extreme rainfall events and human induced land cover changes. Regarding the first aspect, Mouhamed et al. (2013) stated that during the 1960–2000 period, the cumulated rainfall of extremely wet days showed a positive trend in West African Sahel and that extreme rainfalls had become more frequent during the last decade of the 20th century, compared to the 1961–2000 period. Regarding the second factor, human pressure on the environment (significant increase in cropland and disappearance of natural bushes and landscapes, for example) has led to severe soil crusting and desertification. Higher runoff as a long-term consequence of land clearing and soil crusting was described at various scales in the Sahel, from small watersheds of a few tens of square kilometers (Albergel 1987) to larger catchments of up to 21,000 km^2 (Mahé et al. 2005). Leblanc et al. (2008) estimated that in a 50 km^2 study area, at a mean distance of 60 km east of Niamey, $\sim 80\%$ of the natural vegetation had been cleared between 1950 and 1992, firstly to open new areas for agriculture and secondly for firewood supply. As demonstrated by Albergel (1987), the increase of highly eroded areas with high runoff coefficients causes a reduction in the critical amount of rainfall that causes flooding. Descroix et al. (2012) pointed out that the exceptional flood that happened near the city of Niamey in 2010 was provoked by a rainfall that was only average with respect to the long-term record. These modifications, reducing the water amount needed for a critical rainfall, cause an increase in flood hazard frequency.

Moreover the Sahel area is characterized by one of the highest annual population growth rates and new settlements or urban expansion take place in flood prone zones where land is cheap or free. In several cases, houses destroyed by floods were built on valley bottoms or in the pathway of drainage channels exacerbating the damaging effects of rainfall and flooding. The causes of flood damage also include the human effects of inadequate drainage facilities and poor construction materials.

During the 1970s and 1980s many measures (including: grass strips, bench terraces, stone lines, half-moon micro catchments, zai-holes) were applied in order to increase water retention. Some of them, like zai-holes, are traditional water retention measures, while others were specifically developed considering the Sahel climatic and environmental characteristics. These techniques are helpful to rehabilitate degraded lands because on the one hand they slow down or stop surface runoff and consequently reduce soil erosion. On the other, they increase water infiltration that can be exploited for agricultural purposes or to induce natural revegetation. These water retention measures are also effective in reducing flood risk because, by reducing runoff, they increase the critical amount of rain that causes flooding. In the Niger Sahel environment, the appropriate positioning of these measures can be difficult due to the smooth morphology and scarce slopes that characterize the area. Considering the limited funds that are generally available for their realization, the placement optimization of these measures is a key factor. In order to do this, an important resource is provided by Geographic Information Systems (GIS) (Goodchild 2009). These systems, which can manage and analyze different types of spatial or geographical data, are very effective for environmental analysis. Until recently, their use for flood risk reduction plans in developing countries was limited by two main factors: the cost of software and necessary geographic layers (primarily the costs related to satellite imagery), and the technical and scientific skills required to run hydrological models that perform well analyzing the water cycle driver, but very complex to use. This article presents a new method (FREM-Flooding Risk Evaluation Method), based on the Curve Number (CN) runoff estimation method, which uses only free and open-source software and requires only basic-medium GIS skills. The FREM method has been developed in order to provide a simple tool to analyze the flood risk at basin and sub-basin level and to optimize strategies for risk reduction based on water retention measures. A case study is presented for the Ouro Gueladjo basin, located in the Tillabery Region of Niger.

12.2 Materials

12.2.1 Study Area

The study area (Fig. 12.1) is a 141 km^2 catchment located in southwest Niger in the Ouro Gueladjo Rural Community, on the right bank of the Niger River. Its elevation ranges from 185 to 270 m a.s.l. (ASTER DEM) and it is characterized by the following morphological units (Fig. 11.2): plateau, steep edge (talus), gently sloping glacis and valley bottom.

Plateau and talus are composed of clayey sandstone called Continental Terminal (Machens 1967). The upper surface, locally covered by Aeolian sands, represents the remnant of a nearly horizontal laterite plateau. Two levels of cuirass layers are

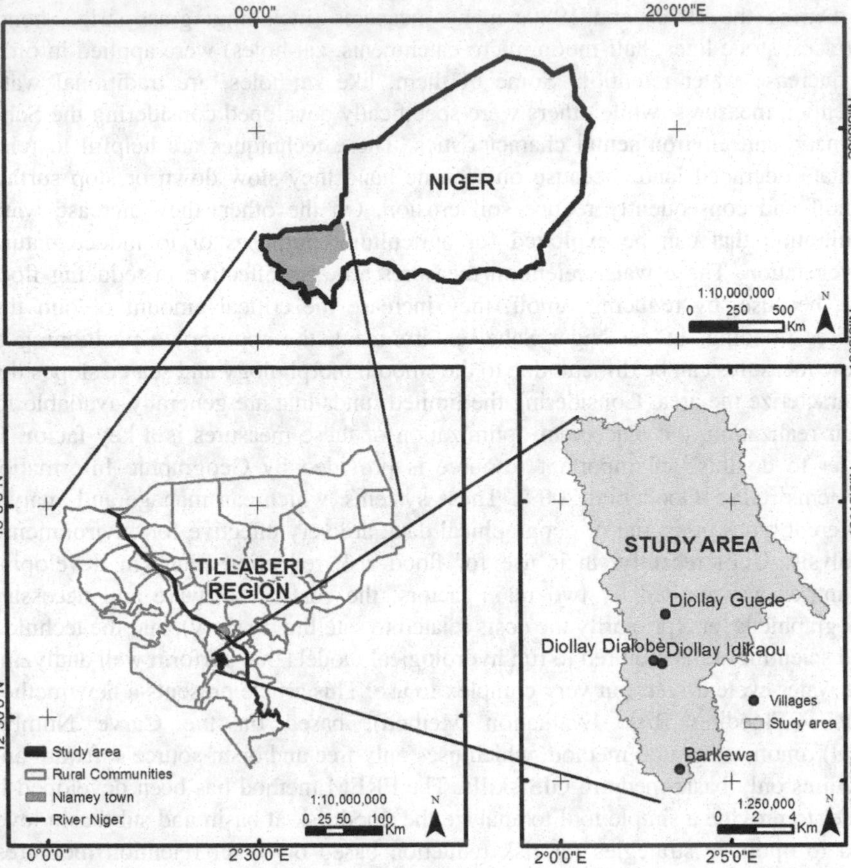

Fig. 12.1 Location of the study area

present in the study area: the higher level at about 260–270 a.s.l. and the lower at 230–225 m a.s.l. Gently sloping glacis is covered in the upper part by crusted bare soils and in the lower by sand layers; it is incised by a network of parallel gullies. In the sandy valley bottom, a large seasonal riverbed ("kori"), called Bukouroul, dug during wetter periods of the Holocene, holds a string of isolated ponds during the rainy season.

Regarding land cover in the study area (Fig. 12.2), the plateau is mainly covered by tiger bush or its remnants (*Combretum micranthum, Guiera senegalensis*, Boscia sp. dominate), a scattered woodland with alternating bare areas and vegetated strips. Due to firewood harvesting, vegetative cover on the plateau is generally degraded and has completely disappeared in some areas, generally near the edges. A few dispersed small millet and/or sorghum fields can be found where residual sand layers are formed and deep enough for cultivation.

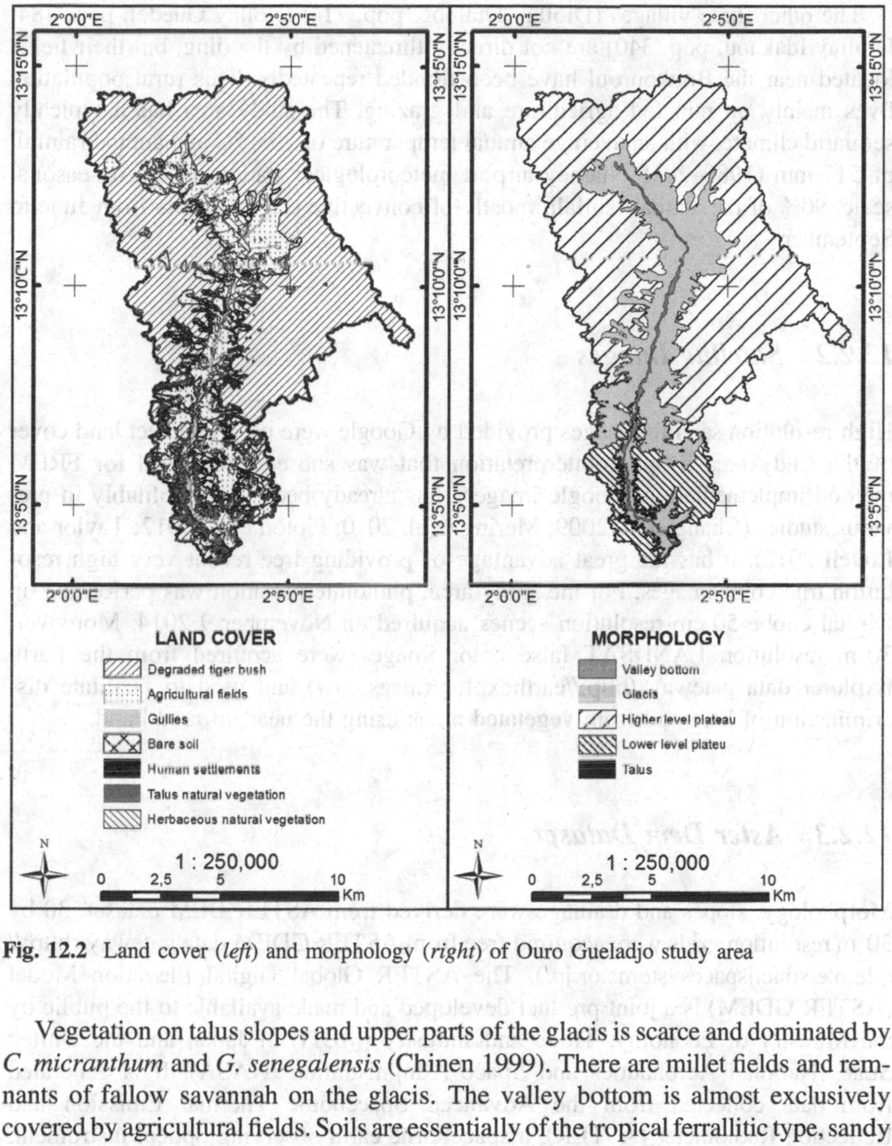

Fig. 12.2 Land cover (*left*) and morphology (*right*) of Ouro Gueladjo study area

Vegetation on talus slopes and upper parts of the glacis is scarce and dominated by *C. micranthum* and *G. senegalensis* (Chinen 1999). There are millet fields and remnants of fallow savannah on the glacis. The valley bottom is almost exclusively covered by agricultural fields. Soils are essentially of the tropical ferrallitic type, sandy and weakly structured (D'Herbès and Valentin 1997). On the generally hardly-cultivable laterite plateaus, thin acidic lithosols directly overlie the ferruginous iron pan. Hill slopes are covered by 0.5–8 m-thick sandy ferruginous soils, sometimes interlaid with other iron pan levels. Valley bottoms have weakly leached sandy ferruginous soils. All are rich in sesquioxides (Al_2O_3 and Fe_2O_3), poor in organic matter (0.5–3%), and have little fertility (low nitrogen and phosphate content).

There are 4 small villages (Fig. 12.1) in the basin. Barkewa village (pop. 314) is the most exposed to flooding being near the catchment outlet on the right bank of the Boukouroul between the glacis and the valley bottom.

The other three villages (Diollay Dialobè, pop. 714; Diollay Guedel, pop. 484; Diollay Idakaou, pop. 340) are not directly threatened by flooding, but their fields located near the Boukouroul have been flooded repeatedly. This rural population lives mainly on rain fed agriculture and grazing. The study area has a typically semiarid climate, with an average annual temperature of 29.8 °C and annual rainfall of 517 mm (1980–2015 Niamey airport meteorological station data). At seasonal scale, 90% of the annual rainfall, mostly of convective origin, occurs from June to September.

12.2.2 Satellite Images

High-resolution satellite images provided by Google were used to detect land cover of the study area by photointerpretation that was subsequently used for FREM method implementation. Google imagery has already been used profitably in previous studies (Chang et al. 2009; Mering et al. 2010; Ploton et al. 2012; Taylor and Lovell 2012): it has the great advantage of providing free recent very high resolution true color images. For the study area, photointerpretation was performed on Digital Globe 50 cm resolution scenes acquired on November 9 2014. Moreover, 30 m resolution LANDSAT false color images were acquired from the Earth Explorer data gateway (http://earthexplorer.usgs.gov/) and used to facilitate discrimination of bare soil from vegetated areas using the near infrared band.

12.2.3 Aster Dem Dataset

Morphology, slopes and drainage were derived from ASTER DEM dataset. 30 by 30 m resolution grids were acquired free from ASTER GDEM data gateway (http://gdem.ersdac.jspacesystems.or.jp/). The ASTER Global Digital Elevation Model (ASTER GDEM) is a joint product developed and made available to the public by the Ministry of Economy, Trade and Industry (METI) of Japan and the United States National Aeronautics and Space Administration (NASA). It is generated from data collected from the Advanced Spaceborne Thermal Emission and Reflection Radiometer (ASTER), a spaceborne earth observing optical instrument. Fujisada et al. (2005) stated that standard deviation is about 10 m, and so the vertical accuracy of DEM is about 20 m with 95% confidence. Hirano et al. (2003) assessed that ASTER DEM is suitable for a range of environmental mapping tasks in landform studies.

12.2.4 Soter Soil Niger DB

Soil data regarding the study area were derived from the SOTER (Soil and Terrain Digital Database) database that can be freely accessed on the internet (https://www.unihohenheim.de/atlas308/b_niger/projects/b2_1_2/html/english/ntext_en_b2_1_2.htm). The South West Niger SOTER database (Oldeman and Van Engelen 1993; Graef et al. 1998) has been specifically developed for land evaluation and management purposes; it contains geomorphological and soil data in GIS or GIS compatible format and offers the possibility of relatively precise spatial estimations of terrain related factors such as soil types and soil factors.

12.2.5 Rainfall Data and Critical Rain

Daily rainfall data of the Torodi meteorological station, the nearest to the study area and located 23 km to the west, were acquired from the National Meteorological Service (DMN) of Niger, in order to detect the critical amount of rain that caused floods in the catchment.

Critical rain is considered as the minimum threshold of rain causing a flood, it is site specific and is identified by the study of rain/flood occurrences. Given that DMN collects daily rain data, the model cannot take into account the rain intensity or basin concentration time. The simplified model is based on a simple relationship between daily rainfall amount and probability of flooding, through the mediation of the structural features of the sub-basin. In the target area, the latest relevant flood was on August 23 2013 due to an 80 mm critical rainfall.

12.3 Method

12.3.1 Theoretical Basis

The method is a simplified version of the Soil Water Assessment Tool (SWAT) (Gassman et al. 2007; Neitsch et al. 2011), the most widely used hydrological simulator. This method provides a simple tool to analyze flood risk at basin and sub-basin level and to optimize strategies for risk reduction. More specifically, the method allows: (i) the optimal placement of water retention structures in the elementary sub-basins that contribute most to the overall risk, (ii) the assessment of the contribution of each hydraulic structure to reduce the total risk, (iii) a priority ranking to be given to these structures identifying which are the most urgent.

The method is physically based, computationally efficient, and is based on Remote Sensing (RS) and GIS analysis integrated with field surveys. FREM methodology is shown in Fig. 12.3; it requires a preliminary watershed and

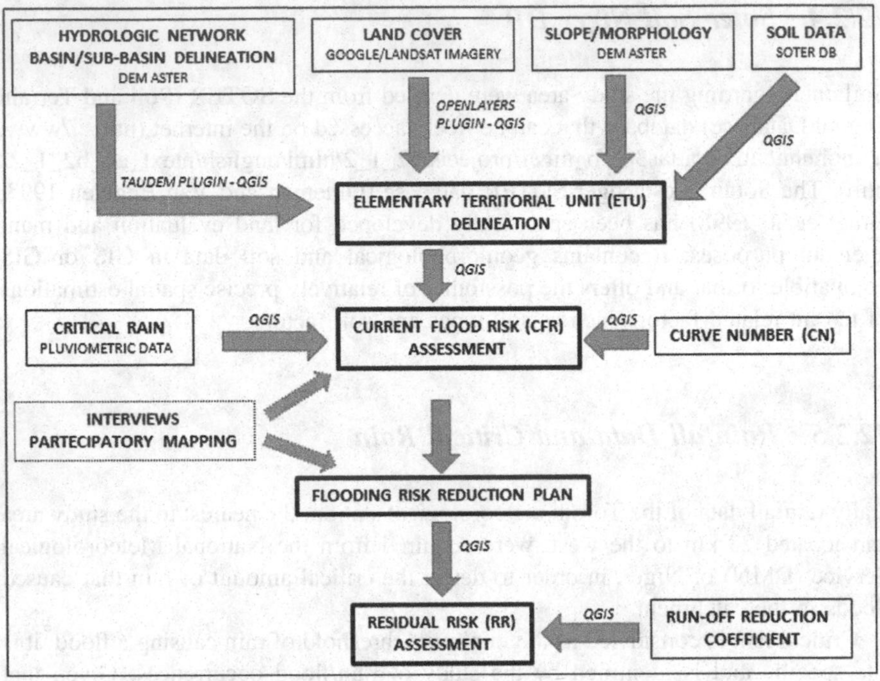

Fig. 12.3 FREM methodology

sub-basin delineation based on DEM data performed using QGIS software and TAUDEM plugin. Areas with the same slope, soil type and land cover form an Elementary Territorial Unit (ETU), a basic computational unit assumed to be homogeneous in hydrologic response to rains.

This preliminary analysis allows the assessment of the Current Flood Risk (CFR) in the study area. The risk is expressed as the amount of water that concentrates in the flooded areas after a critical rain, the minimum daily amount of rain that previously caused flooding in the study area. Therefore, knowing the impact of each area of the catchment, in terms of water that flows away and doesn't infiltrate, water retention measures can be planned. At this stage, two accessory types of maps can be derived to facilitate risk reduction planning. Runoff maps show the areas that have the highest runoff coefficients; priority maps, highlighting the units with higher runoff coefficients and dimensions, indicate the areas where it is more important to take action.

Finally, evaluating the runoff reduction coefficients of each water retention measure, the Residual Risk (RR) is calculated as the amount of water due to a critical rain that would concentrate in the flooded areas after implementation of the planned water retention measures. Areas with the same slope, soil type and land cover form an Elementary Territorial Unit (ETU), a basic computational unit assumed to be homogeneous in hydrologic response to rains.

12.3.2 CN Runoff Estimation

The curve number (CN) method (USDA 1986) allows surface runoff volumes of individual storm events to be estimated; it is widely used because of its simplicity and the limited number of parameters required for runoff prediction (e.g., Graf 1988; Ponce and Hawkins 1996; Bhuyan et al. 2003). It is used in a wide range of design situations and has been integrated into many hydrologic, erosion, and water-quality models such as CREAMS (Knisel 1980), SWRRB (Williams et al. 1985; Arnold et al. 1990), AGNPS (Young et al. 1989), EPIC (Sharpley and Williams 1990), PERFECT (Littleboy et al. 1992), WEPP (Risse et al. 1994) and SWAT.

The CN method is an empirically-based set of relationships between rainfall, land-surface conditions and runoff depth. The basic equations used in the CN method are:

$$R = (P - 0.25)^2/(P + 0.85) \quad R = O\,for\,P \leq 0.2S \tag{1}$$

$$S = (1000/CN) - 10 \tag{2}$$

where R = runoff depth (inches), P = precipitation depth (inches), S = potential maximum retention (inches), and CN is the Curve Number. The coefficients 0.2 and 0.8 in Eq. (1) arise from an assumption that the initial abstraction value (I_a) S is equal to 0.2S.

CN is a function of soil type, land cover, slope and antecedent runoff condition. For each portion of the basin homogeneous in its hydrologic response to rains, a CN value, derived from specific tables, is conferred. CN has a range from 30 to 100; lower numbers indicate low runoff potential while higher numbers are for increasing runoff potential. CN are referred to units with 5% slope and average moisture conditions; specific formulas are provided for slope and moisture adjustments.

12.3.3 Runoff Reduction Coefficients

In order to estimate the impact of water retention measures in the risk reduction plan, runoff reduction coefficients were derived from the available literature and conferred to packages of water retention measures that are specific for land cover types (Table 12.1).

These runoff reduction coefficients, although indicative, are used to comparatively assess the effect of water retention measures on the different hydrological units. They can be used to develop scenarios that indicate the potential effects of risk reduction plans based on different runoff reduction techniques.

Table 12.1 Runoff reduction coefficients of water retention measure packages

Land cover	Water retention measures	Runoff reduction coefficient
Natural vegetation	Stone lines/grass strips	0.6[a]
Croplands	Field fencing with stone lines or grass strips, natural assisted regeneration of tree layers, contour lines ploughing.	0.25[b]
Bare soil, degraded areas	Grass strips/contour bench terraces, reforestation, half-moon micro catchments	0.3[ab]

[a]JGRC (2001)
[b]Roose and Bertrand (1971), Lamachère and Serpantié (1991), Nasri et al. (2004), Al Ali et al. (2008)

12.3.4 GIS Software

Method implementation was performed using QGIS, a free and open-source Geographic Information System software. QGIS was downloaded from the QGIS gateway (http://www.qgis.org/en/site); the software provides a plugin mechanism that can be used to add support for a new data source or to extend the functionality of the main program in a modular way. Watershed delineation and definition of its hydrologic network and sub-basins was performed with Terrain analysis using Digital Elevation Models (TauDEM) QGIS open-source plugin (Tarboton 2005), a set of tools for the extraction and analysis of hydrologic information from topography as represented by a DEM. Land cover photointerpretation was performed using the Open Layers Plugin, which allows a number of image services to be added, including the Google satellite images that were used in this study.

12.3.5 Interviews and Participatory Mapping

A few days of field surveys were conducted during November 2014 for ground truth validation of preliminary ETU delineation and to do a participatory mapping with village inhabitants. Participatory mapping and interviews were carried out in order to appreciate local perception of flood risk and to obtain information about aspects that cannot be recorded through remote sensing analysis. From this point of view, participatory mapping offers a tool to integrate GIS analysis and local perspective; it is a key to integrate top-down and bottom-up approaches and gives a voice to the most marginalized members of the community who are also the most likely to be affected by risk and climate and environmental changes.

12.4 Results

The FREM method allowed a flooding risk reduction plan to be conceived to a critical rain of 80 mm, equal to the amount that caused the last flood in the catchment during 2013; this plan has already been thoroughly explained in Tarchiani and Fiorillo (2016). The analysis indicated that the croplands of the Guellay villages, located on the glacis and in the bottom alluvial zones, are threatened by 42 small sub-basins (Fig. 12.4), covering an area of 112.1 km^2, that discharged 4,413,992 m^3. Most of these sub-basins, corresponding to 96.1 km^2, are covered by degraded or highly degraded tiger bush formations on the plateau that are responsible for 82% of the amount of water that flows on the flooded areas. Barkewa, the village most exposed to floods, is threatened by 6 small sub-basins that surround the village on the south-western and north-eastern sides (Fig. 12.4).

These sub-basins, covering an area of 9.9 km^2, in 2013 discharged 303,596 m^3 to the flooded Barkewa area. A classification of each sub-basin according to runoff coefficient allowed the prioritizing of areas to be treated. Priority areas were identified as large surfaces with high runoff rates, mainly bare soil or degraded bushes. Regarding the Diollay area (Table 12.2), due to the fact that most of the incoming water arrives from the plateau, the proposed measures involve mainly restoration and protection of tiger bush (8501 ha).

The plateau has been heavily exploited in the last decades for firewood harvesting and in the study area tiger bush is heavily degraded or has been totally cleared. On the glacis, proposed water retention measures involve mainly the restoration of marginal areas with natural vegetation by the implementation of

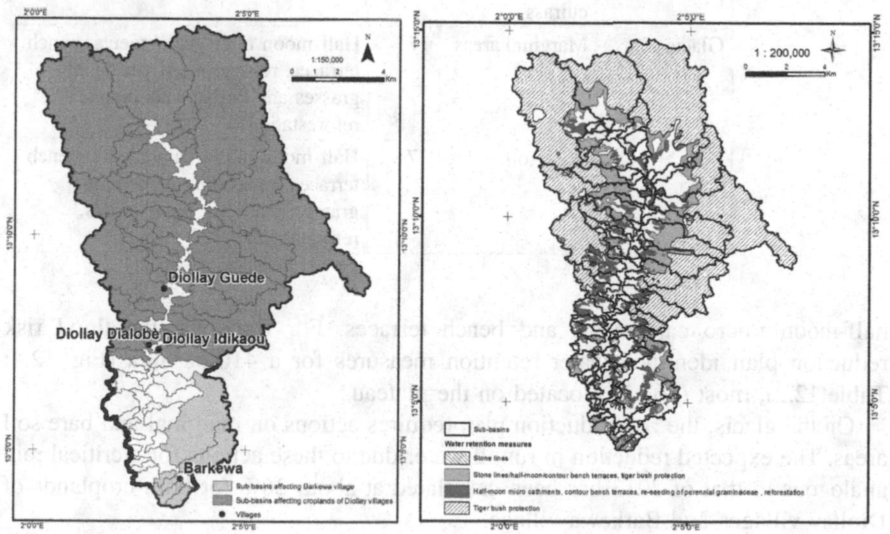

Fig. 12.4 Sub-basins affecting croplands of Diollay and Barkewa village (*left*) and Ouro Gueladjo study area flood risk reduction plan (*right*)

Table 12.2 Flood risk reduction plan for the Ouro Gueladjo case study

Sub-basins	Morphology	Land cover	ha	Water retention measure
Affecting croplands of Diollay villages	Plateau	Degraded tiger bush	8048	Tiger bush protection measures
		Very degraded tiger bush	1453	Stone lines/half-moon micro catchments/bench terraces (according to the soil type), tiger bush protection
	Glacis	Marginal areas with natural vegetation	494	Half-moon micro catchments, bench terraces
		Lateritic cuirass	6	Stone lines
		Crusted bare soil	18	Half-moon micro catchments, bench terraces, reforestation
Affecting Barkewa village	Plateau	Degraded tiger bush	88	Tiger bush protection measures
		Very degraded tiger bush	147	Stone lines/half-moon micro catchments/bench terraces (according to the soil type), tiger bush protection
		Marginal areas with natural vegetation	15	Half-moon micro catchments, bench terraces, re-seeding of perennial grasses and herbaceous plants, reforestation
		Bare soil	41	Half-moon micro catchments, bench terraces, re-seeding of perennial grasses and herbaceous plants, reforestation
		Lateritic cuirass	66	Stone lines
	Glacis	Marginal areas	17	Half-moon micro catchments, bench terraces, re-seeding of perennial grasses and herbaceous plants, reforestation
		Bare soil	37	Half-moon micro catchments, bench terraces, re-seeding of perennial grasses and herbaceous plants, reforestation

half-moon micro catchments and bench terraces. For Barkewa, the flood risk reduction plan identified water retention measures for a 410 ha area (Fig. 12.4; Table 12.2), most (367 ha) located on the plateau.

On the glacis, the risk reduction plan requires actions on marginal and bare soil areas. The expected reduction in runoff water due to these actions for a critical rain analogous to that of 2013 has been calculated at about 26% for both croplands of Diollay villages and Barkewa village.

12.5 Discussion

The Ouro Gueladjo case study demonstrated that the method can be profitably applied in the Sahel environment providing a useful tool for environmental evaluation and to optimize flooding risk reduction plans. The method, compared to other more sophisticated hydrologic models, is much simpler and therefore requires only basic/medium GIS technical skills and can be performed just using QGIS software simply by applying the formulas presented in this paper. Models like SWAT are high-performing and allow many parameters like water, sediment, nutrient, pesticide, and fecal bacteria yields to be evaluated and determined at basin and sub-basin scale, but are much more difficult to implement and require many more data. FREM performs an evaluation focused only on risk reduction planning, but in a much more intuitive and direct way. Moreover it needs less data to be implemented. For example, regarding climatic and meteorological parameters, SWAT requires a large number of climatic inputs (daily precipitation, maximum and minimum temperature, solar radiation data, relative humidity, and wind speed), which complicates model parameterization and calibration, while FREM needs only the amount of critical rain that previously caused flooding in the study area.

The method has the great advantage of using only open-source software and data, a crucial factor for its applicability in developing countries. QGIS has already reached maturity; it is very stable, with high performances analogous to those of commercial software. TauDEM plugin has been demonstrated to be very effective in hydrologic network and basin delineation. The Openlayers plugin allows the very high resolution, and usually very recent free Google images in a real GIS environment to be used, overcoming the limited GIS possibilities provided by Google Earth. A limit that has emerged in the use of Google images is the lack of near infra-red bands, a key factor in environments like the Sahel, where crops and natural vegetation are not easily distinguishable from bare soil or highly degraded areas. This may be compensated by the integration of other free images like the LANDSAT scenes that have been used in this case, despite their limited spatial resolution. The recent availability of 10 m free high-resolution multispectral images provided by the latest generation of satellites, such as the Sentinel, should allow this limitation to be overcome.

However, the case study highlighted some limitations of free available ASTER DEM data. In the South West Niger Sahelian area there is a high presence of endorheic basins. Regarding the Ouro Gueladjo case study, field verifications indicated that Barkewa village is affected only by the 6 nearest surrounding sub-basins due to some ephemeral ponds that impede drainage from northern sub-basins. This aspect was not detected by the GIS hydrologic network analysis due to the not sufficient ASTER DEM spatial resolution, the highest actually available free for the study area, and vertical accuracy. However due to the endorheic nature of many watersheds in the region, the use of this methodology for flood risk assessment must be restricted to small-medium catchments and after field verifications. Moreover, the 30 m resolution is not enough to distinguish micromorphology and in some cases to precisely delineate sub-basins. Therefore actual DEMs don't allow the physical and morphological aspects involved in the water cycle to be defined at a scale comparable to that provided by Google images for land cover delineation. This can reduce accuracy in flood risk reduction plans and must be kept into account.

Soil information is often a critical aspect in environmental studies due to the fact that it can only be detected at a high scale resolution by field surveys and only a few data are usually available free. In the present case study, the availability of the Niger SOTER database helped to overcome this problem. Anyway the CN method requires only one type of soil information, the infiltration capacity that is classified in just four classes. This aspect facilitates implementation of the CN method in the FREM methodology because in the Sahel environment rough infiltration rates, like those required, can easily be identified.

In the FREM model the impact of groundwater level was not considered because soil water balance in the Sahel is controlled more by surface than by deep soil conditions (Collinet and Valentin 1979, 1984; Hoogmoed and Stroosnijder 1984); therefore superficial hydrodynamics are mainly determined by surface conditions that are sufficient to explain significant infiltration (Casenave and Valentin 1992).

Finally, the Ouro Gueladjo case study highlighted that hydrologic modelling for risk reduction planning cannot be performed adequately without considering the opinions and needs of local inhabitants. Remote sensing and GIS are extremely powerful tools but some aspects that must be taken into account in risk reduction planning cannot and will never be understood and considered through them. In this study, a field campaign allowed remotely sensed data and local community perception, needs and knowledge of risk, to be integrated through the use of participatory mapping with the inhabitants of the villages located in the basin.

12.6 Conclusions

A real integration of hydrological risk reduction and adaptation to climate change does not exist, despite the many reports that pledge this integration. Climate change is currently affecting the water cycle and flood process by multiplying flash floods as well as river floods in some areas, (Bates et al. 2008). Since 2009, the World Meteorological Organization has been promoting an integrated approach to flood management: IFM (WMO 2009) that also aims at reducing the risks related to poor management of surface water and adaptation to climate change. At a local level, participatory methodologies for adaptation to climate change focus more on their suitability for the communities rather than appropriate planning of vast rural landscapes. The municipalities have no information on climate change in their territory or any analytical and climate prediction capacity. They therefore cannot set adaptation and risk reduction policies based on the results of robust analysis of local conditions.

There is thus a strong need for new and effective tools for flood related risk reduction planning. Floods are becoming one of the most common disasters even in the Sahel and, due to climate change and environmental degradation, they are going to be increasingly frequent in the next decades. The method has been conceived for the Sahel environment but could also be applied in other environments with only slight changes, mainly related to specific runoff reduction coefficients for different water retention measures.

The application of the methodology in the Municipality of Ouro Gueladjo was actively used in the implementation of an action plan that was incorporated in the Municipal Development Plan in accordance with the resources the Municipal Council is planning to invest. The interest and participation of local stakeholders and populations confirms the importance of continuing and improving approaches integrating local and scientific knowledge to meet the real needs of communities and to ensure sustainable land management.

References

Al Ali, Y., J. Touma, P. Zante, S. Nasri, and J. Albergel. 2008. Water and sediment balances of a contour bench terracing system in a semi-arid cultivated zone. *Hydrological Sciences Journal* 53 (4): 883–892. doi:10.1623/hysj.53.4.883.

Albergel, J. 1987. Sécheresse, désertification et ressources en eau de surface: application aux petits bassins du Burkina Faso. *The influence of climate change and climatic variability on the hydrologic regime and water resources (Proceedings of the Vancouver Symposium, August 1987)*. AHS Publication, vol. 168, 355–365.

Arnold, J.G., J.R. Williams, A.D. Nicks, and N.B. Sammons. 1990. *SWRRB; a basin scale simulation model for soil and water resources management*. Texas: A&M University Press.

Bates, B., Z.W. Kundzewicz, S. Wu, and J. Palutikof. 2008. *Climate change and water. IPCC Technical Paper VI*. Geneva: IPCC Secretariat.

Bhuyan, S.J., K.R. Mankin, and J.K. Koelliker. 2003. Watershed-scale AMC selection for hydrologic modeling. *Transactions of the ASAE* 46 (2): 303. doi:10.13031/2013.12981.

Casenave, A., and C. Valentin. 1992. A runoff capability classification system based on surface features criteria in semi-arid areas of West Africa. *Journal of Hydrology* 130 (1): 231–249.

Chang, A.Y., M.E. Parrales, J. Jimenez, M.E. Sobieszczyk, S.M. Hammer, D.J. Copenhaver, and R.P. Kulkarni. 2009. Combining Google Earth and GIS mapping technologies in a dengue surveillance system for developing countries. *International Journal of Health Geographics* 8: 49. doi:10.1186/1476-072X-8-49.

Chinen, T. 1999. Recent accelerated gully erosion and its effects in dry savanna, southwest of Niger. In *Human response to drastic change of environments in Africa*, ed. N. Hori, 67–102.

Collinet, J., and C. Valentin. 1979. Analyse des différents facteurs intervenant sur l'hydrodynamique superficielle. Nouvelles perspectives. *Applications agronomiques. Cahiers ORSTOM Série pédologie* 17: 283–328.

Collinet, J., and C. Valentin. 1984. Evaluation of factors influencing water erosion in West Africa using rainfall simulation. *Challenges in African Hydrology and Water Resources. Symposium, Harare*, 1984/06: 451–461.

D'Herbès, J.M., and C. Valentin. 1997. Land surface conditions of the Niamey region: ecological and hydrological implications. *Journal of Hydrology* 188–189: 18–42. doi:10.1016/S0022-1694(96)03153-8.

Descroix, L., P. Genthon, O. Amogu, J.L. Rajot, D. Sighomnou, and M. Vauclin. 2012. Change in Sahelian rivers hydrograph: The case of recent red floods of the Niger river in the Niamey region. *Global and Planetary Change* 98: 18–30. doi:10.1016/j.gloplacha.2012.07.009.

Fujisada, H., G.B. Bailey, G.G. Kelly, S. Hara, and M.J. Abrams. 2005. Aster DEM performance. *Geoscience and Remote Sensing, IEEE Transactions* 43 (12): 2707–2714.

Gassman, P.W., M.R. Reyes, C.H. Green, and J.G. Arnold. 2007. The soil and water assessment tool: Historical development, applications, and future research directions. *Transactions of the ASABE* 50 (4): 1211–1250.

Goodchild, M.F. 2009. Geographic information system. In *Encyclopedia of Database Systems*. Springer US, 1231–1236. doi:10.1016/j.proeps.2009.09.160.

Graf, W.L. 1988. *Fluvial processes in dryland rivers*, 1–346. New York: Springer-Verlag.

Graef, F., N. Van Duivenbooden, and K. Stahr. 1998. Remote sensing and transect-based retrieval of spatial soil and terrain (SOTER) information in semi-arid Niger. *Journal of Arid Environments* 39 (4): 631–644. doi:10.1006/jare.1998.0423.

Hirano, A., R. Welch, and H. Lang. 2003. Mapping from ASTER stereo image data: DEM validation and accuracy assessment. *ISPRS Journal of Photogrammetry and Remote Sensing* 57 (5): 356–370. doi:10.1016/S0924-2716(02)00164-8.

Hoogmoed, W.B., and L. Stroosnijder. 1984. Crust formation on sandy soils in the Sahel. I. Rainfall and infiltration. *Soil and Tillage Research* 4: 5–23. doi:10.1016/0167-1987(84)90013-8.

JGRC. 2001. Guide technique de la conservation des terres agricoles. *Documentation technique de la JGRC Vol. 5*. Société Japonaise des Ressources Vertes. http://www.reca-niger.org/IMG/pdf/Guide_technique_conservation_terres_agricoles.pdf.

Knisel, W.G. 1980. *CREAMS: A field scale model for Chemicals, Runoff, and Erosion from Agricultural Management Systems [USA]. United States*. Department of Agriculture. Conservation research report (USA).

Lamachère, J.M., and G. Serpantié. 1991. Valorisation agricole des eaux de ruissellement et lutte contre l'érosion sur champs cultives en mil en zone soudano-sahelienne. In *Utilisation rationnelle de l'eau des petits bassins versants en zone aride*, 165–178. AUPELF-UREF.

Leblanc, M.J., G. Favreau, S. Massuel, S.O. Tweed, M. Loireau, and B. Cappelaere. 2008. Land clearance and hydrological change in the Sahel: SW Niger. *Global and Planetary Change* 61 (3): 135–150. doi:10.1016/j.gloplacha.2007.08.011.

Littleboy, M., D.M. Silburn, D.M. Freebairn, D.R. Woodruff, G.L. Hammer, and J.K. Leslie. 1992. Impact of soil erosion on production in cropping systems. I. Development and validation of a simulation model. *Soil Research* 30 (5): 757–774.

Machens, E. 1967. Notice explicative sur la carte géologique du Niger occidental, a l'échelle du 1/200 000. *Editions du Bureau de recherches géologiques et minières*.

Mahe, G., J.E. Paturel, E. Servat, D. Conway, and A. Dezetter. 2005. The impact of land use change on soil water holding capacity and river flow modelling in the Nakambe River. *Burkina-Faso. Journal of Hydrology* 300 (1): 33–43. doi:10.1016/j.jhydrol.2004.04.028.

Mering, C., J. Baro, and E. Upegui. 2010. Retrieving urban areas on Google Earth images: Application to towns of West Africa. *International Journal of Remote Sensing* 31 (22): 5867–5877. doi:10.1080/01431161.2010.512311.

Mouhamed, L., S.B. Traore, A. Alhassane, and B. Sarr. 2013. Evolution of some observed climate extremes in the West African Sahel. *Weather and Climate Extremes* 1: 19–25. doi:10.1016/j.wace.2013.07.005.

Nasri, S., J.M. Lamachere, and J. Albergel. 2004. The impact of contour ridges on runoff from a small catchment. *Revue des Sciences de l'Eau* 17 (2): 265–289.

Neitsch, S.L., J.R. Williams, J.G. Arnold, and J.R. Kiniry. 2011. *Soil and water assessment tool theoretical documentation version 2009*. Texas Water Resources Institute.

Oldeman, L.R., and V.W.P. Van Engelen. 1993. A world soils and terrain digital database (SOTER)-An improved assessment of land resources. *Geoderma* 60 (1): 309–325. doi:10.1016/0016-7061(93)90033-H.

Ponce, V.M., and R.H. Hawkins. 1996. Runoff curve number: Has it reached maturity? *Journal of Hydrologic Engineering* 1 (1): 11–19. doi:10.1061/(ASCE)1084-0699(1996)1:1(11).

Ploton, P., R. Pélissier, C. Proisy, T. Flavenot, N. Barbier, S.N. Rai, and P. Couteron. 2012. Assessing aboveground tropical forest biomass using Google Earth canopy images. *Ecological Applications* 22 (3): 993–1003. doi:10.1890/11-1606.1.

Risse, L.M., M.A. Nearing, and M.R. Savabi. 1994. Determining the Green-Ampt effective hydraulic conductivity from rainfall-runoff data for the WEPP model. *Transactions of the ASAE* 37 (2): 411–418. doi:10.13031/2013.28092.

Roose, E.J., and R. Bertrand. 1971. Contribution à l'étude de la méthode des bandes d'arrêt pour lutter contre l'érosion hydrique en Afrique de l'Ouest. Résultats expérimentaux et observations sur le terrain. *L'agronomie Tropicale* 26 (11): 1270–1283.

Sharpley, A.N., and J.R. Williams. 1990. EPIC-erosion/productivity impact calculator: 1. Model documentation. *Technical Bulletin-United States Department of Agriculture* 1–1768.

Taylor, J.R., and S.T. Lovell. 2012. Mapping public and private spaces of urban agriculture in Chicago through the analysis of high-resolution aerial images in Google Earth. *Landscape and Urban Planning* 108 (1): 57–70. doi:10.1016/j.landurbplan.2012.08.001.

Tarboton, D.G. 2005. *Terrain analysis using digital elevation models (TauDEM)*. Utah State University, Logan. http://hydrology.usu.edu/taudem/taudem3.1/. Accessed 26 June 2016.

Tarchiani, V., and E. Fiorillo. 2016. Plan villageois de réduction du risque d'inondation et de sècheresse dans la commune d'Ouro Gueladjo au Niger. In *Risque et adaptation climatique dans la région Tillabéri, Niger*, ed. V. Tarchiani, and M. Tiepolo, 177–203. Paris: L'Harmattan. ISBN 978-2-343-08493-0.

USDA. 1986. Urban hydrology for small watersheds. *Technical Release* 55: 2–6.

Williams, J.R., A.D. Nicks, and J.G. Arnold. 1985. Simulator for water resources in rural basins. *Journal of Hydraulic Engineering* 111 (6): 970–986.

WMO. 2009. *Integrated flood management*. Report n. 1047. http://www.apfm.info/publications/concept_paper_e.pdf. Accessed 26 June 2016.

Young, R.A., C.A. Onstad, D.D. Bosch, and W.P. Anderson. 1989. AGNPS: A nonpoint-source pollution model for evaluating agricultural watersheds. *Journal of Soil and Water Conservation* 44 (2): 168–173.

Chapter 13
Knowledge for Transformational Adaptation Planning: Comparing the Potential of Forecasting and Backcasting Methods for Assessing People's Vulnerability

Giuseppe Faldi and Silvia Macchi

Abstract In recent years there has been growing recognition that people's vulnerability is not just as an outcome of possible climate impacts, but rather a dynamic contextual characteristic of a socio-ecological system. Accordingly, along with the acknowledgement of the close connection between climate change adaptation and sustainable development, the scientific debate on adaptation has increasingly focused on the issue of transformation of current systems in response to a changing environment. The need for transformational adaptation, especially in high vulnerability contexts, induces planners to broaden and diversify both knowledge and methods to deal with the growing uncertainty and complexity of socio-ecological systems. By focusing on future studies, this chapter aims to explore the transformational knowledge contribution of forecasting and participatory backcasting methods in assessing people's vulnerability using the case study of coastal Dar es Salaam (Tanzania). Participatory backcasting helps understanding contextual vulnerability and community aspirations, and defining shared adaptation goals and action, Conversely, forecasting proves to be fundamental for the identification of boundary conditions and system thresholds relevant to a specific problem, thus providing knowledge that is valuable for integrating global and local perspectives.

Keywords Transformational adaptation · Vulnerability · Backcasting · Forecasting · Dar es salaam

G. Faldi (✉) · S. Macchi
DICEA—Department of Civil Constructional and Environmental Engineering, Sapienza University of Rome, Via Eudossiana 18, 00184 Rome, Italy
e-mail: giuseppe.faldi@yahoo.com

M. Tiepolo et al. (eds.), *Renewing Local Planning to Face Climate Change in the Tropics*, Green Energy and Technology,
DOI 10.1007/978-3-319-59096-7_13

13.1 Toward Transformational Adaptation: The Need to Innovate Vulnerability Assessment Methods

In recent years, there has been a recognition of the close connection between Climate Change (CC) adaptation and sustainable development, as well as the consolidation of an interpretation of vulnerability as not only the result of possible impacts of CC but also as a contextual characteristic of a Socio-Ecological System (SES) (*contextual vulnerability*) (O'Brien et al. 2007). The debate on adaptation has become increasingly focused on the transformation of current systems in response to CC, and the multitude of economic, political, environmental, and social changes that many communities, cities and regions must address.

Making reference mainly to Resilience and Transition theories (Martens and Rotmans 2005; Folke 2006; Olsson et al. 2006; Van der Brugge and Rotmans 2007), the literature on CC adaptation understands transformation as a structural modification of an SES deriving from a series of specific interacting changes that occur in different domains (economic, cultural, technological, ecological, and institutional), and which may occur as a response to a specific spontaneous or deliberate event or action, or develop gradually over time (IPCC 2014).

According to this perspective, adaptation is no longer understood as simply securing a city with respect to a possible climate impact (*incremental adaptation*), but also as an opportunity to favour the deployment of transformative actions oriented towards sustainability and equity goals (*transformational adaptation and change*) (O'Brien 2012; Park et al. 2012; IPCC 2014). More specifically, incremental adaptation is considered the reproduction of actions and behaviours that have already been tested in the past, or that are currently underway, and aimed at maintaining the integrity and essence of a system with respect to the impacts of CC (Park et al. 2012). Vice versa, transformational adaptation includes the actions and behaviours that modify the fundamental attributes of a system in response to a determined impact of CC, and transformational change is understood as a modality of reducing risk and the vulnerability, not only by adapting to CC, but also by challenging those social and economic structures that have contributed to CC, and to social vulnerability more generally (IPCC 2014).

In particular, the need for transformational adaptation and change has been emphasized (contemporaneously with incremental action) especially in contexts like Sub-Saharan urban environments characterized by high degrees of social vulnerability and exposure to severe climatic stress (Kates et al. 2012).

This emerging theoretical perspective on adaptation led the scientific community to identify a broad knowledge gap as regards how to plan in practice in a way that favours transformational adaptation and change in an SES, specifically for contexts of high social vulnerability (IPCC 2014).

In order to contribute to filling this gap, we first assume that promoting the conditions to facilitate a system transformation begins with changing the way we represent and understand a given question upon which to intervene ("transformation of understandings" for "transformation of practices") (Dente 2014). The type of knowledge upon which the decision-making process is based as well as the way in which that knowledge is constructed, therefore, become fundamental elements in favouring a transformative process.

The transformational perspective on adaptation therefore induces planners to broaden and diversify both knowledge and methods to deal with the growing uncertainty and complexity of socio-economic, environmental, and climatic systems. In particular, planners need to update the most widely-used climate proofing methods for vulnerability assessment (such as impact/risk analyses, developed predominantly for the management of natural disasters), as they are unable to fully grasp the future uncertainty and the multidimensional vulnerability processes occurring in an SES, and, therefore, favour an incremental vision of adaptation. Planners also need to research alternative or complementary methods for supporting decision-making that focus the assessment process directly on the most vulnerable actors and their experiential knowledge, and that are able to embrace the uncertainty and complexity of the multi-stress context in which vulnerability dynamics are produced, and to propose long-term future projects that consider people's legitimate expectations of change (Eakin and Luers 2006; Pelling 2011; O'Brien 2012; IPCC 2014).

In pursuit of that goal, the present work focuses on the broad field of future studies in which scenario methods are increasingly used in planning for adaptation to CC (Börjeson et al. 2006), and seeks to explore the knowledge contribution towards transformational adaptation on a local scale that arises from the use of the two main scenario building approaches for assessing people's vulnerability: forecasting and backcasting.

The argument is explored through the case study of Dar es Salaam (Tanzania), where a vulnerability assessment to access to water of the coastal peri-urban population was developed combining forecasting and participatory backcasting in the framework of the Project "Adapting to Climate Change in Coastal Dar es Salaam" (ACC Dar) (www.planning4adaptation.eu).

After illustrating the use of the two scenario building approaches in the vulnerability assessment process, the chapter presents the case study, the scenario methodology, the research method, the results and emerging knowledge contribution, ending with the potential use of forecasting and participatory backcasting methods for transformational adaptation planning.

13.2 Scenario Analysis for Vulnerability Assessment

The growing interest in adaptation has been translated into progressively more complex methodologies to support the definition of adaptation strategies: from the "classic" environmental impact studies to the more detailed "first" and "second generation" climate change vulnerability studies that encompass social parameters in the assessment process (Füssel and Klein 2006). Among the various methodological instruments applied to adaptation planning and vulnerability assessment in particular, scenario analysis methods are increasingly used because they are

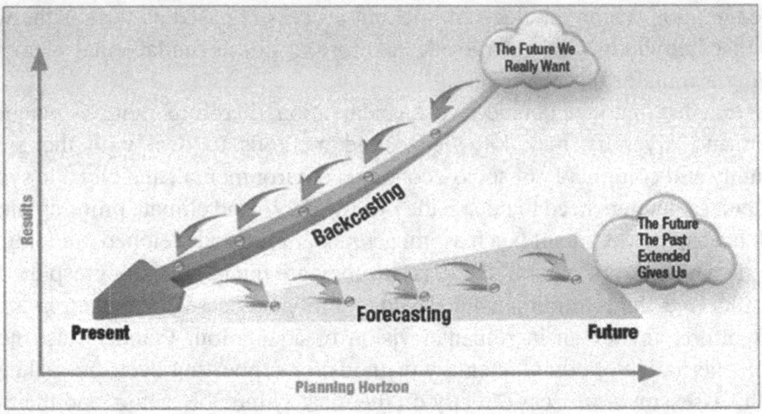

Fig. 13.1 Different perspectives of the forecasting and backcasting approaches (mod. from Roche 2012)

considered useful in addressing situations of high uncertainty characterized by low levels of control (Peterson et al. 2003; Börjeson et al. 2006).

Within the field of future studies, two main schools of thought on scenario analysis can be distinguished that correspond to two different approaches and perspectives on scenario building (Fig. 13.1) (Dreborg 1996; Van Notten et al. 2003).

The first is the forecasting approach (exploratory scenario: what could possibly happen?), developed within the broad school of strategic planning. In the strategic vision, scenario analysis represents a method that, through the construction and exploration of possible stories and paths that move from the present to the future, can be used as a learning process aimed at understanding the relations between different drivers and variables in a system, thus facilitating the development of a certain degree of flexibility in the system in question as regards potential future events (Van der Heijden 2005; Lindgren and Bandhold 2009).

The second is the backcasting approach (normative scenario: what would be preferable to have happen?), which is framed in the broader field of sustainability planning (Robinson 1990; Dreborg 1996), where is it used in a variety of spheres, such as studies on the "transition to sustainability" for socio-technological (Rotmans et al. 2001) and urban (Hopkins 2008) transition. According to that approach, scenario analysis is used primarily to define one or more desirable visions and images of the future that present a solution to a given problem, and secondarily, moving the perspective from the future to the present for the definition of the actions and changes necessary in order for such future images to emerge (Vergragt and Quist 2011). A recent evolution of this approach is so-called participatory backcasting, which is based on the involvement of different actors (experts and non-experts) in the creation of a future vision and the development of future-present paths with the goal of promoting a social learning process for the actors involved (Robinson 2003; Quist and Vergragt 2006; Quist 2007).

As regards the use of scenario analysis in CC adaptation studies, the scenario methods that draw from the forecasting approach, albeit with substantial differences in their scenario building techniques (top-down vs. bottom-up, quantitative vs. qualitative, analytic vs. participatory), have played and continue to play a dominant role in impact studies and CC vulnerability studies and are thus the most widely used in adaptation planning, especially at the regional and supra-regional levels (e.g. SRES scenarios, socio-economic scenarios, down-scaled climatic scenarios) (IPCC 2000; Swart et al. 2004). In particular, forecasting scenario analysis is especially used in CC studies to explore specific environmental and socio-economic questions or phenomena that have a direct correlation to CC, and to understand, visualise, and analyse the relative main future trends, impacts, and climate risks, in order to identify the vulnerability of the people exposed and to thus facilitate the identification and prioritization of various adaptation policy responses (IPCC 2014).

In any case, despite their wide use in adaptation planning, various authors (Gydley et al. 2009, O'Brien 2012) have emphasized how the use of the "more consolidated" forecasting methods in the vulnerability assessment process cannot help but support an incremental view of adaptation, especially on a local scale, because forecasting, which is based on the exploration of dominant trends and is incapable of taking contextual vulnerability dynamics and their future uncertainty into account, tends to consider the future as a restatement of the mechanisms of the present, thus favouring a conservative perspective on adaptation.

Contrariwise, backcasting is suggested in the literature (Giddens 2009; Van der Voorn et al. 2012; IPCC 2012) as theoretically able to favour the unfolding of transformative processes in that, by moving from a utopian and desirable future, it tends to detach from present dynamics and, especially if applied in a participatory way, can introduce certain elements into the adaptation planning process that can promote transformation (shared vision and social learning). In any case, despite the hypothetical recognition of this potential, its use at the practical level for informing local adaptation planning has not yet been widely experimented.

13.3 Levels of Knowledge for Planning Adaptation Locally: Understanding the Access to Water in Coastal Dar es Salaam

13.3.1 Study Area and Methods

The case study was carried out in Dar es Salaam (Fig. 13.2), the largest city in Tanzania, and one of the most important metropolises of eastern Sub-Saharan Africa. In the first phase of the developed vulnerability assessment process (Fig. 13.3a), a forecasting method was applied analytically to study a biophysical phenomenon at the urban level that could be influenced by CC, namely seawater intrusion into the coastal aquifer.

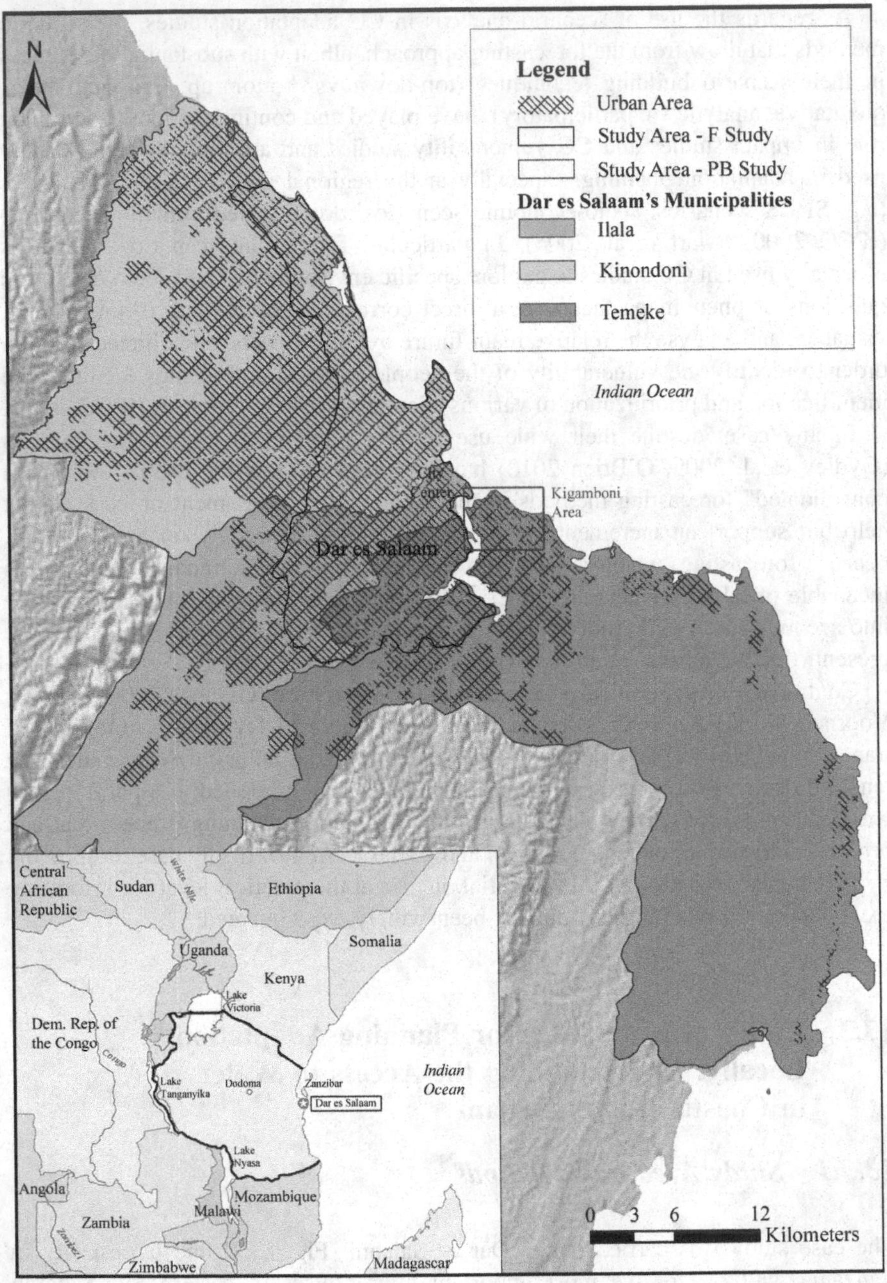

Fig. 13.2 Dar es Salaam Region and the areas of the Forecasting (F) and Participatory Backcasting (PB) studies

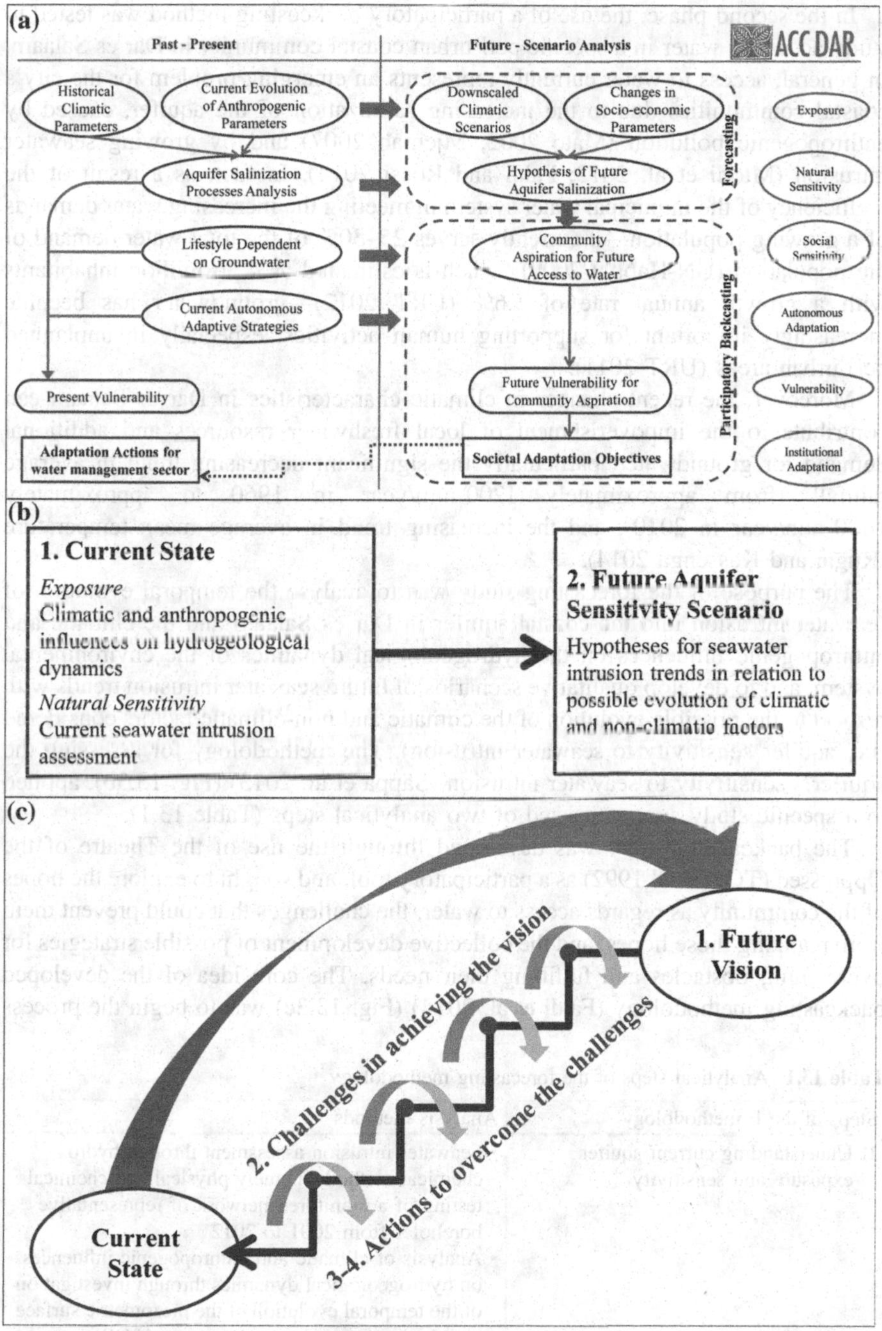

Fig. 13.3 The different phases of the developed vulnerability assessment process (**a**), and the conceptual frameworks of the forecasting (**b**) and participatory backcasting (**c**) methodologies

In the second phase, the use of a participatory backcasting method was tested to study access to water in a specific peri-urban coastal community in Dar es Salaam. In general, access to water currently represents an emerging problem for the city's coastal communities due to the increasing salinization of the aquifer, caused by anthropogenic pollution (Mato 2002; Mjemah 2007) and by growing seawater intrusion (Mtoni et al. 2012; Faldi and Rossi 2014). In fact, as a result of the inefficiency of the municipal water system in meeting the increasing water demands of a growing population—it currently serves 25–30% of the total water demand of the population (UN-Habitat 2009), which is estimated as 4.36 million inhabitants with a growth annual rate of 5.6% (URT 2012)—groundwater has become increasingly important for supporting human activities, especially in unplanned peri-urban areas (URT 2011).

Moreover, the recent changes in climatic characteristics in Dar es Salaam can contribute to the impoverishment of local freshwater resources and additional demand for groundwater, particularly the significant decreasing trend in average rainfall (from approximately 1200 mm/year in 1960 to approximately 1000 mm/year in 2010), and the increasing trend in average mean temperature (Rugai and Kassenga 2014).

The purpose of the forecasting study was to analyse the temporal evolution of seawater intrusion into the coastal aquifer in Dar es Salaam and the climatic and anthropogenic influences on the hydrogeological dynamics of the environmental system, and to develop qualitative scenarios of future seawater intrusion trends with respect to the possible evolution of the climatic and non-climatic factors considered (i.e. aquifer sensitivity to seawater intrusion). The methodology for assessing the aquifer's sensitivity to seawater intrusion (Sappa et al. 2013) (Fig. 13.3b), applied to a specific study area, consisted of two analytical steps (Table 13.1).

The backcasting study was developed through the use of the Theatre of the Oppressed (TO) (Boal 1992) as a participatory tool, and sought to explore the hopes of the community as regards access to water, the challenges that could prevent them from realizing those hopes, and the collective development of possible strategies for overcoming obstacles and fulfilling their needs. The core idea of the developed backcasting methodology (Faldi et al. 2014) (Fig. 13.3c) was to begin the process

Table 13.1 Analytical steps of the forecasting methodology

Steps of the F methodology	Analysis methods
1. Understanding current aquifer exposure and sensitivity	• Seawater intrusion assessment through hydro chemical methods, namely physical and chemical testing of a monitored network of representative boreholes from 2001 to 2012 • Analysis of climatic and anthropogenic influences on hydrogeological dynamics through investigation of the temporal evolution of the piezometric surface and Active Groundwater Recharge (AGR)
2. Exploring future aquifer sensitivity scenario to seawater intrusion	Development of qualitative hypotheses for seawater intrusion trends related to the possible future evolution of climatic and non-climatic factors

Table 13.2 Steps of the participatory backcasting methodology

Backcasting phases	Steps of the PB-TO methodology
Community workshop	1. Development of a shared vision of future access to water 2. Identification of the challenges in achieving that vision
Various public Forum Theatre sessions	3. Preparation of a theatrical representation staging the vision and related challenges that emerged during the workshop 4. Collective exploration of alternative actions and strategies to overcome challenges

by defining the community's shared future vision regarding access to water, to then look backwards from that future to the present situation in order to identify the challenges that might arise as they work towards achieving that vision, and finally to identify strategies and actions for overcoming those challenges and achieving the desired future. The methodology is organized in 4 consecutive steps, carried out during a two-phase exercise (Table 13.2). By using a variety of creative TO techniques, active community participation is fostered throughout the entire process. In fact, participants' interaction represents the core of the process, which relies heavily on a qualitative traditional knowledge system, made up of words, images, stories, and scenes. The developed methodology was tested though a backcasting exercise carried out in Kigamboni ward, a peri-urban coastal area located in the south of Dar es Salaam. The first phase of the exercise involved 24 young people, while in the second phase 11 public Forum Theatre events were performed in various areas within the Kigamboni ward, involving in total more than 2000 people.

The results that emerged from the forecasting and backcasting studies were evaluated with respect to the knowledge contribution provided, identified in terms of:

- understanding the problem of the access to water induced by the use of a given method (way of addressing the future uncertainty and complexity of the context)
- objectives resulting from that understanding (consideration on people's expectations of change)
- action resulting from that understanding.

13.4 Results of the Forecasting and Participatory Backcasting Studies

In the first phase of the vulnerability assessment, the fundamental knowledge contribution of the use of forecasting applied to the study of environmental systems consisted mainly in its ability to provide a better understanding of pressure factors determining the seawater intrusion. This allowed for identification of the boundary conditions of the disruption and the thresholds of the SES in relation to the evolution of the disruption and, consequently, the determination of general principles identifying the supra-local limits of local adaptation initiatives. In particular, the forecasting study shows an increase in the areas affected by seawater intrusion in the period 2001–2012, and how this increase is correlated to the dropping piezometric level, the over-exploitation of the aquifer, and the decrease in average AGR (20%) due to the combined effects of CC (rainfall decrease and mean temperature rise) and anthropogenic causes (increasing soil sealing due to an increase in urban areas from 40 to 65%, and a decrease in soil and vegetation areas from 55 to 32%). The analysis of water demand over the 10-year period also provided an estimate of the exploitation of groundwater by the population.

The combined decrease in resource availability and increase in the estimated groundwater withdrawal suggest that unplanned and uncontrolled groundwater exploitation is a significant factor of hydrogeological imbalance, one that can be related to a general increase of the aquifer's sensitivity to seawater intrusion (Sappa et al. 2013). The qualitative hypotheses regarding the future development of climatic and anthropogenic variables, considering the persistence of current trends, indicate a drastic decline in annual AGR in the future.

At the same time, the rate of groundwater extraction is expected to grow, due to increasing population. The comparison of the AGR and water demand values, estimated according to different water demand scenarios, indicate that the water demand is already equal to the AGR value, with possible negative consequences for the population (Fig. 13.4).

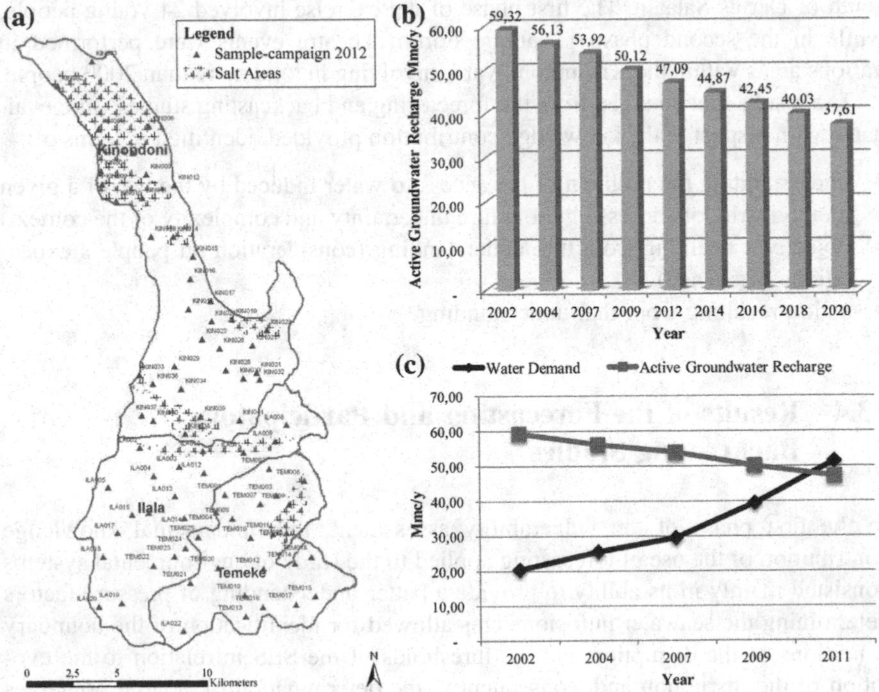

Fig. 13.4 Some specific results of the forecasting study (mod. from Sappa et al. 2013): **a** Sectors affected by seawater intrusion (2012), **b** Estimation of future evolution of AGR, **c** AGR versus water demand (2001–2012)

The forecasting study has therefore demonstrated that the problem has already reached a limit state, providing in addition an order of relevance for the drivers of this disruption. In this specific case, CC represents a possible multiplier of the effects of the disruption, but cannot be considered a primary stress factor in light of the weight of over-exploitation of the aquifer and soil sealing in recent years.

Thus, with respect to the identification of boundary conditions and system thresholds, the forecasting study has proven able to identify those determinative factors that are currently exceeded, or whose future development according to current trends would entail scenarios and tendencies to be avoided (business as usual scenario). This knowledge is fundamental to a vulnerability assessment aimed at adaptation in transformational way, in that it offers the possibility of identifying whether and where a change in the SES is necessary with respect to a specific disruption. Through the use of forecasting in studying specific disruptions to the environmental system, it is therefore possible to define adaptation objectives beyond the local level, which adaptation initiatives should always follow in order to limit an already excessive disturbance or to recognize one that has reached critical levels.

In the case study, the conservation and sustainable use of groundwater resources is therefore framed as a priority general principle with fundamental significance for the coastal communities of Dar es Salaam.

Despite the emergence of such strata, forecasting study was unable to understand the contextual dynamics affected by the disturbance and the eventual existence of cross-scalar influences, namely the internal factors and variables of the context that could constitute additional drivers of the disturbance. This contributes to an increase in uncertainty when predicting what the future dynamics of an SES could be with regard to the evolution of the disturbance.

In the second part of the assessment process, analysis of the results of the practical experimentation of participatory backcasting allowed for the definition of a framework of knowledge elements that participatory backcasting, when applied at the community level, can bring to a vulnerability assessment aimed at transformational adaptation.

A first knowledge element that emerged from the use of participatory backcasting in the vulnerability assessment is strictly linked to the visioning process. The construction of a shared future vision highlighted in a clear and detailed way the desires and hopes of the community as regards access to water: from the legitimate expectation of stable access to good quality water at a reasonable price, which would be guaranteed by having access to multiple water sources, to the desire to generate one or more socio-economic activities at both the individual and community level. In particular, within the desired vision (Fig. 13.5a) every family in the community has access to a sufficient amount of water for their domestic and productive purposes (2000 L/day per household). Water supply should also be ensured by one or more well-constructed, deep wells and various public standpipes connected to the municipal water system. The desire for alternative water sources derives from need to guarantee a continuous and stable supply, so as to relieve people from the need to spend a considerable part of their day obtaining water. In addition, adequate access to water would allow people to develop various socio-economic activities on a broader scale, both at the household level (such as family agriculture) and the community level (such as larger-scale agriculture, livestock grazing, ice production for consumption, and fish farming in artificial ponds). Understanding community aspirations therefore provides several kinds of information that is valid at the community level, and the consideration of which in the vulnerability assessment process could facilitate identification of possible specific adaptation objectives that are socially shared and more detailed, and vice versa, not extrapolated from an external vulnerability assessment that is neutral with respect to the context.

Secondly, the use of participatory backcasting for vulnerability assessment allowed for recognition of the multidimensionality of the problem being studied, i.e. by identifying the various vulnerability mechanisms existing in the context and the connections that arise between them. In particular, during practical experimentation it became clear that the question of access to water is very complex and involves a series of interrelated dynamics and multiple transversal aspects of community life. The question is correlated to a series of technical/environmental problems (qualitative degradation of the resource, management and maintenance of water systems to be installed in the area), however limited access to quality water resources also depends on other problematic and interrelated social, economic, and political conditions. In fact, the definition of challenges in achieving the vision facilitated the immediate emergence of a multitude of critical situations linked to access to water: from establishing a common objective within the community as regards the development of possible shared projects to improve their access to water, to obtaining and eventually managing the funds needed to develop such community projects, to the risk that political authorities or individuals representing the community might use their power to obtain personal economic benefits from the failure of community projects (Fig. 13.5a).

This capacity of backcasting allowed for a reduction in the level of uncertainty in understanding the possible evolution of an SES in response to a given environmental or socio-economic pressure factor that directly affects the context, or to a specific already planned (or to be planned) adaptation action. In addition backcasting demonstrates the main causal processes and the level of entry points upon which it is important to focus in order to reduce/transform vulnerability conditions.

Lastly, in the collective exploration of possible actions for overcoming the challenges and achieving the shared vision, the knowledge contribution provided by participatory backcasting demonstrates a more operational character. The public events carried out produced various practices directly proposed by the community. This included actions involving political authorities, proposed in a critical but collaborative way, such as the creation of independent community groups for gathering funds (from various sources) or for gathering, sharing, and distributing information (plans, regulations, laws), as well as the creation of a community water association with a clearly defined political strategy and an economic and technical plan, and finally actions that assumed in some cases the occurrence of illegal acts (breaking of pipes to directly access water or forcing political leaders to step down) (Fig. 13.5b).

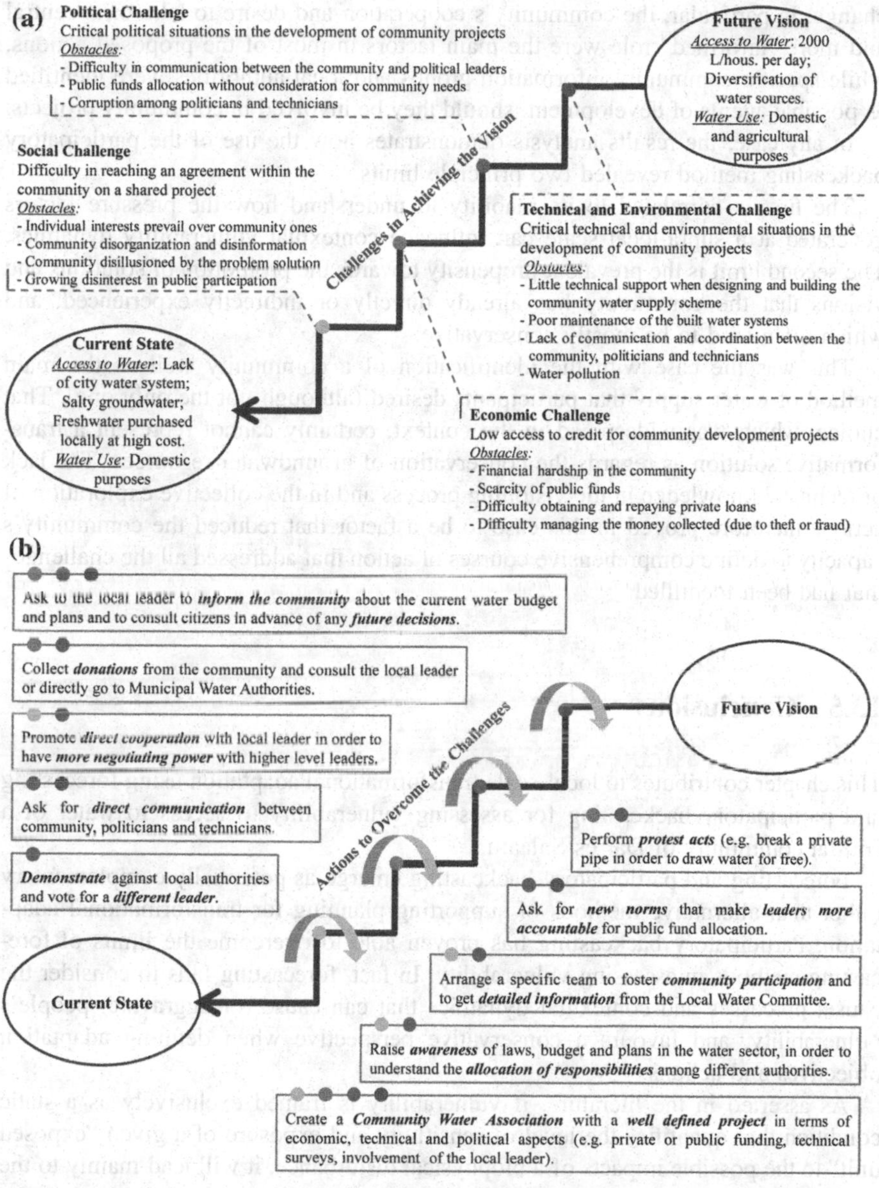

(a)
Political Challenge
Critical political situations in the development of community projects
Obstacles:
- Difficulty in communication between the community and political leaders
- Public funds allocation without consideration for community needs
- Corruption among politicians and technicians
Future Vision
Access to Water: 2000 L/hous. per day; Diversification of water sources.
Water Use: Domestic and agricultural purposes
Social Challenge
Difficulty in reaching an agreement within the community on a shared project
Obstacles:
- Individual interests prevail over community ones
- Community disorganization and disinformation
- Community disillusioned by the problem solution
- Growing disinterest in public participation
Challenges in Achieving the Vision
Technical and Environmental Challenge
Critical technical and environmental situations in the development of community projects
Obstacles:
- Little technical support when designing and building the community water supply scheme
- Poor maintenance of the built water systems
- Lack of communication and coordination between the community, politicians and technicians
- Water pollution
Current State
Access to Water: Lack of city water system; Salty groundwater; Freshwater purchased locally at high cost.
Water Use: Domestic purposes
Economic Challenge
Low access to credit for community development projects
Obstacles:
- Financial hardship in the community
- Scarcity of public funds
- Difficulty obtaining and repaying private loans
- Difficulty managing the money collected (due to theft or fraud)
(b)
Ask to the local leader to inform the community about the current water budget and plans and to consult citizens in advance of any future decisions.
Collect donations from the community and consult the local leader or directly go to Municipal Water Authorities.
Future Vision
Promote direct cooperation with local leader in order to have more negotiating power with higher level leaders.
Actions to Overcome the Challenges
Ask for direct communication between community, politicians and technicians.
Perform protest acts (e.g. break a private pipe in order to draw water for free).
Demonstrate against local authorities and vote for a different leader.
Ask for new norms that make leaders more accountable for public fund allocation.
Arrange a specific team to foster community participation and to get detailed information from the Local Water Committee.
Current State
Raise awareness of laws, budget and plans in the water sector, in order to understand the allocation of responsibilities among different authorities.
Create a Community Water Association with a well defined project in terms of economic, technical and political aspects (e.g. private or public funding, technical surveys, involvement of the local leader).

Fig. 13.5 Results of the participatory backcasting (Faldi et al. 2014): Vision and challenges/obstacles in achieving the vision (**a**), main actions identified during Forum Theatre (**b**)

The main knowledge contribution that emerges in this phase of the process consists in the capacity of participatory backcasting to demonstrate modalities for carrying out an action, which identify the factors (to consider) and the agents of development (to mobilize) that the community considers potentially able to favour change. In particular, the community's cooperation and desire to take on a central and more "informed" role were the main factors in most of the proposed actions, while specific community information groups and local authorities were identified as possible agents of development, should they be involved in community projects.

In any case, the results analysis demonstrates how the use of the participatory backcasting method revealed two principle limits.

The first is correlated to its inability to understand how the pressure factors generated at a supra-local scale can influence contextual vulnerability dynamics. The second limit is the prevalent propensity towards the promotion of solutions and visions that the community has already directly or indirectly experienced, and which often tend to be mostly conservative.

This was the case with the identification of a community well as the main method of water supply that participants desired (although not the only one). That choice, which was widespread in the context, certainly cannot represent a transformative solution as regards the conservation of groundwater resources. The lack of technical knowledge in the visioning process and in the collective exploration of actions therefore proved in this case to be a factor that reduced the community's capacity to define comprehensive courses of action that addressed all the challenges that had been identified.

13.5 Conclusions

This chapter contributes to local-scale transformational adaptation using forecasting and participatory backcasting for assessing vulnerability to access to water of a coastal community of Dar es Salaam.

Forecasting and participatory backcasting emerge as potentially complementary rather than alternative methods of supporting planning for transformational adaptation. Participatory backcasting has proven able to overcome the limits of forecasting methods in assessing vulnerability. In fact, forecasting fails to consider the causal processes and contextual dynamics that can cause (or aggravate) people's vulnerability, and favours a conservative perspective when defining adaptation objectives and actions.

As asserted in the literature, if vulnerability is framed exclusively as a static condition that quantifies the level of sensitivity and exposure of a given "exposed unit" to the possible impacts of a biophysical disturbance, it will lead mainly to the identification of incremental adaptation measures. This is evident within the case study: the understanding of the potential evolution and relevance of the seawater intrusion problem leads to the definition of primarily technical-infrastructural or regulatory actions aimed at managing or reducing the disturbance itself. Although the conservation of water resources emerged as a general principle in the urban context of Dar es Salaam, that type of action cannot always be considered effective or "no regret" in that the proposed representation is inevitably missing a part of the problem, namely the understanding of how such initiatives can address contextual mechanisms of vulnerability.

The use of participatory backcasting in the vulnerability assessment can therefore fill such gaps, because the alternative representation of the problem that it proposes allows the adaptation process to focus on the dynamics of contextual vulnerability. Participatory backcasting in this case study shows that groundwater salinization represents only a part of the problem. People's vulnerability is characterized as a dynamic condition that results from different and interconnected socio-economic, environmental, and political processes that reveal the reasons why people are unable to access quality water at a reasonable price.

In addition, participatory backcasting has proven potentially able to fill the gaps in forecasting in terms of defining the objectives and actions that come up against the future uncertainty of contextual vulnerability. The identification of shared community objectives, entry points and possible modalities of action provides greater operability in the definition of medium-term adaptation options. This in turn allows for a reduction of uncertainty with respect to the efficacy of such options in the specific context, and optimization of available financial resources, which can be scarce in general, and especially in Sub-Saharan contexts.

According to the representation of the problem proposed in the participatory backcasting, new infrastructure are desired and preferred in the Kigamboni community, but to be an effective and optimal option it would have to be preceded by actions that address various social, economic, and political problems

From this point of view, participatory backcasting proposes a change in the representation of the problem, which can thus create the conditions for defining local adaptation strategies that are transformative, i.e. oriented not only to the definition of actions for counteracting the effects of CC, but also towards the development of future projects for sustainability and equity.

In any case, the complementarity of the two methods is also clear in the opposite direction. Indeed, the fundamental knowledge contribution that emerged from the forecasting study can potentially compensate the limited technical knowledge within participatory backcasting. In fact, the ability of forecasting to identify, on a supra-local level, the boundary conditions and upper limits beyond which a system change becomes necessary allows for the definition of general objectives that a local adaptation action should seek to meet. Such objectives therefore constitute exclusionary principles for the eventual range of actions defined at the local level.

From the perspective of transformational adaptation, forecasting can therefore bring to the vulnerability assessment process the type of technical knowledge that is necessary to integrate the global into the local perspective. Such knowledge, if introduced into the participatory process, can also allow for an expansion of the horizon of visions and possible courses of action that are available to the community. In the present case study, this translated into the possibility of allowing people to recognize that a community well may not be the most desirable option, thus motivating them to search for alternative solutions that could be equally desirable.

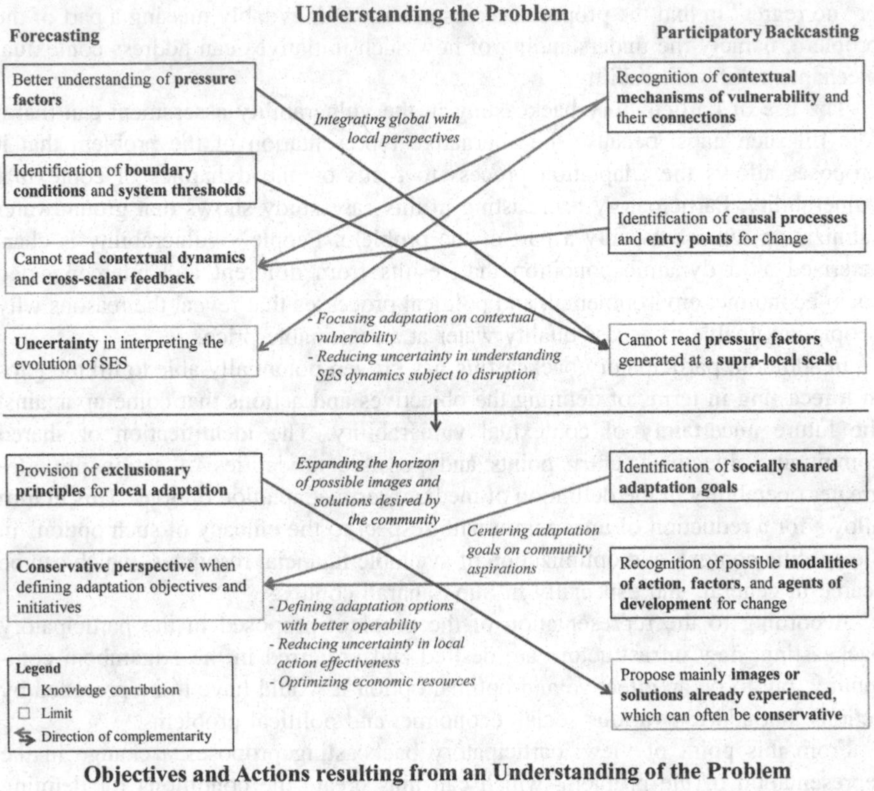

Fig. 13.6 Knowledge contribution and limits of forecasting and participatory backcasting: framework of complementarity for transformational local adaptation planning

The framework of the elements of complementarity in the use of forecasting and participatory backcasting in planning transformational adaptation is outlined in Fig. 13.6. On the basis of the experiment results, the configuration of a platform of action for transformational local adaptation in a high vulnerability context involves a change in the representation of contextual vulnerability, and the promotion of a co-production of knowledge. In fact, local traditional knowledge, as mobilized in the backcasting exercise, proved determinative in understanding how vulnerability was formed and "experienced" in the context, and how people wanted to transform it, while technical knowledge, if introduced and "reworked" within the participatory process, is important in defining and potentially expanding the conditions surrounding desires and consequent actions.

In any case, the combination at the community level of the two forms of knowledge and, in this specific case, the two different methods experimented, would require the definition of specific instruments and techniques that introduce technical knowledge into the participatory process, while preventing dynamics of subjugation within the community. How to configure that transdisciplinary dialogue of methods at the community level in a way that optimizes potential and limits shortcomings for transformational local adaptation therefore requires further experimentation and study.

References

Boal, A. 1992. *Games for actors and non-actors*. Oxford: Routledge.

Börjeson, L., M. Höjer, K.-H. Dreborg, T. Ekvall, and G. Finnveden. 2006. Scenario types and techniques: Towards a user's guide. *Futures* 38 (7): 723–739. doi:10.1016/j.futures.2005.12.002.

Dente, B. 2014. *Understanding policy decisions*. Cham: Springer. doi:10.1007/978-3-319-02520-9.

Dreborg, K.-H. 1996. Essence of backcasting. *Futures* 28 (9): 813–828. doi:10.1016/S0016-3287(96)00044-4.

Eakin, H., and A.L. Luers. 2006. Assessing the vulnerability of social-envoronmental systems. *Annual Review of Environment and Resources* 31: 365–394. doi:10.1146/annurev.energy.30.050504.144352.

Faldi, G., S. Macchi, O. Malcor, and M. Montemurro. 2014. Participatory backcasting for climate change adaptation: supporting community reasoning on access to water in Dar es Salaam. ACCDar Project. http://www.planning4adaptation.eu/Docs/newsInfoMaterial/07 2014/Backcasting%20Working%20Paper_Faldi_Macchi_Malcor_Montemurro.pdf. Accessed 28 November 2016.

Faldi, G., and M. Rossi. 2014. Climate change effects on seawater intrusion in coastal Dar es Salaam: Developing exposure scenarios for vulnerability assessment. In: *Climate Change Vulnerability in Southern African Cities*, eds. S. Macchi and M. Tiepolo, 57–72. Cham: Springer. doi:10.1007/978-3-319-00672-7_4 .

Folke, C. 2006. Resilience: The emergence of a perspective for social-ecological systems analyses. *Global Environmental Change* 16: 253–267.

Füssel, H.-M., and R.J.T. Klein. 2006. Climate change vulnerability assessments: an evolution of conceptual thinking. *Climatic Change* 75 (3): 301–329. doi:10.1007/s10584 006-0329-3.

Giddens, A. 2009. *The politics of climate change*. Cambridge: Polity Press.

Gidley, J.M., J. Fien, J.-A. Smith, D.C. Thomsen, and T.F. Smith. 2009. Participatory futures methods: Towards adaptability and resilience in climate-vulnerable communities. *Environmental Policy and Governance* 19 (6): 427–440. doi:10.1002/eet.524.

Hopkins, R. 2008. *The transition handbook. From oil dependency to local resilience*. White river junction: Chelsea Green Publishing.

IPCC. 2000. *Special report on emissions scenarios*. A special report of working group III of the Intergovernmental Panel of Climate Change. Cambridge: Cambridge University Press.

IPCC. 2012. *Managing the risks of extreme events and disasters to advance climate change adaptation*. A special report of working groups I and II of the Intergovernmental Panel on Climate Change. Cambridge: Cambridge University Press.

IPCC. 2014. *Climate change 2014: Impacts, adaptation, and vulnerability*. Contribution of working group II to the fifth assessment report of the IPCC. Cambridge: Cambridge University Press.

Kates, R.W., W.R. Travis, and T.J. Wilbanks. 2012. Transformational adaptation when incremental adaptations to climate change are insufficient. *PNAS-Proceedings of the National Academy of Sciences of the United Stated of America* 109 (19): 7156–7161. doi:10.1073/pnas.1115521109.

Lindgren, M., and H. Bandhold. 2009. *Scenario planning: The link between future and strategy*, 2nd ed. Basingstoke: Palgrave Macmillan.

Mato, R.R.A.M. 2002. *Groundwater pollution in urban Dar es Salaam, Tanzania: Assessing vulnerability and protection priorities*. Eindhoven: University Press, Eindhoven University of Technology.

Martens, P., and J. Rotmans. 2005. Transitions in a globalizing world. *Futures* 37 (10): 1133–1144. doi:10.1016/j.futures.2005.02.010.

Mjemah, I.C. 2007. Hydrogeological and hydrogeochemical investigation of a coastal aquifer in Dar es Salaam, Tanzania. Gent: Ph.D. dissertation, Ghent University.

Mtoni, Y., I.C. Mjemah, K. Msindai, M. Van Camp, and K. Walraevens. 2012. Saltwater intrusion in the Quaternary aquifers of Dar es Salaam region, Tanzania. *Geologica Belgica* 15 (1-2): 16–25.

O'Brien, K. 2012. Global environmental change II: From adaptation to deliberate transformation. *Progress in Human Geography* 36 (5): 667–676.

O'Brien, K., S. Eriksen, L.P. Nygaard, and A. Schjolden. 2007. Why different interpretations of vulnerability matter in climate change discourses. *Climate Policy* 7 (1): 73–88.

Olsson, P., L.H. Gunderson, S.R. Carpenter, P. Ryan, L. Lebel, C. Folke, and.S. Holling. 2006. Shooting the rapids: navigating transitions to adaptive governance of social-ecological systems. *Ecology and Society* 11 (1): 18.

Park, S.E., N.A. Marshall, E. Jakku, A.M. Dowd, S.M. Howden, E. Mendham, and A. Fleming. 2012. Informing adaptation responses to climate change through theories of transformation. *Global Environmental Change* 22: 115–126.

Pelling, M. 2011. *Adaptation to climate change: From resilience to transformation*. Abingdon: Routledge.

Peterson, G.D., G.S. Cumming, and S.R. Carpenter. 2003. Scenario planning: a tool for conservation in an uncertain world. *Conservation Biology* 17 (2): 358–366.

Quist, J. 2007. *Backcasting for a sustainable future: the impact after ten years*. Delft: Eburon.

Quist, J., and P. Vergragt. 2006. Past and future of backcasting: The shift to stakeholder participation and a proposal for a methodological framework. *Futures* 38: 1027–1045.

Robinson, J. 1990. Futures under glass: A recipe for people who hate to predict. *Futures* 22 (8): 820–843. doi:10.1016/0016-3287(90)90018-D.

Robinson, J. 2003. Future subjunctive: Backcasting as social learning. *Futures* 35: 839–856.

Roche, P. 2012. A new possibility - An exciting future shaping our present. http://www.executivecoachingwithlpr.com/2012_08_01_archive.html. Accessed 5 December 2016.

Rotmans, J., R. Kemp, and M. van Asselt. 2001. More evolution than revolution: Transition management in public policy. *Foresight* 3 (1): 15–31. doi:10.1108/14636680110803003.

Rugai, D., and G.R. Kassenga. 2014. Climate change impacts and institutional response capacity in Dar es Salaam, Tanzania. In: *Climate Change Vulnerability in Southern African cities*, eds. S. Macchi, and M. Tiepolo, 39–55. Cham: Springer. doi:10.1007/978-3-319-00672-7_3.

Sappa, G., M.T. Coviello, G. Faldi, M. Rossi, A. Trotta, and S. Vitale. 2013. Analysis of the sensitivity to seawater intrusion of Dar es Salaam's coastal aquifer with regard to climate change. ACCDar Project. http://www.planning4adaptation.eu/Docs/events/WorkShopII/WP_Activity_2.2_def.pdf. Accessed 25 November 2016.

Swart, R.J., P. Raskin, and J. Robinson. 2004. The problem of future: sustainability science and scenario analysis. *Global Environmental Change* 14 (2): 137–146. doi:10.1016/j.gloenvcha.2003.10.002.

UN-Habitat. 2009. *Tanzania: Dar es Salaam City Profile*. Nairobi: UNION, Publishing Services Section.

URT. 2011. *Dar e Salaam City Environment Outlook 2011*. Dar es Salaam: Vice-President's Office, Division of Environment, United Republic of Tanzania.

URT. 2012. *Population and housing census: Population distribution by administrative areas*. Dar es Salaam: Ministry of Finance, National Bureau of Statistics, United Republic of Tanzania.

Van der Brugge, R., and J. Rotmans. 2007. Towards transition management of European water resources. *Water Resource Management* 21 (1): 249–267. doi:10.1007/s11269-006-9052-0.

Van der Heijden, K. 2005. *Scenarios: The art of strategic conversation*, 2nd ed. Chichester: Wiley.

Van der Voorn, T., C. Pahl-Wostl, and J. Quist. 2012. Combining backcasting and adaptive management for climate adaptation in coastal regions: A methodology and a South African case study. *Futures* 44: 346–364.
Van Notten, P.W.F., J. Rotmans, M.B.A. van Asselt, and D.S. Rothman. 2003. An updated scenario typology. *Futures* 35 (5): 423–443. doi:10.1016/S0016-3287(02)00090-3 .
Vergragt, P.J., and J. Quist. 2011. Backcasting for sustainability: Introduction to the special issue. *Technological Forecasting and Social Change* 78 (5): 747–755. doi:10.1016/j.techfore.2011.03.010.

Chapter 14
An Effective Approach to Mainstreaming DRR and Resilience in La Paz, Mexico

Juan Carlos Vargas Moreno, Enrico Ponte, Sophia Emperador and Marcela Orozco Noriega

Abstract This chapter proposes a methodology to delineate strategic priorities and identify specific locations for disaster risk management tactics that integrate a geospatial approach with qualitative and quantitative analysis. This methodology was applied to a case study developed in the mid-sized city of La Paz, capital of Baja California Sur, Mexico and serves as an example of an integrative mainstreaming disaster risk management approach. The mainstreaming process was developed in four distinct phases (technical, institutional, social and financial) that engage a diverse range of stakeholders, representing a broad range of professional knowledge and experience. This study presents the first spatial resilience evaluation in Mexico and details how a mainstreaming approach can be integrated with technocratic methods to better dimension risk and provide decision makers with a versatile and effective multifaceted tool tailored to local needs.

Keywords Mainstreaming · Geospatial analysis · Resilience strategies · Decision makers support · Disaster risk reduction

14.1 Introduction

The economic, social and environmental impacts of disasters will increase in the coming decades as the effects of climate change further exacerbate this trend. Global economic losses due to adverse natural events between 1980 and 2014 have

J.C. Vargas Moreno · E. Ponte (✉) · S. Emperador · M. Orozco Noriega
GeoAdaptive LLC, 250 Summer Street. First Floor, 02210 Boston, MA, USA
e-mail: eponte@geoadaptive.com

J.C. Vargas Moreno
e-mail: jcvargas@geoadaptive.com

S. Emperador
e-mail: semperador@geoadaptive.com

M. Orozco Noriega
e-mail: arqmarcelaorozco@gmail.com

M. Tiepolo et al. (eds.), *Renewing Local Planning to Face Climate Change in the Tropics*, Green Energy and Technology,
DOI 10.1007/978-3-319-59096-7_14

been estimated at US$4.2 trillion (Munich Re 2015). The occurrence of extreme weather patterns such El Niño of 2015/2016, one of the strongest recorded, and the World Bank's estimate of 100 million people in extreme poverty by 2030 (World Bank 2016) highlight the dangerous confluence of conditions that are predicted to take place in the near term.

These circumstances underscore why the integration of risk reduction, climate change adaptation and resilience strategies must become an urgent priority as a means to address a growing urban risk in a more effective and efficient manner.

In 2012, the World Bank (WB) presented the 'Sendai Report: Managing Disaster Risks for a Resilient Future' outlining 4 "Priorities for action" to mainstream or integrate Disaster Risk Management (DRM) and disaster and climate risk across WB activities in developing countries. This report was produced using data from 2012–2015 to assess the general progress in mainstreaming activities across WB efforts, such as: (i) in lending operations; (ii) in strategies; (iii) across global practices; and (iv) through investment in key dimensions of DRM; and (v) through partnerships.

The WB has made encouraging progress on the mainstreaming of DRM, both in terms of the quantity and type of operations. For example, annual financial contributions to DRM increased from US$3.7 billion in 2012 to US$5.7 billion in 2015. The mainstreaming approach has promoted a more favorable balance between ex ante risk management and post-disaster response financing.

Effective DRM requires attention at every step of the cycle: from disaster preparedness to resilient recovery. The WB has invested in developing multi-disciplinary expertise in key dimensions of DRM, from risk modeling and assessment, early warning and weather and climate services, and risk financing and insurance. Under the Global Practice for Social, Urban and Rural Development, and Resilience's (GSURR) leadership, the WB has advanced specialized expertise and operational capacity to offer services in these areas to member countries. The Global Facility for Disaster Reduction and Recovery (GFDRR) has supported progress through specialized thematic technical teams, as well as grant funding that has informed the design of projects and provided support for investment operations.

New tools and methods are improving how disaster risks are identified and addressed. For example, the Innovation Lab program, hosted by GFDRR, has promoted open data platforms, citizen-led data collection with smart-phones, and other innovative approaches utilizing remote sensing and geospatial hazard datasets. Such tools have increasingly been used to support policy decisions and investment allocation. The incorporation of hazard data, as well as the consideration of future climate change impacts, has increasingly informed investment projects that might otherwise fail to account for extreme events.

This chapter presents an innovative methodology that uses strong quantitative and qualitative approaches, which were informed by a robust mainstreaming process and organized in four steps. As part of this process, a participatory phase was developed involving a variety of stakeholders representing a range of professional expertise, interests and local knowledge, comprised of representatives from: multilevel government agencies, NGOs, members from the private sector, academia, and civil society. The application of the methodology is presented as a case study developed in the city of La Paz (Baja California Sur, Mexico).

14.1.1 Mainstreaming: A State of the Art Approach

With the increase of disaster related risk, particularly in urban areas, there is growing consensus among experts that the key to the lessening of risk lies in the 'mainstreaming' of disaster risk reduction into development planning and policy. The Urban Agenda recently presented at the HABITAT III Conference in Quito underlined the relevance and crucial nature of such approach, as a means "… to strengthen the resilience of cities and human settlements, using mainstreaming holistic and data-informed disaster risk reduction and management at all levels, reducing vulnerabilities and risk, especially in risk-prone areas of formal and informal settlements" (UN-Habitat 2016).

The term "mainstreaming" is derived from the concept of how small, isolated tributaries flow into the larger main-stream of a river, a seamless integration of disparate flows into a larger whole (Pelling and Holloway 2006: 16). Hence "mainstreaming risk reduction" describes a process that fully incorporates and integrates the efforts of disaster risk reduction (DRR) into larger relief efforts and development policy (La Trobe 2005). This approach aims to radically expand and enhance DRR so that it is incorporated into normal practice, and fully institutionalized within an agency's relief and development agenda. Essentially, this process merges the key principles of DRR with development goals, governance arrangements, institutional policies and practices.

When put into practice, mainstreaming requires an analysis of how potential hazard events could affect the performance of policies, programs and projects, on the other hand, it should also consider the impact of these on the vulnerable landscapes. Results from these analyses can help inform risk sensitive development, which is now widely recognized as a critical component to achieve sustainable development.

The mainstreaming process must also include or consider the effects of climate change, and incorporate additional priority interventions as appropriate. This process requires investing in climate change monitoring and forecasting (both science and policy related) as part of broader national efforts. Additional interventions such as budgeting and financing for adaptation include the integration of adaptation into national systems and the leveraging of special funding sources and modalities. Policy interventions can intervene at different levels including general measures that are reviewed with a climate lens, as well as adaptation-specific measures (UNDP-UNEP 2011).

Effective mainstreaming of DRM across multiple key sectors is vital for resilient development "because DRM is not a sector in itself, but a process to protect development progress, reduce losses and support growth" (Mackay and Bilton 2003). This process is embedded in the day-to-day operations of national and local organizations, across strategic sectors, and completed with specific resources (human, financial, technical, material, knowledge) allocated to manage potential risks.

The introduction of mainstreaming efforts can occur at all levels of governance, from its insertion into development plans, processes and initiatives at the local level (e.g. city master plans or individual infrastructure projects), to efforts at the sub-national and national levels. Risk management principals can also be mainstreamed within decision-making processes, legislation, regulations, organizational protocols and subsidy regimes. Regardless of the level of governance or administrative avenue, mainstreaming largely takes place within the core sectors related to development activities used by most governments worldwide. These include agriculture, transportation, utilities, housing, health and education, amongst others.

As a social justice-led approach to policy making, mainstreaming allows for the integration of equal opportunity principles, strategies and practices into the everyday workings of the government and other public bodies. It has three key characteristics: it is a deliberate process; there are multiple routes and/or outputs that can be targeted (e.g. policies, plans and legislation); and it should take place across multiple levels of government.

According to this definition and characterization of mainstreaming, the following three purposes can be defined:

- To make certain that all the development programs and projects are designed with evident consideration for potential disaster risks (La Trobe 2005). Currently, future risks are evaluated through separate processes, when it should be done concurrently. Therefore, there is often a basic lack of technical capacity for understanding, evaluating and acting on risk information (Oates et al. 2011).
- To make certain that all the development programs and projects do not inadvertently increase vulnerability to disaster in all sectors (social, physical, economic and environment), influence the existing development agenda (Jahan 1995) or modify existing laws and other legal provisions, as defined in the Hyogo framework of action in 2005.
- To make certain that all disaster relief and rehabilitation programs and projects are designed through a deliberate process that takes place across multiple levels of government (Pervin et al. 2013; Bettencourt et al. 2006). Currently, DRM is seen as being 'outside' of sectoral development, having low priority in core sectors, and lacking the appropriate national legislation (Schipper and Pelling 2006; Pelling and Wisner 2009).

The decentralization of DRM for local authorities is a critical element of good governance. However this is often undermined by constraints due to local institutional capacity, financing difficulties, challenges with cross-government coordination and low levels of citizen participation in risk-management activities (William 2011). Experience has shown that in order for decentralization to be effective other supporting aspects are needed, such as: "strong national leadership on DRM; mechanisms to enforce DRM policies; high levels of public awareness about risks and risk reduction; adequate technical capacity to undertake risk-reduction actions; and incentives that create strong political interest in risk-informed planning" (Scott and Tarazona 2011).

14.2 Methodology

Creating a clear 'vision' of mainstreaming approach has proved to be difficult for governments and other agencies around the World. Mainstreaming is a complex process that requires guidance by technical specialists and leaders. This can be hampered by capacity of local sectoral staff, weak institutional mandates for mainstreaming, and political pitfalls (Bahadur and Tanner 2013). In addition, the process can be obstructed by the grouping and compartmentalization of funding streams for DRM and climate change development, inadequate interest from the private sector, and the scarcity of estimates for the cost of mainstreaming. These conditions make it difficult to obtain financing for the application of mainstreaming (Becker-Birck et al. 2013). In the end, solutions and strategies are needed to ensure that sectors no longer operate without a clear acknowledgement of risk and its impact on development. This requires strong political leadership and agreement between the different stakeholders (Harris and Bahadur 2014).

These challenges serve as the starting point from which GeoAdaptive has developed its methodology.

As a means to reduce risk and encourage the development of prosperous, equitable and sustainable societies, GeoAdaptive has applied a specific mainstreaming methodology which refers to processes that include the results of risk assessments and climate change adaptation practices within current planning documents at all levels of government (e.g. local: city master plans; regional: State development plans; and national: five-year plans). Through this approach diverse risk profiles are developed, which indicate specific areas for investment and help define strategic priorities.

GeoAdaptive incorporates a robust approach that divides mainstreaming into four phases providing planners and decision makers a tool that allows them to understand climate risk and to run a cost-benefit analysis of the potential implementation of specific measures and strategies. These four phases are described as follows:

- Technical mainstreaming. The goal of this phase is to validate the results obtained during the previous risk analysis and to train local experts to best use the information that has been collected. Through this phase different risk profiles are considered and merged with accessible and timely information on the economic returns of risk-informed decisions and risk-reducing investments. Within this phase GeoAdaptive works with the technical team from the local government and other key stakeholders to interpret the information collected and the analytical results. As part of these efforts training is provided in disaster related topics, as well as other skills related to the improvement of communications within the local administration (ADPC 2010). In addition, officials are provided training in public participatory processes and multi-stakeholder dialogue that help resolve conflict over development priorities. Training programs provide the relevant risk information for local needs, encourage the respect and utilization of indigenous knowledge and promote traditional practices of coping with disaster risks (UNISDR 2015). A capacity assessment helps formulate a development strategy that will be necessary to address deficiencies, and optimize existing capacities. Within this phase GeoAdaptive also defines a first list of priority actions and recommendations with the purpose of identifying structural and non-structural resilience measures, the economic viability of future investments and the specific risks and benefits related to the implementation of these measures. This process is conducted by GeoAdaptive's technical team of experts, in coordination with local or national task groups, or units for disaster risk reduction (DRR).

- Institutional mainstreaming. Through this phase GeoAdaptive develops a detailed analysis to prioritize the most powerful actions and strategies and defines their implementation. The methodology supports the development of government-based decision processes that are based on local laws, consensus oriented, participatory, efficient, equitable, transparent and accountable. GeoAdaptive aims to increase the culture and effective practice of DRR amongst local institutional stakeholders ensuring that the local and national authorities can function adequately during risk situations. Considering that a lack of coordination and cooperation is still one of the key obstacles to achieving an effective mainstreaming process, GeoAdaptive's team encourages the active involvement of different stakeholders in the internal mainstreaming process, including the private sector and non-governmental groups (e.g. environmental or community groups). National legislatures can provide an overarching framework for risk reduction and can enable risk reduction strategies in the key ministries (ADPC 2010). Regional governments with powers to draw up locally enforceable legislation can often issue appropriate executive orders, ordinances and other directives to require departments and agencies, private companies, voluntary groups and citizens in its jurisdiction to carry out certain risk reduction actions (ADPC 2010). Aligning local and national legislation, policies and practices with global frameworks for DRR and sustainable development (e.g. HFA, MDGs, and Habitat Agenda) can generate support from international agencies for local DRR initiatives, in the form of technical advice and/or financial resources. The institutional structure should strengthen the horizontal and vertical integration of DRR between different levels of government, between various key agencies, between other stakeholders (e.g. civil society, private sector, academia, etc.) and between neighboring localities (Benson 2009). Multi-sector and multi-level communication and cooperation seems more likely to happen in many countries when the highest level of executive power, such as the Prime Minister or President oversees the coordination of DRM (Benson 2009). Many countries have established disaster management committees, with members representing different economic sectors and interests, to facilitate the mainstreaming process.
- Educational mainstreaming. GeoAdaptive's approach aims to improve the integration of science policy with risk reduction and adaptation strategies. The participatory nature of our approach amongst universities, the private sector and representatives from community organizations provides GeoAdaptive with

expert knowledge on local conditions, the opportunity to validate analytical results, to identify site-based adaptation strategies and to raise local awareness. Overall this phase begins with stakeholder validation of the results from the organizational and institutional mainstreaming phases (e.g. scenario results, forecasts from prior assessments, etc.). This mainstreaming phase is developed through different workshop, training activities and meetings, which are facilitated across the all four phases of a project.

- Financial mainstreaming. Most local governments have yet to incorporate DRR and resilience strategies into their yearly budget, although some do have contingency or calamity fund for emergency response (ADPC 2010). Moreover, in most cases, local governments are expected to dedicate budget line for DRR and resilience from existing budgets (ADPC 2010). Regardless, local government also need to explore other sources of disaster-related funding. The Sendai Framework encourages countries to mainstreaming disaster risk into their macroeconomic strategies and public finances. Integrating the mainstreaming process is a central part of building resilience, as a means of protecting and guaranteeing the livelihood of individuals and businesses, as well as the operation of the public sector institutions. To strengthen the country's economic and fiscal resilience, governments should integrate disaster risk in public finance, including the development of risk-sensitive investments and risk financing mechanisms (UNISDR 2016). The capacity of countries to develop a sound economic analysis on the benefits of DRR is a central support for investment and policy decisions conducting to resilience. GeoAdaptive's process is developed to help decision-makers better identify the best funding mechanisms for the different actions and strategies prioritized in the institutional mainstreaming phase.

Departing from the results obtained by the four different mainstreaming phases, GeoAdaptive's team can evaluate the benefits of the strategies and measures developed in the prior phases. The resulting recommendations must be complemented with appropriate monitoring systems and should enable the development of strategies at the planning, execution and evaluation stages. However, monitoring mechanisms (particularly in developing countries) are often ad hoc, inconsistent, lacking transparency, or non-existent. This presents a challenge to policy-makers seeking to institutionalize risk-informed planning through national legislation. Therefore, GeoAdaptive's final validation is obtained through the identification of goals from local authorities, an evaluation of the impacts and the development of a monitoring and evaluation process. These goals are organized in four groups:

- Implementing supportive policies: Within institutional mainstreaming GeoAdaptive recognizes the importance of strengthening political commitments, the implementation of policy and institutional frameworks that incorporate disaster risk assessment, facilitate the investment of requisite resources, and strengthen the social capital of vulnerable communities.

- Utilizing local experience and wisdom in risk assessment: GeoAdaptive recognizes the importance of citizen's experience of disasters and mitigation actions. Hence, the incorporation of citizen's experience of disaster and mitigation actions. Hence, the incorporation of traditional and local knowledge during our participative approach ensures that the process is responsive to the local conditions of the communities at risk, thereby enhancing its effectiveness.
- Basing risk assessment on information management: Starting with a risk assessment process that considers local information and data, we recognize the basic governance responsibility of providing information on potential and actual risks.
- Ensuring professional management of risk assessment systems: Ensuring professionalism in risk assessments requires investment in data, information and communication systems and in the development the requisite institutional and human capability to manage the process.

14.3 La Paz, Case Study

The following case study presents a pilot project that provided a rapid spatial analysis of the impact (both damages and losses) of hurricane Odile (September 15, 2014) on the Mexican municipality of La Paz. Hurricane Odile was the most intense tropical cyclone to affect the Baja California since 1976 (Fig. 14.1).

According to the National Center for Disaster Prevention (CENAPRED 2015) Odile caused six deaths, stranded 32,000 tourists and generated pecuniary damages close to MXN $12 billion pesos (state-wide).

The study area focused on the urban extent of the municipality of La Paz, the capital of the state of Baja California Sur. Situated northwest of Mexico City, La Paz is situated in a bay on the southeastern portion of the Baja California Peninsula with an urban population of 215,178 residents (INEGI 2010). This coastal city is located in an arid and semi-desert landscape and is the civic, educational and commercial center of the region (Fig. 14.2). Although it's the primary center of government for the state, the economic drivers of the region are focused on tourism and commerce.

The proximity, geographic isolation, and economic and infrastructural interdependencies amongst the municipalities of La Paz and Los Cabos generate an intrinsic relationship. This inter-reliant relationship was taken into consideration during the analytical phase and these results were integrated into the study as a means to identify resilience strategies at the regional level.

The analysis and final results summarized here form part of a larger effort: "Intervenciones estratégicas hacia un futuro resiliente en La Paz, Baja California Sur", developed with the support of the International Community Foundation (ICF) and the Inter-American Development Bank (IDB).

Fig. 14.1 Damage caused by Hurricane Odile in La Paz, Mexico (*Source* www.lacrónica.com, mmspress.com.mx)

14.3.1 Project Need

The impact of Hurricane Odile underscored the need for a holistic and systematic review of the effects a large natural hazard could have upon the region and how the city could become more resilient from a physical, social, economic, and institutional perspective. The development of such assessment allowed city officials to rethink how they can adapt and prepare for the effects of climate change and natural hazards. This additional information can help move towards a new model of sustainable urban development that increases the city's regional competitiveness, while maintaining a high quality of life, for which La Paz is known. Within this context, the study evaluated the impacts caused by Hurricane Odile and provided an analysis of the capacities to prevent and respond to extreme events in La Paz, as a means to formulate strategies of resilience.

The study aims to: (1) assess the impact of Hurricane Odile and the city's capacity to cope with and respond to disasters, and (2) identify a series of strategic interventions that increase the long term resilience of the city.

Completed in a two part structure, the first part relied on geospatial assessment that identified the intensity of natural hazards (flooding, hurricanes, earthquakes,

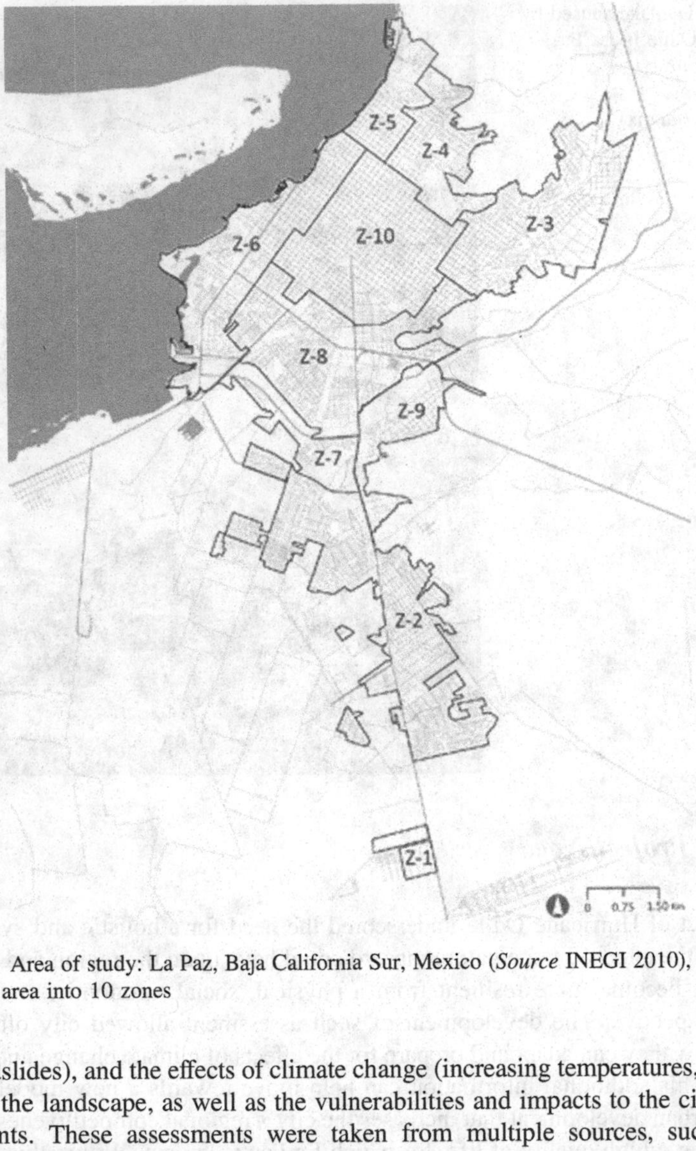

Fig. 14.2 Area of study: La Paz, Baja California Sur, Mexico (*Source* INEGI 2010), division of the study area into 10 zones

and landslides), and the effects of climate change (increasing temperatures, sea level rise) on the landscape, as well as the vulnerabilities and impacts to the city and its inhabitants. These assessments were taken from multiple sources, such as the studies conducted for the city of La Paz, as part of the IDB's Emerging and Sustainable Cities Initiative (ESCI), the risk atlas, as well as other official and academic reports. The corresponding results served as a comprehensive base for further analysis aimed to understand the underlying issues that exacerbate the vulnerability of the city and guide the delineation of actions to reduce these, while increasing the long-term welfare and well-being of the population.

The second part of the project focused on the spatial evaluation of the urban resilience post-natural disaster. As a pilot project the study lacked comparable references with respect to the evaluation method or with the development of recommendations. Therefore added value of this study lies in the introduction of the spatial dimensioning of the analysis and in the evaluation of both the constraints of social resilience and of the municipal infrastructure (Fig. 14.3).

The spatial evaluation allowed stakeholders to visualize and geographically identify the extent and intensity of multiple hazards, the vulnerability of the city's social institutions and infrastructure, as well as the overall impacts. Through this process, a first evaluation of spatial resilience was developed that specifically considered the city's management and adaptation capacity, the response actions and the reconstruction of diverse sectors. The resulting analysis allowed the definition of priorities and specific actions that have the potential to increase resilience in the city and generate a more sustainable future. The development of the resilience assessment was highly dependent on a participatory process, with contributions from local experts (national, state and local level). As a means to organize this component a "Resilience Committee" was formed, composed of key representatives from municipal, state and federal agencies, civil society, academia and the private sector. These select participants contributed with their knowledge, insight and experience throughout the project to validate data, evaluations and proposals prepared by the project team.

This chapter focuses on the second part of the study and the application of the geospatial resilience assessment and the participatory process that guided the identification and refinement of actions and resilience strategies. The overall process was guided by the following steps:

1. Evaluation (14.3.2). Resilience Analysis. The baseline evaluation of the city is the first step, comprised of a series of geospatial assessments that include: a

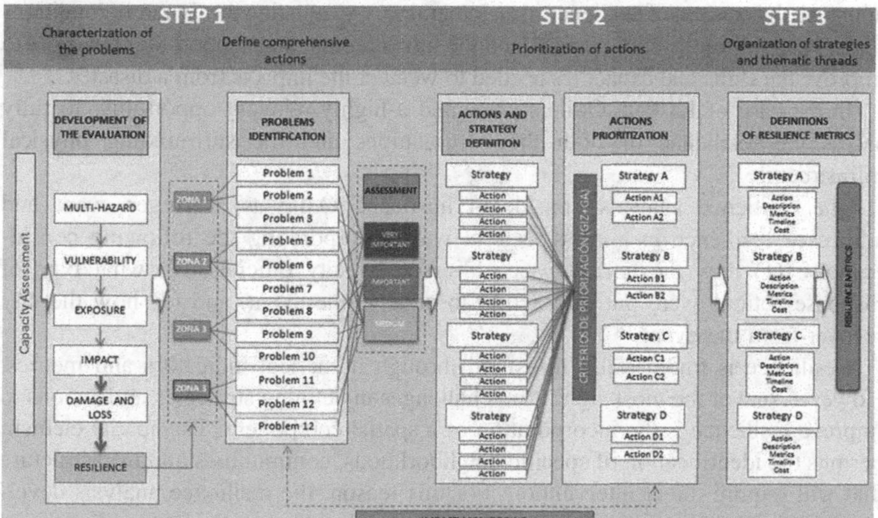

Fig. 14.3 Methodological framework for the definition of actions and strategies

multi-hazard assessment generated by different sources of information to identify areas of greatest latent danger; a vulnerability assessment (physical and social); an exposure assessment (physical and social); the impact analysis of Hurricane Odile, including the identification of damages and economic losses; and a resilience evaluation, including disaster prevention and response capacities in the city.

2. Identification and appraisal of challenges (14.3.3). Technical Mainstreaming. Having identified the baseline conditions the key challenges are identified based on the predetermined geographic organization of the city (zones).
3. Definition and prioritization of actions (14.3.3). Institutional Mainstreaming. Specific actions are defined for each challenge and each is evaluated through a series of filters and prioritization criteria. The objective is to identify the key, priority actions that will contribute the most to the development of integral, multi-sector resilient strategies.
4. Organization of actions by thematic grouping and definition of strategies (14.3.3). Institutional and Educational Mainstreaming. Once all the actions have been evaluated by the prioritized they are organized into specific thematic groupings based on implementation synergies, interdependencies and other interrelated benefits.

14.3.2 Resilience Analysis

Exposure and impact analysis allow the dimensioning of natural hazard effects. However, they do not indicate whether a city has the capacity to resist, adapt, cope and recover from the negative effects of that hazard. For this reason, it is necessary to approach the concept of resilience in an integral way, evaluating the dynamic conditions of the behavior of the communities and the infrastructure that support them in order to improve the skills and capacities needed to weather the impacts from a disaster.

In the case of La Paz, Odile's presented a highly relevant opportunity to fully assess the resilience of both the communities and the surrounding physical infrastructure.

The resilience analysis considered hurricane Odile as a trigger event and addressed resilience in a cross-sectoral way, incorporating the following considerations: (1) how the hazard was addressed by city officials, (2) what type of response actions were put into action to address the event, and (3) how the city recover from the event.

Resilience is traditionally measured through numerical indicators and metrics. However, one of the most significant challenges in defining strategies and actions to improve resilience is the incorporation of a spatial component. The spatial element permits the identification of specific neighborhoods, communities and infrastructure that will require direct intervention. For this reason, the resilience analysis developed for La Paz has the distinctiveness of being designed to explore resilience not only through quantitative methods but also through the incorporation of a geographic component. Through this analysis, specific areas of high or low resilience can be identified and directly associated with specific conditions or triggers. This particular approach, was shaped on the resilience evaluation by Cutter et al. (2008), present resilience as a dynamic process dependent on antecedent conditions, the disaster's severity, time between hazard events, and influences from exogenous factors.

The analysis developed was based on a geographical database generated from information obtained from official sources (INEGI) and supplemented with data obtained from secondary sources, prior evaluations and interviews with local officials. The data collected was systematized in a Geographic Information System (GIS), creating a digital environment for the assessment. The geostatistical evaluation was elaborated through an analysis of geographical spatial statistics, which required the segmentation of the city in a grid of 50 × 50 m cells. Each cell was assigned a numerical value that represented the factor being evaluated. The use of geostatistical analysis permitted the identification of a resilience value and therefore pinpoint areas of lower resilience, as well as the factors or sectors that structurally contributed to the inherent resilience of the city and its communities. Additionally this approach allowed for the creation of a general spatial index of resilience capacities for each sector, allowing the visualization of the behavior and spatial distribution of the resilience dynamics.

Each zone in the city was further explored based on the population's access to health, emergency and support services (e.g. hospitals, shelters, etc.). The accessibility assessment was developed through the analysis and quantification of spatial networks, which analyzed the route and travel time from each housing area in the city to the specific points identified for emergency and support services. The final product identified the residential areas with greater difficulty to access support and emergency services during an emergency. For example, the results in La Paz show that the southern communities have the greatest difficulty accessing or receiving emergency support.

When comparing the maximum and minimum values from the results of the resilience capacities spatial analysis, the emergency infrastructure and the health service sector resulted in the lowest values for the following reasons:

- Medical centers have a low capacity (number of beds and doctors per capita).
- Shelters lack sufficient medical capabilities.
- Access to health services and shelters are limited in flood scenarios.

Although the city provides ample coverage of basic services (77.91% of the population has access to electricity, water and sewer service according to INEGI), accessibility and coverage in the southern communities of La Paz is a priority issue. The access to health and emergency services is dire especially in flood scenarios due in most part from the potential isolation of the zone due to its geographic location south of the city center and the lack of services to support this part of the city.

A map for each of the services analyzed was created to visualize the results of the spatial-sectoral resilience capacity analysis: spatial patterns clearly show that the

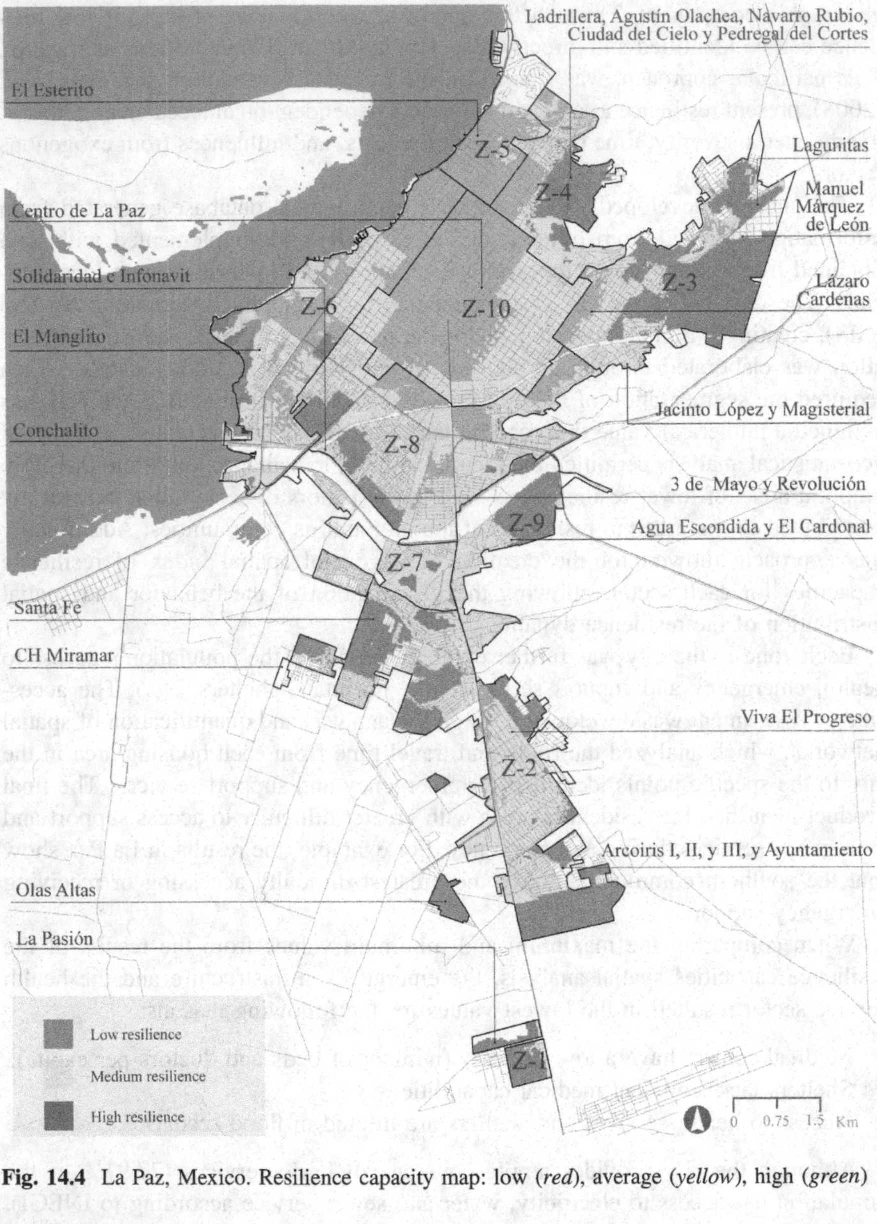

Fig. 14.4 La Paz, Mexico. Resilience capacity map: low (*red*), average (*yellow*), high (*green*)

locations with the least capacity for resilience are located in peripheral areas of the city. The city center (Z-10) presents the conditions of greatest resilience for urban development (Fig. 14.4). In contrast the peripheral areas (Z-1, Z-3, Z-4, Z-9) have the lowest resilience values. It is important to consider this evaluation, given the urban growth pattern of the city, which has been expanding during the last twenty years, from a compact urban model to one of urban sprawl.

Authorities should consider these resilience patterns, in addition to other factors (e.g. provision of services, environmental impact, etc.) and the consequences that arise if they continue to allow the city to grow in this manner. On the other hand, they must also comply and enforce adequate construction standards and environmental measures necessary to maintain and raise the level of resilience. For example Zones 1, 3, 4 and 9 are shown to have the lowest capacity to prevent and respond to emergencies. During hurricane Odile zone 9 (Agua Escondida) experienced few impacts however, it received low resilience value for it mostly comprised of informal and marginalized communities. Zones 1, 3 and 4 registered the highest number of impacts to housing, as they are in multi-hazard areas and lack access to emergency services, which means these communities are in high impact areas and have little capacity to cope.

The city center (Z-10) has a high capacity to cope with events caused by natural hazards. Despite the damages to the road networkin this zone during Odile, its access to basic and emergency services remained intact, for it is not in a high multi-hazard area and all of its sectors have an integrated prevention and response actions plan. On the other hand, zones 5 and 6, which include areas with the majority of the hotels and services of the city, proved to have a medium to high response capacity. It should be noted that the study is based on the impacts and response actions to hurricane Odile, and in this case, there was no significant damage in these zones. However, this area is highly vulnerable to both storm surge and coastal flooding.

14.3.3 Mainstreaming Analysis

The main challenges identified for the city were organized and evaluated considering the results of the resilience analysis, which included: the multi-risk analysis, the resilience analysis, the previous works completed for the ESCI initiative and the impact assessment for hurricane Odile. Additional analysis was done to evaluate the priority areas identified as part of the resilience analysis and the priority areas identified as part of the resilience analysis and the priority areas identified by the local government and civil society groups. Different layers of information were analyzed to observe the spatial overlaps that occur between the different perspectives and to identify the neighborhoods (*colonias*) that should be considered priorities, having incorporated broader perspectives from different local constituencies. Once identified, these prioritized areas were linked to the geo referenced damages caused by hurricane Odile in order to verify the most prone areas. Of the 109 specific impacts reported to road network, 2 occurred in vulnerable areas mapped by the city council, and 8 occurred in vulnerable areas reported by civil society.

The subsequent actions and strategies developed were identified through the previous mentioned methodological framework.

Identification and Appraisal of Challenges-Technical Mainstreaming
In order to better understand and explore the role of resilience planning can have upon the city the most important and critical challenges were identified. These challenges were taken from the different components from the resilience analysis as well as directly from city officials and key civil society groups. As a result 22 challenges were identified in addition to the geographic reach of each issue. The challenges were categorized into two main groups: problems specific to a zone or area of the city (e.g., zone 10, or neighborhood X), or problems that affect the entire city (e.g. water shortage). Through the examination and analysis of these spatial patterns, the relationships with each geostatistical zone of the city were explored.

Definition and Prioritization of Actions—Institutional Mainstreaming
For each of the 22 challenges specific actions were developed to address the issues at hand. In preparation for this exercise, an extensive review of projects and prior evaluations of the related issues was carried out to generate a list of actions or interventions vetted by previous studies. A total of 81 strategies, actions or interventions were noted, organized by sector and reviewed with the Resilience Committee for further verification.

Using the information gathered for the challenges, geographic and quantitative, as well as the corresponding strategic interventions, a total of 29 actions were generated in response to the 22 challenges. During this identification and evaluation process, the team used the information collected during multiple workshops with local stakeholders and meetings held with the Resilience Committee. The final actions were identified based on the following set of considerations described below, which would facilitate the implementation and feasibility of the proposed actions.

- the level of vulnerability in the location where the action is being proposed and its relationship to other hazards
- the effects of climate change and other hazards on the action
- clear identification of the stakeholders that should be involved or the relationships they have with other implementing organizations
- the monitoring and evaluation processes for the identified actions
- direct relationship or link with other identified actions or current efforts
- geographic scale of the action (some measures need to be applied at the city scale, others at a site scale)
- length and duration of the implementation phase of the action

The list (Table 14.1) represents an example of the challenges and corresponding actions, however the final list was further refined through a set of filters in order to identify the most relevant actions to help improve the resilience of the city.

Once the actions were defined, characterized by sectors geographical located, a prioritization scheme based on a multi-criteria analysis was developed to identify the most relevant actions to implement as part of the resilience strategy. A total of 13 criteria were used for this adaptation prioritization phase. The first 10 key criteria from the criteria of the prioritization framework used by GIZ (2015) called "Methodology of prioritization of adaptation measures" the team selected 10 key

Table 14.1 List of problems organized by thematic axis with their corresponding actions

Category	Challenge	Action
Energy	15 Electric distribution system is isolated from national lines, its design makes it vulnerable to hydro-meteo events	Upgrading electric transmission and distribution system
	16 Lack of alternative energy sources	Promote and implement alternative energy generation in safe areas
Communication and transportation	17 Primary road in flood prone areas	Causeway construction
	18 Fiber-optic cable prone to hurricanes and floods	New fiber-optic undersea cable from Topolobampo to La Paz
Natural areas	19 Lack of regulation for urban forestry location and maintenance	Landscaping strategy to reduce the heat island effect and power demand
	20 Tree canopy loss increase heat island effect and vulnerability to natural hazards	Green corridors incorporating greenway
	21 Coastal dunes alteration increase flood and erosion threats	Coastal dunes restoration/stabilization, and prevention of salt water intrusion
	22 Mangrooves and wetlands loss expose the city to storm surge and wind	Coastal protection plan incorporating mangroves/wetlands protection, flood, erosion and salt water intrusion prevention

criteria were adapted from the prioritization framework used by the German Agency for International Cooperation (GIZ) "Methodology of prioritization of adaptation measures" (2015). The rating scale for each criterion was taken from the descriptions outlined in the GIZ's methodological documentation.

The base evaluation criteria were supplemented with three additional filters developed by the GeoAdaptive (GA) team in order to further enrich the evaluation, described as follows:

(1) Resilience filter

- Criteria: The action increases the resilience of the city and minimizes its vulnerability
- Evaluation: Contributes slightly (1), medium (2) or substantially (3)

(2) Financial feasibility filter

- Criteria: The general estimated implementation cost of the action, taking into account projects of a similar scale and complexity.
- Evaluation: $ (1), $$ (2), $$$ (3)

(3) Technical and political viability filter

- Criteria: The action is viable taking into consideration the technical knowledge available in the city and the support of politicians, based on similar projects in the city.
- Evaluation: not viable (1), moderately viable with additional support (2), viable with the technical knowledge available to the city (3)

This final list of 13 criteria was discussed and validated with the Resilience Committee to ensure that the most vital factors were taken into consideration in light of the committee's perspective regarding the importance and applicability of the actions.

All filters, both GIZ and supplementary formulations, were evaluated with an equitable weight.

For the final evaluation, the values of the 13 filters were weighted using the following equation:

$$A = \frac{1}{n} \times \sum_{t=1}^{n} x_{\mathrm{i}}$$

Where:

A is the average value of the performance evaluation
Σ is the sum of all values
n is the number of variables
X_i is the value of each individual variable in the criteria list
n is the number of terms or items evaluated

The evaluation ranks were as follows:

- Between 0 and 1 = Low priority
- Between 1 and 2: Medium priority
- Between 2 and 3: High priority

Several of the final actions are related to urban planning initiatives such as water resource planning, coastal protection and disaster risk reduction strategies. Therefore, through this process, several actions were designed and prioritized to ensure the distribution amongst key sectors, as well as the responsible entities of implementation. Due to this objective, the responsibility for implementing the proposed actions does not lie exclusively with government entities, but involves other stakeholders, including civil society, the private sector and academia.

The application of the filters yielded a total of 20 prioritized actions from of the original group of 29, with actions prioritized for each of the key sectors. The top 20 actions scored evaluations ranging from 2 to 3, or high priority. These were further analyzed to explore their potential distribution with respect to the sectorial categories associated with each of the problems. It was encouraged that decision makers in La Paz evaluate the incorporation of these actions into existing efforts, given their preventative nature and high prioritization score (Table 14.2).

Organization of Actions by Thematic Grouping and Definition of Strategies-Institutional and Educational Mainstreaming

The delineation of multi-sectoral strategies aimed to group or organize the prioritized actions into larger thematic groups that allowed the identification of synergies between the actions, sectors and stakeholders that can facilitate their implementation in a more efficient manner. Each strategy for La Paz was developed considering the involvement of various stakeholders and the linking of different institutions that would contribute to the achievement of the objectives set out by these thematic groups. Likewise, the strategies ensured that there were measures and activities at a variety of scales that assured that there were challenges identified at the local, urban and regional scale. This aspect was particularly important in La Paz because of its location on a peninsula, making it vulnerable to large hydro-meteorological events since it relies on basic infrastructure that is regionally operated (energy, water, transportation).

It was equally important that the multi-sectoral model explicitly indicate the responsibilities of each stakeholder involved. In order to achieve this, each action was linked not only to the various municipal, state and federal government institutions and private and academic stakeholders but also to the tools, actions and projects already proposed or underway, such as those indicated by the PIMUS (Plan Integral de Mobilidad Urbana Sustentable), the PACC-LAP (Plan de Acción de Cambio Climático - La Paz) or proposed by the GIZ.

The prioritized actions were grouped around specific thematic groups by adhering to the following the principles:

- synergy among the actions within the same thematic group
- collaboration between stakeholders and institutions
- sectoral interdependence within the same thematic group
- capture and distribution of shared benefits
- extent or duration of implementation within the same thematic group (some actions are short term, while others are long term)

As a result, 5 key thematic groups were defined, providing a comprehensive base to organize the prioritized 20 actions. The 5 thematic groups are:

- 1: territorial management of natural hazards,
- 2: provision of urban infrastructure and services
- 3: strengthening social and business participation
- 4: institutional risk management
- 5: Integrated management of natural resources

14.3.4 Results

The Roadmap and the Resilience metrics were two tools developed as part of the study with the aim to guide and assist the implementation of the prioritized actions.

Table 14.2 Example of the application of filters to the actions identified for the city of La Paz

		Multi-criteria													
		GIZ										GA			
		Transversality	Coordination: stakeholders/sectors	Feasibility	Flex. /"no regrets"	Coordination: ecosystems	Sustainability	Attention to vulnerable population	Active participation	Strengthening adaptive capacity	Evaluation and Feedback	F. Technical viability	F. Financial viability	F. Resilient	Total
#	Actions	1	2	3	4	5	6	7	8	9	10	A	B	C	
	URBAN PLANNING AND RISK MANAGEMENT														
1	Include technical standards that integrate hazard, vulnerability and risk maps into plans in order to restrict development in high risk areas and identify alternative programmatic or land uses.	3	3	2	2	3	3	3	1	3	3	3	3	3	3
2	Relocate housing that is located in unmitigable high risk areas to low risk zones, with access to basic services.	2	3	3	3	1	2	3	2	2	3	3	2	3	3
3	Channelize and install flood protection infrastructure to prevent the flooding of nearby housing and key infrastructure.	3	3	3	3	1	2	3	3	3	3	3	2	3	3
4	Build stabilization projects on the edges of the El Trinfo and El Cajoncito streams to reduce erosion and protect the streambeds.	1	3	3	2	1	1	3	3	2	2	2	1	1	2
5	Develop a public awareness and education campaigns to deal with prevention and preparation when faced with the threat of natural hazards.	3	3	2	3	3	3	3	2	3	3	3	2	3	3

They contributed to the implementation by defining the general scope of the prioritized actions within a strategy, the dimensioning of activities, approximate costs, and the potential impacts of each action.

The implementation success of these actions will depend on the coordination and collaboration between stakeholders and the potential need to create temporary alliances to facilitate the process.

Given that resilience depends on a continuous process between government, private sector and civil society, it will require the development and sustainment of a continuous dialogue where adjustments and improvements in the prioritized actions are discussed and explored given the current political, financial, and temporal context as well as environmental conditions.

The Road Map

The Road Map outlines the prioritized interventions or actions, which were organized into five thematic strategies (Table 14.3). Specific resilience metrics were identified for each action that can be used to monitor the implementation of the action, the period of implementation, the expected impact, and the approximate implementation cost for each project.

As a complement to the Road map, a locator map was included that spatially identified the prioritized actions. However, not all actions were identified geographically because some are programmatic in nature or refers to training and should be applied or available city-wide and were not specific to a particular area or geographic location.

The approximate projected costs were provided for each action as a reference. These were determined through research of other similar projects in the region or Mexico and by interviews with professionals or government representatives. However, looking at the distribution of the financial burdens and responsibilities of the programmed actions, it is noticeable that the urban management strategies represent the most relevant and widespread of the projected investments.

The average execution period for the actions varies; however, most fall within the medium term (approximately 1 year), providing an opportunity for the identification of financing mechanism, and could be initiated soon after the completion of the study in order to take advantage of the momentum generated from the project, maximizing its potential for implementation.

It should be noted that the distribution of the financial costs associated with the execution of the actions was also carefully considered in this study.

32% of the final actions were identified as requiring Municipal level intervention. More than 50% of the 5 actions prioritized under the "Urban Planning and Risk Management" strategy also required financial support at the municipal level. This cost, although a large undertaking for the Municipal government, represents an opportunity for local authorities to demonstrate leadership in the public sector and improve the internal capacities of the municipal staff.

Given the fact that the initial stages of the proposed projects will represent a greater use of resources, the municipal government may consider requesting assistance in terms of management of the respective federal authorities, or request technical and economic assistance through international cooperation. The lowest

Table 14.3 Road Map–Summary table of the costs for the 5th thematic axis: "Integral management of natural resources"

	STRATEGY 5	**Theme: Integrated management of natural resources** Promote the management of natural resources for the mitigation of hazards and the long-term sustainability of the region.					
	Action	Resilience Measure	Date: 6 months	Date: 1 year	Date: 5 years	Impacts	Aproximate Cost (MXN pesos)
A18	Develop a water resource management plan to manage drought conditions for the agricultural sector and coordinate additional activities, recommendations and training for farmers.	M3	X	X		Increased efficiency of water consumption for agricultural activities. Diversification of agricultural products based on water consumption.	$24,000,000
	Establish a training program for farmers to implement best management practices (BMPs).						
A19	Develop an Urban Landscape Ecology Strategy integrating ecology with urban planning.	M3, M6, M14 M15	X	X		Reduction of heat island effect. Increased infiltration of runoff.	$2,400,000
	Identify species of vegetation and trees that are resistant to winds and drought conditions to be more resilient to climatic changes						
A20	Design and implement a series of green corridors.	M3, M6, M14 M15	X	X		Minimizing dependence on air conditioning, affecting overall energy consumption.	$ 30,000,000
	A green infrastructure system across the city that can improve the urban environment.						

economic costs are represented by the "Strengthening social and business participation in emergency management" with MXN$598,856 pesos and the "Provision of urban infrastructure and services" with 2,940,000 pesos. These two categories represent an area of opportunity to increase the city's resilience, since they can help create change in the behavior and culture of La Paz's residents through education and training programs. Other strategies such as the "Integrated management of natural resources" (MXN$56,400,000 pesos), have a much higher implementation costs due to the complexity and scale of the efforts required.

Resilience Metrics

It is fundamental to improve and guide the efforts of decision-makers and leaders of civil society to manage and assess the implementation and impact of resilience focused projects. Monitoring and evaluation systems are essential elements for assessing the performance of actions and interventions in the light of minimizing risk and improving resilience. As a response, a series of resilience metrics were designed to evaluate of the state of the city in reference to resilience and measure the effect of current and future interventions.

More than 300 existing indicators for resilience, vulnerability and risk were examined through a literature review (including global systems and tools available to decision makers, for example indicators from UNISDR, IDB, World Bank and the Rockefeller Foundation). Indicators were selected and developed with quantifiable and evaluable metrics from easily accessible and periodically generated data. The final set of indicators were selected considering expert technical knowledge and the following characteristics:

- Applicability to local conditions in La Paz and other similar cities in the region
- Prevalence in the literature
- Indicators that address multiple components of urban resilience (including risk)
- Ability to be supervised by the Resilience committee
- Relevance to infrastructure and social capacities to address threats
- Availability of data needed to support indicators in La Paz.

A total of twenty-one resilience metrics were developed to form the resilience assessment matrix. These indicators cover both the socio-organizational capacities and the functioning and performance of infrastructure and urban systems, in relation to the occurrence of natural hazards.

The metrics are divided by topics to facilitate its application:

- 1: Access and availability to basic services and infrastructure.
- 2: Health and social welfare.
- 3: Risk management and emergency management.
- 4: Damage and impacts in emergencies.

In order to facilitate the indicator evaluation, a traffic light assessment was developed, using a benchmark scale generated from global (World Bank) and local standards (reference indexes in Mexico and in Baja California Sur). The proposed metrics allow the use of both statistical and geographical information periodically provided by official institutions such as INEGI or the city government, which generate the necessary information for evaluation in a reoccurring and financial sustainable way. The traffic light assessment facilitates the evaluation through a clear and simple system (light grey, grey, black) that can be easily interpreted and would provide a baseline or reference for future evaluations (Tables 14.4 and 14.5).

Table 14.4 Classification of the traffic light assessment

Meets or exceeds resilience and sustainability standards	(light grey)
Has gaps in its resilience measure	(grey)
Below the minimum standard	(black)

Table 14.5 Resilience metrics focused on the access and availability to basic services and infrastructure

No.	Resilience Measure	Description	Thresholds A			La Paz BCS
			Light Grey	Grey	Black	
M1	Potable water coverage	Percent of houses with access to potable water.	80–100%	60–80%	<60%	97.9% ICES 2012
M2	Efficiency in water use	Water consumption per capita	55–90	37–55 y 90-110	<37 y >110	82.54 m3 / person / year ICES 2012
M3	Availability of water resources	Net water balance (internal and external resources)	>0 MMC/ year	<0 MMC / year, with conservation plans	<0 MMC / year, without conservation plans	<0 million m3/year
M4	Electricity coverage	Percent of houses with electric service.	90–100%	70–90%	<70%	98% ICES 2012
M5	Technology and Communication Coverage	Percentage of houses with cell phones.	90–100%	80–90%	<80%	74% ICES 2010
		Percent of houses with internet.	50–100%	40–50%	<40%	36% ICES 2010
M6	Percent of managed runoff.	Percent of vegetation with respect to the urban footprint	>40%	30–40%	<30%	N/D

14.4 Conclusions

The planning and development of a resilient society requires a complex and highly integrated approach. Establishing processes and analysis that identify timely and tangible strategies that respond to the challenges posed by climate change and natural hazards continue to be an unknown issue for many cities.

Multiple discussion points emerge when carrying out a study of this nature. The methods, approaches, and evaluation instruments used are rapidly evolving. On the other hand, studying resilience in the context of the recent impact of a hurricane such as Odile, poses even more significant challenges because of the availability of information and data that is sensitive to social, environmental and economic constraints.

The participation of local stakeholders is fundamental to the success of this process, particularly in the institutional and educational mainstreaming phases. The development of a participatory and knowledge-generating process rather than an advisory one is a critical distinction that must be highlighted.

A stakeholder analysis is a crucial first step to identify influential participants and key stakeholders of the city, who can form part of a working consultative group (Resilience Committee). The formation of the group can help ensure the early involvement and support from key members of the city, as well as a better transition between recommendations and the implementation of projects to ensure a "live" final product. The members of the group should be key members representing the different levels of government, the institutional organizations from each sector involved, as well as from civil society and the private sector. It is necessary to emphasize the level of commitment required by the Committee at the beginning, throughout and after the study to ensure the validity and follow-up of the implementation and evaluation processes. The formation of this group is essential to the development and improvement of the local capacity of local resilience experts in the city (technical mainstreaming).

In addition to local stakeholder knowledge it is fundamental to include and build upon the knowledge and results from prior studies and efforts. The incorporation of existing technical information and data helped the alignment and contextualization of the results and the formulation of actions that are verified with previously validated information (technical mainstreaming).

The compilation of georeferenced data (GIS, .kml, etc.) was essential for the spatial assessment, the visualization of challenges and proposals, and the analysis of resilience capabilities. It is recommended to coordinate and train collaborators from various local institutions in the collection of information through the use of free platforms such as Open Street Maps and Quantum GIS. This will allow their involvement from the beginning of the study, as well as the generation of local capacities.

The experiences gained in the area of resilience and risk management reveal that no sector or entity is capable of addressing all aspects of response and adaptation. Resilience and risk management are multi-sectoral issues that require a systemic and holistic perspective. This requires the development of processes and analyzes that recognize the interdependencies between sectors and formulate actions that encompass and motivate a multi-sector, transdisciplinary and multi-institutional collaboration. Key to achieving this goal is to have representatives from different sectors in discussions and analysis that recognize their interactions and synergies.

References

ADPC. 2010. Mainstreaming disaster risk reduction. Book 4, urban governance and community resilience guides. Bangkok: ADPC.

Bahadur, A., and T. Tanner. 2013. Policy climates and climate policies: Analysing the politics of building urban climate change resilience. *Urban Climate* 7: 20–32.

Becker-Birck, C., J. Crowe, J. Lee, and S. Jackson. 2013. *Resilience in action: Lessons from public-private collaborations around the world.* Boston, US: Meister Consultants Group Inc.

Benson, C. 2009. *Mainstreaming disaster risk reduction into development: Challenges and experience in the Philippines*. Geneva: ProVention.

Bettencourt, S. et al. 2006. Not if, but when: Adapting to natural hazard in the Pacific Islands region: A policy note. *World Bank Policy Note*. Washington, D.C.: World Bank.

CENAPRED. 2015. Ciclones tropicales (Huracanes). http://www.atlasnacionalderiesgos.gob.mx/index.php/riesgos-hidrometeorologicos/ciclones-tropicales-huracanes.

Cutter, S., L. Barnes, M. Berry, C. Burton, E. Evans, E. Tate, and J. Webb. 2008. A place-based model for understanding community resilience to natural disasters. *Global Environmental Change* 18 (4): 598–606. doi:10.1016/j.gloenvcha.2008.07.013.

GIZ. 2015. Propuestas de temas y proyectos GAU prioritarios. Anexo 6.

Harris, K., and A. Bahadur. 2014. The case of Vanuatu: Integration of disaster risk management, environment and climate change. In *Vanuatu: Integration of disaster risk management, environment and climate change in Vanuatu*. London: ODI.

INEGI. 2010. Censo de población y vivienda 2010 del Instituto Nacional de Estadística y Geografia.
Jahan, R. 1995. *The elusive agenda: Mainstreaming women in development*. London: ZED Books.
La Trobe, S. 2005. *Mainstreaming disaster risk reduction: A tool for development organisations*. Australia: TEARFUND.
Mackay, F., and Bilton, K. 2003. Learning from experience: Lessons in mainstreaming equal opportunities. Scottish Executive Social Research.
Munich Re. 2015. NatCat SERVICE Loss database for natural catastrophes worldwide.
Oates, N., et al. 2011. *The mainstreaming approach to climate change adaptation: Insights from Ethiopia's water sector. ODI Background Note*. London: Overseas Development Institute.
Pelling, M. and A. Holloway. 2006. Legislation for mainstreaming disaster risk reduction. Middlesex: Tearfund archive www.preventionweb.net/english/professional/publications/v.php?id=1198. Accessed: September 27 2013.
Pelling, M., and B. Wisner. 2009. Reducing urban disaster risk in Africa. In *Disaster risk reduction: Cases from urban Africa*, ed. M. Pelling, and B. Wisner. London: Routledge.
Pervin, M., et al. 2013. *A framework for mainstreaming climate resilience into development planning*. London: IIED.
Schipper, L., and M. Pelling. 2006. Disaster risk, climate change and international development: Scope for and challenges to integration. *Disasters* 30 (1): 19–38. doi:10.1111/j.1467-9523.2006.00304.x.
Scott, Z., and M. Tarazona. 2011. Study on disaster risk reduction, decentralization and political economy: Decentralization and disaster risk reduction. Global Assessment Report 2011. Geneva: UN Office for DRR.
UNDP-UNEP. 2011. *Mainstreaming climate change adaptation into development planning: A guide for practitioners*.
UN-Habitat. 2016. Urbanization and development: Emerging futures. In *World cities report 2016*. Nairobi: UN-Habitat.
UNISDR. 2015. *Sendai framework for disaster risk reduction 2015-2030*. Geneva: UNISDR.
UNISDR. 2016. Panel 3 Disaster risk reduction in public finance planning and investment, Reunión ministerial y de autoridades de alto nivel sobre la implementación del marco de Sendai para la reducción del riesgo de desastres 2015–2030 en las Américas. http://eird.org/ran-sendai-2016/eng/sessions/session-1/S1P3-Session-1-Panel-3.pdf.
Williams, G. 2011. Study on disaster risk reduction, decentralization and political economy: The political economy of disaster risk reduction. In *Global assessment report 2011*. Geneva: UN Office for DRR.
World Bank. 2016. *Shock waves, managing the impacts of climate change on poverty*. Washington DC: The World Bank.

Chapter 15
Possible Impact of Pelletised Crop Residues Use as a Fuel for Cooking in Niger

Stefano Bechis

Abstract One of the main causes of deforestation in Africa is the cutting of trees for cooking purposes. Even though gas and other fuels are available, in theory, their diffusion is slower than the increase of the population that continues to use wood as a fuel. The trends for the near future show that, in many areas, the forests will disappear and expose the soil to desertification. An intermediate solution to address the deforestation problem can be the use of other biomass sources, different from the traditional wood sticks, for cooking. In this paper, a system that uses agricultural residues and tree prunings, instead of wood, as a fuel, is introduced. The work done consisted in setting up a pellet making facility, and a gasifier stove. The results obtained by the pellet fuelled stove, show that its efficiency is higher than that of the improved cook stoves that use wood, and that, in addition, the stove does not produce smoke during normal cooking operation, making its use safer for the operators and other persons close to the fire.

Keywords Woodfuel · Deforestation · Gasification · Residues biomass

15.1 Introduction

During the last three decades, worldwide, the total number of people using biomass for cooking, around 2.7 billion, remained the same. The percentage of the world population that uses biomass for cooking purpose has been decreasing, and this might be regarded as a fairly positive datum, but this reduction is only the result of the same number of people relying on firewood, divided by the rising total number of inhabitants on planet Earth (Fig. 15.1). As a matter of fact, the expected switch to fuels other than wood, in rural areas of Developing Countries, is still not happening, on a global scale. At continental scale, it can be noted that in Africa there is a rising number of people relying on biomass (Fig. 15.2), while in the Western Pacific

S. Bechis (✉)
DIST, Politecnico and University of Turin, Viale Mattioli 39, 10125 Turin, Italy
e-mail: stefano.bechis@unito.it

M. Tiepolo et al. (eds.), *Renewing Local Planning to Face Climate Change in the Tropics*, Green Energy and Technology,
DOI 10.1007/978-3-319-59096-7_15

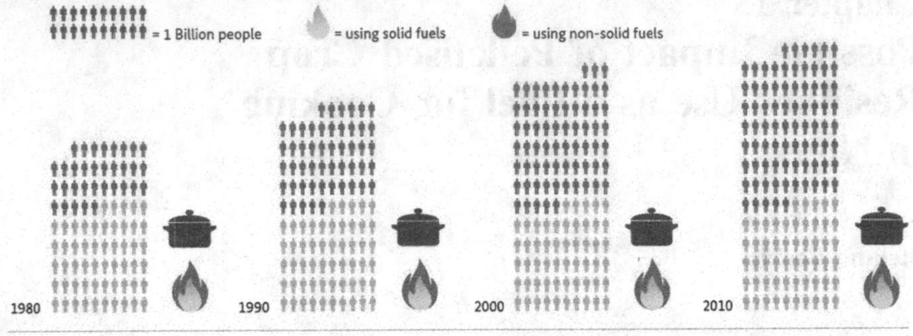

Fig. 15.1 Number of people using solid and non-solid fuels (Roth 2014 elaborated from Bonjour et al. 2013)

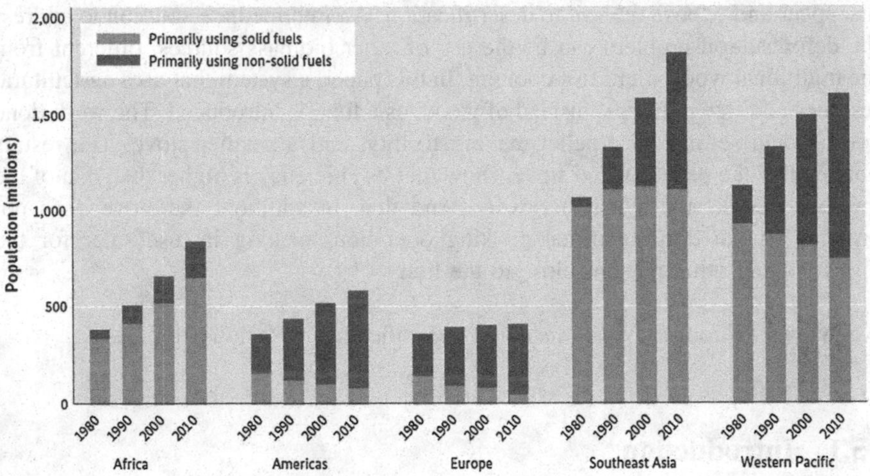

Fig. 15.2 Trend in the use of solid and non-solid fuels divided by geographical area (Roth 2014)

states, as well as in Europe and South America, the absolute number of people using biomass as a fuel is declining.

The strong increasing number of people using wood fuel in Africa, represents a serious sustainability problem over this issue in that specific continent.

The present deforestation process observed in Africa derives, in some sense, from decades of abundance of firewood. It has been reported (Ozer 2004) that before the great droughts of late 1960s, no shortage of firewood had been recorded. The supply of wood fuel exceeded the demand, and it was regarded as an infinite resource. For this reason, no measures to save firewood, using it in an efficient way, were studied and adopted. The firewood supply was limited to the collection of dead wood, even in small towns. Since no alive trees were cut for this purpose, the regeneration of the forest was assured.

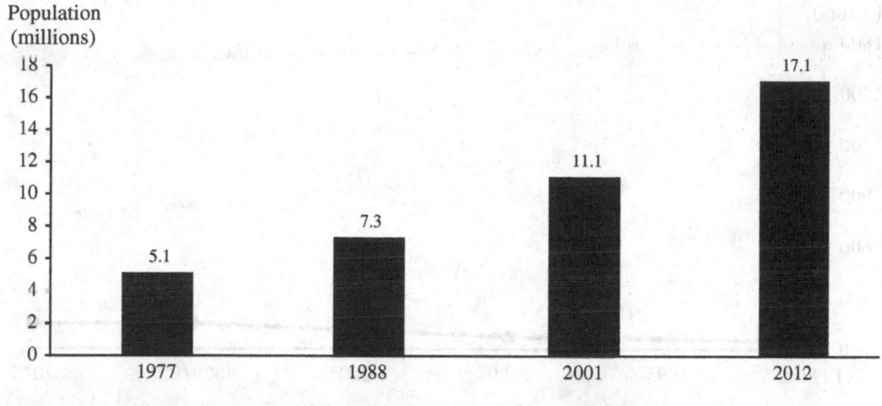

Fig. 15.3 Total population in Niger according to censuses from 1977 to 2012

The main driver in the increasing wood fuel consumption is the population increase in the countries that have the higher percentages of population using biomass as a source of energy. In Niger, where 96.4% of families use biomass for cooking purposes (RdN 2014), the yearly population growth rate is the highest in the world, 3.9% in 2014. Niger population increased from around 5 million in 1977 to about 11 million in 2001, and to more than 17 million in 2012 (Gouro et al. 2014; RdN 2014) (Fig. 15.3).

Last estimations (http://countrymeter.info/en/Niger) indicate that population in Niger already exceeds 20 million in 2016. From the 1977 census, to the year 2016, the population quadrupled, in only 39 years.

It has to be pointed out that the rural population continues to grow, at least at the same rate as the urban population. Under this aspect, there is not, in Niger, a general migration towards cities, but a proportional growth of the population both in towns and in rural areas (Tarchiani and Tiepolo 2016).

In this same country, starting from 1990, more than the half of the natural forest has gone lost, and a recent survey (FAO 2015) indicates in less than 800,000 ha the remaining natural forest area (Fig. 15.4). Similar deforestation patterns are observed in several other African countries. Deforestation changes local climate, and is one of the main contributor to global warming (Watson et al. 2000).

To address the deforestation problem, the use of forest resources as wood fuel for cooking must be considered, together with the other causes.

The traditional three stone fire is still widespread in the whole country, and with an overall efficiency around only 15%, it requires for a remarkable quantity of wood (Fig. 15.5). The reason why it is so popular, is that it is simply set up arranging three stones or bricks on the ground, with some wood pieces in the middle.

A number of improved cook stoves has been set up during the last decades, to improve the efficiency of cooking with wood. Unfortunately, those stoves are still not used in a diffuse way, and they still rely on wood sticks to work. To mark a considerable change from the present situation, a change of fuel is needed.

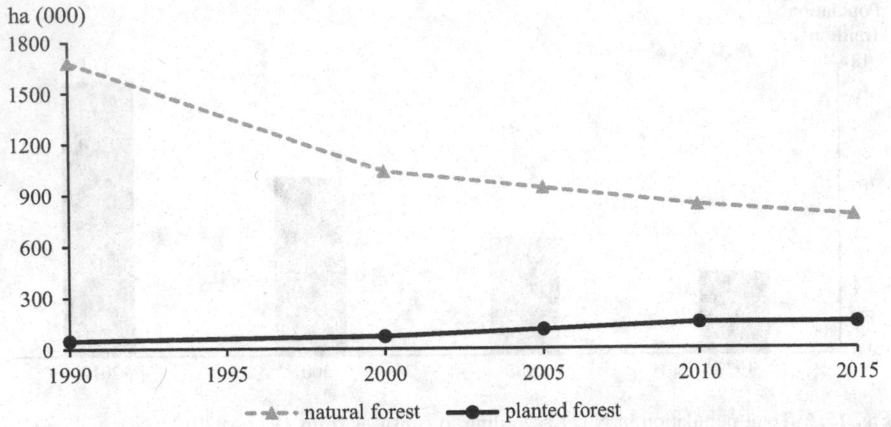

Fig. 15.4 Trends of forest land availability in Niger between 1990 and 2015

Fig. 15.5 A three stone fire made with bricks

Looking into the biomass produced in Niger each year, besides wood, there is a considerable quantity of unused agricultural residues. Those residues can be used as a source of heat for cooking, if appropriately prepared to be burned, or gasified.

Taking into consideration the agricultural production (FAO 2013), it is possible to estimate the quantity of agricultural residues applying the RPR-Residue to Product Ratio conversion index (Lal 2005; Ryan and Openshaw 1991), that indicates the residual biomass produced for each ton of crop production (Table 15.1). A quantity up to 30% of crop residues removal is considered safe under the point of view of possible soil loss (Lindstrom 1986). Some differences in crop residues removal have been observed, with stress on variables like the kind of tillage, soil moisture, and temperature. For these reasons, residue removal effect is site-specific, and cannot be generalized.

Table 15.1 Niger main agricultural crops production and estimate of residual mass

Crop	Production mass T 000	RPR	Residue mass T 000
Millet	2995	1.51[a]	4544
Cow peas, dry	1300	2.90[b]	3770
Sorghum	1287	1.50[a]	1931
Groundnuts, with shell	280	1.00[a]	280
Sugar cane	190	0.25[a]	48
Cassava	150	0.20[b]	30
Sweet potatoes	97	0.25[a]	24
Potatoes	88	0.25[a]	22
Rice, paddy	40	1.50[a]	60
Oilseeds	35	1.37[a]	48
Total	6462		10,756

[a]Lal (2005)
[b]Ryan and Openshaw (1991)

15.1.1 Present Use of Agricultural Residues in Niger

The agricultural residues have different uses.

- some residues are used to feed animals. In this case, livestock is left free to access the fields where the residues are, and the animals directly feed on the edible part of the residues.
- some residues are simply left on the ground, and they degrade under the action of environmental factors.
- some residues are burned directly in the fields, before seeding or transplanting a new crop, if their mass is seen as an obstacle to the agricultural operations.
- in some areas where wood fuel is particularly scarce, some kind of residues, as the biggest millet stalks, are already used by the local population to be burned for cooking purposes.

The use to feed animals is particularly important in Niger, as it is reported that already 37.8% of cereals residues, and 85.7% of legumes residues, are already used, for this purpose (Karimou and Atikou 1998).

When using agricultural residues, a special attention has to be given to the preservation of soil quality, to prevent an excessive removal of crop residues to depress physical and nutritional characteristics of the land (Lal 2005). Where the removal of residues is already high, further removal should be avoided, in order to protect the soil characteristics.

Other residues that can be used as a raw material to produce pellets are tree pruning and dry leaves. Those residues are abundant in urban areas, and, after they are gathered in piles, they are often set in fire directly in the streets, in order to reduce their mass.

15.2 Methodology

The present work intends to contribute to reduce the use of wood fuel for cooking purposes, in order to reverse the tendency in deforestation. To achieve the objective of the work, two different tasks have been accomplished:

1. setting up of a gasification stove specifically designed for the Nigerian traditional cooking exigencies
2. start and optimization of a pellet production chain from agricultural residues and wooden pruning

These two tasks have been carried out in the framework of a EU funded Energy facility project, Renewable energies in the Agadez et Tillabéri regions, under the direction of the Italian NGO Terresolidali (Bechis et al. 2014). The setting up of the gasification stove was carried out starting from an existing model (Brace Research Institute 1999). This P-1 gasification stove (Figs. 15.6 and 15.7) has been adapted to the pots largely used in Niger, and to the requested cooking conditions of the average family.

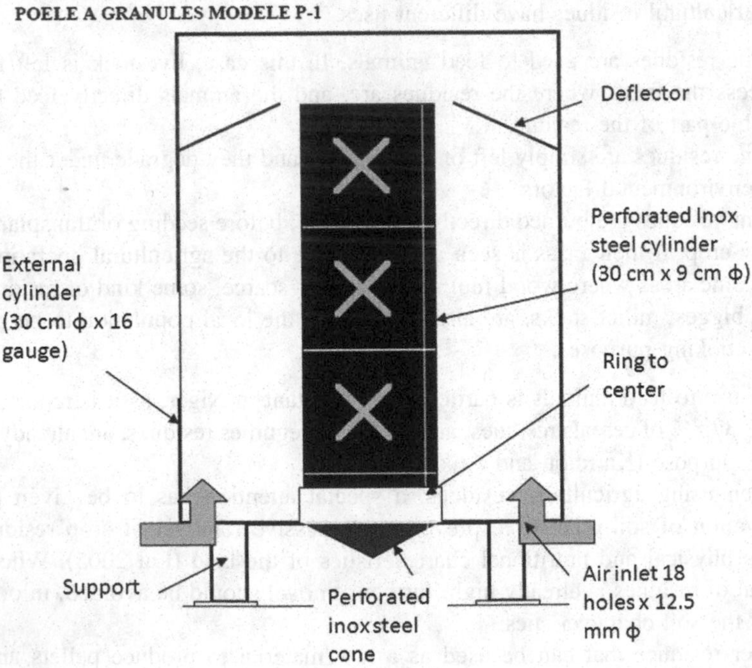

Fig. 15.6 The project of the P-1 gasification stove produced by the Brace Research Institute of Montréal, Canada

Fig. 15.7 The gasification stove developed

The project requirements for the stove can be summarized as:

1. perform the gasification of fuel
2. reach a high overall efficiency
3. be able to cook a complete meal for the average Niger family (7 persons)
4. be stable, to allow stirring
5. be safe for users and children
6. be economic.

The gasification process is necessary to obtain better efficiencies, and a more complete combustion of the biomass used as a fuel.

The gasification process assures less smoke development, and consequently less health risks for the operator. To achieve a stable and steady gasification process, it was necessary to use the feedstock under the form of pellets. High efficiency in thermal conversion allows to perform the cooking of meals with the smallest possible quantity of fuel. As a consequence, high efficiency directly involves fuel savings, with positive effects on economy and sustainability.

To find the right proportions among the different parts of the stove, a first set of around 50 tests was carried out, modifying one parameter at a time. These first tests were performed in Italy, using commercial wooden pellets as a fuel, and 3 dm^3 of water as the heat receiver to assess the heat transfer.

After this first provisional set up, a series of tests were performed in Niger, including some tests where real cooking was performed, to collect the impressions and the suggestions of the potential users on the stove under development. The feedbacks of the users were particularly useful to assess the stove dimensions and stability, as well as the power and lasting of the flame for each fuel charge.

The involvement of users in the early stages of development of the stove, was expressly done in order to take the users exigencies into account from the beginning, to develop a stove with the most suitable, or user friendly, design.

Besides the exigencies of the users, specific attention was paid to the practical building of the stove, in order to allow the local craftsmen to easily build it. Under this point of view, the choice was to make the production of the stove as simple as possible, with dimensions of the stove taking into account the dimensions of the iron sheets available on the Niger market, in order to produce the parts with fewer hours of work, and fewer material scraps.

In parallel with the development of the stove, tests were performed to produce pellets with agricultural residues collected in Niger.

The setting up of the pellet making process passed through a sequence of actions. First of all millet stalks were sent from Niger to Europe to perform the first pellet making tests on small Diesel powered machines. Second, the machines were exported to Niger and were tested on other different raw materials in various proportions. Third, the pellets produced in Niger were analysed at the University of Turin, to assess their physical characteristics. Finally, the produced pellets were used in the Aaron stove to assess their performance and in cooking tests (Fig. 15.8).

After the first experimental phase, and the construction of 350 stoves, a larger facility for the production of pellets, powered by electricity, has been established in the city of Niamey. Tests were also carried out on pellet making for fodder production. The advantages in pelletizing hay and the most edible residues are in a

Fig. 15.8 The machines in the pellet making facility. The electric mill (*left*) and the pellet machine (*right*)

higher product density and a better preservation of the nutritive value. The Aaron stove has been officially tested by the independent body CNES, the National Solar Energy Centre of Niger, and its results compared with those of other stoves (Boubacar 2013). To assess the economic sustainability of the entire pellets cycle, an economic analysis has been carried out by two researchers of the University of Zinder. The cost of pellets has been compared with the reference price for wood fuel in urban areas.

The analysis was completed by technical and economic observations on the feedstock supply chain for pellet making, with the aim to identify possible modifications, that would reduce the cost of the pellet making activity.

15.3 Results and Discussion

The proposed technology proved to be adapted to the local technical level. Craftsmen in Niamey and Agadez could build the stove, with the use of the tools normally available in the average Niger workshop, and a welder. After an investigation it was possible to identify a workshop where to perform the maintenance of the pellet making machines, working on the parts subject to wearing, and even manufacture some of them, as dies and rollers. The people trained in the use of the Aaron stove, quickly learned how to use it, and appreciated the possibility to stay comfortably close to the fire, breathing clean air, as the stove does not produce smoke.

The independent tests carried out by CNES certified that cooking with pellets in the Aaron stove saves 75% of biomass, in comparison with the consumption of the conventional, open 3–stone fire.

The cooking tests indicated that it is possible to cook a complete meal for the average Niger family, with 7 members, with about 1.4 kg of pellets (0.20 kg/day per person). With the traditional three-stone fire still largely used both in rural and urban areas, the quantity of wood fuel necessary to accomplish the same task is one bundle of wood sticks. A series of measures indicated that the average weight of a bundle is approximately 5.8 kg. This fixes the quantity necessary per person in 0.83 kg/day.

Starting from these figures, taking into account the latest indications on Niger population, slightly over 20 million people in 2016, and the percentage of the total population using biomass as a source for cooking, 96.4%, the need for cooking with woodfuel in traditional fires, for the whole Niger population, can be estimated in 5.83 million t per year, while the same projection, in case of a general use of the system developed in this project, is of 1.4 million t of pellets per year.

Under the economic point of view, the study carried out by at the University of Zinder focussed on the production cost of pellets, and on the possibility for the pellets to compete with the traditional wood sticks on the market (Middah et al. 2015).

A comparison of the costs of cooking with wood sticks in the three stone fire and with pellets in the Aaron stove, needs to consider the cost per unit and the number of units necessary to perform the cooking of one meal. The cost of wood sticks varies widely, depending on the site, and on the season. In the capital Niamey, on average, this cost is around 230 FCFA per bundle (0.35 Euros/bundle). Divided by the weight of the bundle, the cost results approx. equal to 40 FCFA per kg (0.06 Euro/kg). The cost of one kg of pellets is higher, around 100 FCFA (0.15 Euro/kg). Taking into account the quantity needed to cook one meal, 1.4 kg of pellets, they cost about 140 FCFA (0.21 Euro per meal) while to cook one meal with the three stone fire an entire bundle of wood sticks is needed. Then, even though the unit cost of pellets is higher, as a small mass of pellets is needed in comparison with wood sticks, to cook a meal, the use of pellets involves an economy of 90 FCFA (around 0.14 Euro) for the preparation of each meal. The costs described, are for the present production rate of pellets, in case of an increase in pellet production, an economy of scale is expected (Fig. 15.9). The two researchers suggested improvements for the organisation of the residues supply chain, with the use of a motorised cart, that would reduce the specific production cost.

The measures taken on pellets made with hay, indicated that the pelletized fodder had seven times higher density than the hay stored in the traditional way. A more dense fodder is easier to store and transport. This development of the project will be object of further investigation.

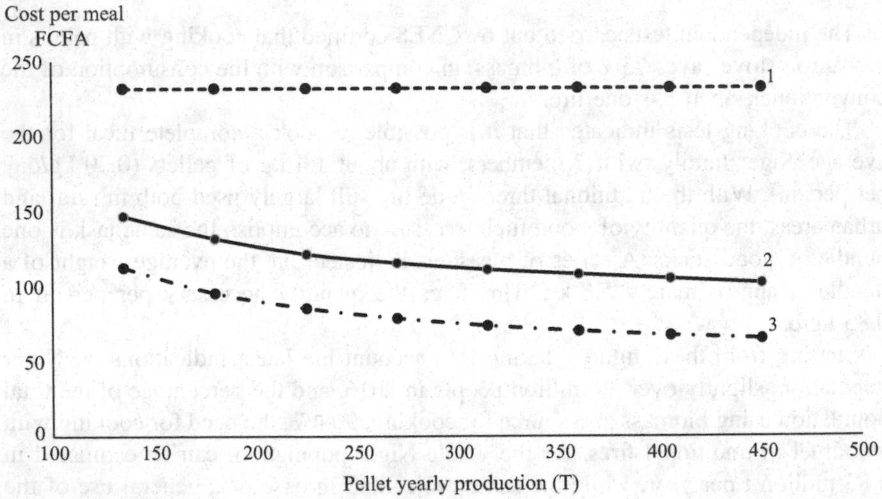

Fig. 15.9 The resulting price of pellet for increasing production quantities at the Niamey pellet production facility. *1*. the wood fuel reference price, *2*. the present price for pellets as a function of the yearly production at the plant, *3*. the calculated price after an enhancement of the feedstock delivery chain

15.4 Conclusions

The achievements of the present work demonstrated that agricultural residues are suitable as a cook fuel in Niger. Their adoption on a large scale mainly depends on the organization of a supply chain of pelletized fuel to the consumers, and on public policies, as the cost of the gasification stove is important when compared to the revenue of the average rural Niger family.

Under the technical point of view, the system developed with pellet making and the gasification stove to use it, can potentially have a significant impact on forests conservation, if it is implemented on a large scale.

A second important aspect of this system is linked to the gasification process, that produces much less harmful smoke than direct burning of biomass. The Aaron stove and its gasification system is likely to have a positive impact on health issues linked to the use of firewood.

Attention has to be paid to the amount of crop residues removed from the fields. In areas where the removal of agricultural residues is already high, due to the use of agricultural residues as fodder for animals, further removal should be avoided, in order to protect the soil quality.

References

Bechis, S. et al. 2014. The Aaron stove, switching to agricultural residues to save firewood. ASABE and CSBE Annual International Meeting. Montréal.

Bonjour, S., et al. 2013. Solid fuel use for household cooking: country and regional estimates for 1980–2010. *Environmental Health Perspectives* 121 (7): 784–790. doi:10.1289/ehp.1205987.

Boubacar, S. 2013. *Résultats des tests sur le foyer Aaron*. Niamey: CNES, Centre National d'Energie Solaire.

Brace Research Institute. 1999. Conception de poêles multifonction à biomasse densifiée en granules pour usages domestiques et communautaires. Montréal: Faculty of Engineering, McGill University, Report No. I-378-P-267.

FAO. 2015. FAO Global forest resources assessment. http://www.fao.org/3/a-i4793e.pdf.

FAO. 2013. *FAO statistical yearbook*. http://www.fao.org/docrep/018/i3107e/i3107e00.htm.

Gouro, A.S., C. Ly, H. Makkar. 2014. Résidus agricoles et sous-produits agro-industriels en Afrique de l'ouest. État des lieux et perspectives pour l'élevage. Accra: Bureau régional pour l'Afrique de la FAO.

Karimou, M., and A. Atikou. 1998. Les systèmes agriculture-élevage au Niger. In *The dry savannas of West and Central Africa*, ed. G. Tarawali, and P. Hiernaux, 78–97. Ibadan: IITA.

Lal, R. 2005. World crop residues production and implications of its use as a biofuel. *Environment International* 32 (4): 575–584. doi:10.1016/j.envint.2004.09.005.

Lindstrom, M.J. 1986. Effects of residue harvesting on water runoff, soil erosion and nutrient loss. *Agriculture, Ecosystems & Environment* 16 (2): 103–112. doi:10.1016/0167-8809(86)90097-6.

Middah, D., C. Harira. 2015. Production des pellets Ong Terre Solidali Niger. Institut Universitaire de Technologie, University of Zinder and PREDAS Programme Régional de promotion des energies domestiques et alternatives au Sahel, Burkina Faso.

Ozer, P. 2004. Bois de feu et déboisement au Sahel: mise au point. *Sécheresse* 15 (3): 243–251.

RdN-République du Niger, Institut National de Statistique du Niger. 2014. Le Niger en chiffres. http://www.stat-niger.org/statistique/file/Affiches_Depliants/Nigerenchiffres2014def.pdf.
Roth, C. 2014. Micro gasification 2.0 cooking with gas from dry biomass. Eschborn: GIZ.
Ryan, P., K. Openshaw. 1991. Assessment of biomass energy resources: a discussion on it's needs and methodology. World Bank.
Soil Quality National Technology Development Team. 2006. Crop residue removal for biomass energy production: effects on soils and recommendations. Technical note No. 19. USDA United States Department of Agriculture—NRCS National Resources Conservation Service soils.usda.gov/sqi.
Tarchiani, V., and M. Tiepolo (eds.). 2016. *Risque et adaptation climatique dans la Région Tillabéri, Niger. Pour renforcer les capacités d'analyse et d'évaluation*. Paris: L'Harmattan.
Watson, R.T., et al. 2000. *Land use, land-use change and forestry*. Cambridge, UK: Cambridge University Press and IPCC.

Chapter 16
Review of Pilot Projects on Index-Based Insurance in Africa: Insights and Lessons Learned

Federica Di Marcantonio and François Kayitakire

Abstract Agricultural risk management involves a portfolio of strategies that can, to different extents, prevent, reduce, and/or properly transfer the impact of a shock. Adaptation, mitigation and coping strategies provide a range of complementary approaches for managing risks resulting from adverse weather events. This paper discusses how the issue of adverse weather events is a challenge for agriculture, and offers an overview of climate risk management strategies and an in-depth examination of one of the most tested risk transfer tools—index-based insurance. It describes the development of the insurance market in Africa, and analyses the major challenges and contributions made by weather index-based insurance.

Keywords Index-based insurance · Risk management · Sub-Saharan Africa

16.1 Introduction

Although agriculture is Africa's largest economic sector, it only generates 10% of its total agricultural output. Mainly rain-fed, agriculture remains highly dependent on weather and sensitive to extreme weather patterns such as erratic rainfall (AGRA 2014). Exposure and vulnerability to adverse events are key determinants for assessing the impacts of shocks that, in the majority of developing countries, still represent the main causes of losses. According to the data of the Centre for Research on the Epidemiology of Disasters-CRED (http://www.emdat.be), approximately 11 million people in developing countries were affected by natural hazards in 2013, 50% of whom were affected by weather-related events. Whilst floods were the most frequent type of natural disaster events, droughts were the

F. Di Marcantonio (✉) · F. Kayitakire
Joint Research Center, European Commission, 21027 Ispra, Italy
e-mail: federica.di-marcantonio@ec.europa.eu

F. Kayitakire
e-mail: francois.kayitakire@ec.europa.eu

M. Tiepolo et al. (eds.), *Renewing Local Planning to Face Climate Change in the Tropics*, Green Energy and Technology,
DOI 10.1007/978-3-319-59096-7_16

most important events in terms of people affected and large economic damages caused (Carter et al. 2014).

The frequency of extreme climate events is generally increasing all over the world (IPCC 2012), but the devastating effects are mainly recorded in those areas where there are high rates of poverty and limited resources and capacity for disaster response. This is true for half of Sub-Saharan countries that are hit by at least one drought every 7.5 years, and by at least one flooding event every three years (Dilley et al. 2005).

In this context, prevention, adaptation and mitigation strategies provide a range of complementary approaches for managing risks that arise from adverse weather events. Effective risk management responses should involve a portfolio of strategies to reduce the impact of and properly transfer the residual part of the risks (the part not covered by other mechanisms). As a part of these mechanisms, a particular form of risk transfer, known as index-based insurance, has received increased attention from a number of academic researchers, international multilateral and non-governmental organisations, and national governments (Miranda and Farrin 2012). The interest shown in the use of this particular tool translated into a number of agricultural insurance pilots that, with the exception of a few cases outside Africa, still suffer from substantial limitations. Besides the numerous advantages of index-based insurance over conventional insurance products, a number of technical and socio-economic challenges have prevented its scalability at a commercial level. These unsatisfactory results are generated by both demand and supply. In this paper, we provide some insights into the reasons behind the difficulties in scaling-up agricultural insurance and, in particular, index-based insurance schemes.

By reviewing pilot projects in Africa and the current literature, this paper also aims to introduce the key concepts and definitions behind risk management; provide background information on the agricultural insurance market in Africa, discussing both its development as well as the different types of insurance products, particularly index-based insurance; and highlight the challenging factors that undermine product upscaling.

16.2 Agricultural Risk Management and Strategies

Agriculture is the largest economic sector of many African countries, employing 65% of the African labour force and accounting for about a third of its gross domestic product (World Bank 2008). Eighty percent of all farms in Sub-Saharan Africa (SSA) are smallholder farmers, who contribute up to 90% of the production in some SSA countries (AGRA 2014). However, production remains primarily at the subsistence farming level (70%), with only a residual part generally commercialised (McIntyre et al. 2009).

Because of its intrinsic nature, the agricultural output remains sensitive to climate variability (IPCC 2012). In addition, the increasing number of catastrophic events and other extreme natural resource challenges and constraints weaken the recovery process or worsen the long-term process of accumulating assets (Carter et al. 2007). These combined factors affect the livelihood of large parts of the population that are vulnerable to weather shocks (Gautam 2006), and whose level of preparedness and ability to properly respond to risks need to be improved.

Over time, individuals involved in the agricultural sector have developed a range of risk-management practices. Rural communities, financial institutions, traders, private insurers, relief agencies and governments all use a variety of both ex-ante and ex-post measures to reduce risk exposure and cope with losses. In some cases, and in the absence of formal mechanisms, rural households have developed individual or collective ex-ante actions for managing risks. In anticipating the negative effects of shocks, the most straightforward decision that a risk-averse farmer makes is to avoid profitable, but risky, activities (Elabed and Carter 2014; Hill 2011).

In the same vein, other informal arrangements, either in the form of ex-ante or ex-post strategies, though effective means for offsetting the negative impact of idiosyncratic shocks, have proven inadequate to protect people from destructive events that impact a large number of individuals simultaneously (Hazell 1992). For instance, Awel et al. (2014) highlight the ineffectiveness of informal risk sharing group arrangements, arguing that this mechanism cannot cope with spatial covariate shocks. Similarly, Dercon et al. (2014), showed that group risk-sharing mechanisms are very strong among households in Ethiopia, but tend to offer only a partial form of insurance, as they are characterised by limited commitment. This does not guarantee full insurance against covariate risks.

Another informal and ex post strategy used by poor farmers and pastoralists is the depletion of productive assets to offset income shocks and stabilise consumption (Carter et al. 2011). This strategy, frequently used by farmers to cope with shocks (Janzen and Carter 2013), has been found to have pernicious effects on household welfare (Hill 2011) and lower households' ability to escape poverty (Lunde 2009).

Whilst moving from informal to formal sharing arrangements appears in theory to be advantageous for rural community members, evidence from a rural village in the Borana area of Ethiopia shows that, due to the complementarity of the two forms of risk arrangement as well as the same selection and monitoring processes, "the formal credit service does not seem to outperform in terms of outreach the informal risk sharing arrangements" (Castellani 2010).

Whilst most ex-ante and ex-post mechanisms implemented as formal or informal mitigation/coping strategies (Fig. 16.1) are in place in many developing countries (albeit to different extents and in different combinations), a comprehensive framework that facilitates multidisciplinary risk evaluation and strategy implementation is commonly lacking.[1] The best way to efficiently combine a variety of instruments is also not yet completely clear. Jaffee et al. (2010) state that 'all these

[1]The methodology for a Rapid agricultural supply chain risk assessment (RapAgRisk), developed by the Agricultural risk management team (ARMT) of the World Bank, provides a system-wide approach for identifying risks, risk exposure, the severity of potential losses, and options for risk management either by supply-chain participants (individually or collectively) or by third parties (e.g. the Government). It is designed to provide a first approximation of major risks,

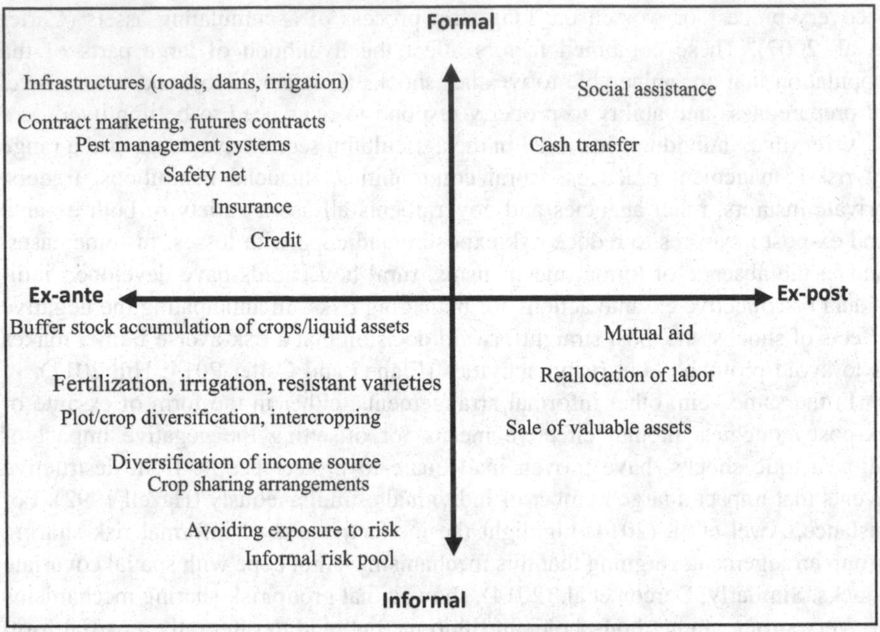

Fig. 16.1 Risk management strategies in agriculture (author elaboration based on World Bank 2005)

instruments have different private and public costs and benefits, which might either increase or decrease the vulnerability of individual participants and the supply chain. When selecting a mix of risk responses, supply chain participants take account of the many inter-linkages among the different types of risk management strategies and instruments'.

There is, however, only fragmented information for some countries as India (Venton and Venton 2004) or Nepal (Dixit et al. 2008), but there is as yet no national information system that can estimate the cost-benefit ratio of disaster management and preparedness programmes.

Hence, while the process for recognising the value of integrated and multi-layer strategies is still far from being widely implemented, countries have focused on the analysis of the potential benefits of specific tools designed to provide protection from one or more agricultural risks. One of these instruments is agricultural insurance.

(Footnote 1 continued)

vulnerabilities, and areas that require priority attention for investment and capacity building (Jaffee et al. 2010).

Table 16.1 Geographic distribution of insurance premiums (Mahul and Stutley 2010)

Region	Global agricultural premium (%)	Estimated crop premium ($ million)	Estimated livestock premium ($ million)	Estimated agricultural premium ($ million)	Agriculture insurance penetration (% 2007 GDP)
Africa	0.4	58.5	5	63.5	0.13
Asia	15.3	1265.9	1047.1	2313.0	0.31
Europe	16.8	2102.6	434.8	2537.4	0.64
LAC	3.2	461.3	26.3	487.8	0.24
North America	63.6	9597.2	3.2	9600.4	5.01
Oceania	0.7	45.6	54.9	100.5	0.38

16.3 Agricultural Insurance

Agricultural insurance has a long history in many countries, and has been largely successful in China and other developed countries (Sandmark et al. 2013). The first agricultural insurance product was developed in Germany in 1700 (Sandmark et al. 2013). It later emerged in the United States, Japan and Canada, and today different types of this product are common in most parts of Europe. Despite heavy government subsidies, insurance penetration remains low even in developed countries, where it never exceeds two percent (Mahul and Stutley 2010). The development of the market is even lower in developing countries, with market penetration in Africa generally being the lowest. A study carried out on a sample of 65 countries (including seven of the eleven countries offering insurance in Africa) by Mahul and Stutley (2010) concluded that agricultural insurance penetration was mostly low in large parts of the surveyed countries, particularly in low- and middle-income countries, where it was less than 0.3%.

Although the estimated global agricultural insurance premium volume almost doubled in the period 2004–2007, it remained low especially in African countries where it roughly reached 63.5 USD million, equivalent to an average of 0.13% of the 2007 agricultural GDP (Table 16.1). Despite the recent relative growth of the insurance industry in Africa, the premium volume generated by the agriculture sector remains marginal (Asseldonk 2013).

Additional insights come from further splitting crop insurance products into the two major groups: traditional indemnity-based products[2] and index-based products (Table 16.2). These results depict a low level of development of the market, and particularly any relevant move of the unconventional products. Figures also suggest

[2]Indemnity-based insurance is generally divided into two categories: (1) named peril, and (2) multiple peril. Named peril crop insurance (NPCI) involves assessing losses based upon a specific risk or peril (hail insurance is the most common type of named peril insurance). Multiple peril crop insurance (MPCI) provides cover for more than one peril, but has never been successfully offered by the private sector on purely commercial terms (Sina 2012).

that the marginal growth of this sector remains predominantly anchored in traditional areas such as indemnity-based crops. This trend is also confirmed by other studies, which report that indemnity-based crops and livestock insurance account for almost 70% of all policies (McCord et al. 2013).

16.4 Penetration of Agricultural Insurance in Africa

The low expansion depicted above could be seen as a snapshot of past market development, which may not reflect current trends. However, the findings of other studies do not diverge significantly from these results. The study of the landscape of microfinance in Africa conducted by the MicroInsurance Centre (Matul et al. 2010) offers some valuable insights to help understand the dynamic of microinsurance markets.[3] In line with the previous findings, the study shows that while life insurance products dominate the insurance market, considerable regional differences remain in product outreach. Indeed, excluding Southern African countries (mainly South Africa), market development is quite unchanged, and agricultural microinsurance is almost inexistent (Fig. 16.2).

Compared to other developed and developing countries, African countries have very limited experience in the agricultural insurance sector. Information collected in 2008 on microinsurance in Africa identified fewer than 80,000 farmers benefiting from agricultural (crop and livestock) insurance (Matul et al. 2010). Agricultural coverage increased to approximately 220,000 people in 2011, although this growth was mainly concentrated in East Africa. In the same year, an average of 8000 policies were issued for each of the 30 different products identified in the region (McCord et al. 2013). In 2014, the number of total polices sold in Africa more than doubled, mainly as a result of the introduction of a significant number of parametric products that were still in a pilot stage. Although the insurance market in Africa has registered an increase in the past ten years, in terms of number of countries entering the market and number of policies sold, the overall outreach is still too small (Fig. 16.3).

From 2011 to 2014, the average agricultural coverage ratio (defined as a percentage of the country's total population covered by agricultural microinsurance) grew from 0.01 to 0.05. This increase was mainly driven by Algeria, Nigeria and Kenya, with an average agricultural coverage ratio of 0.33. Compared to the other two countries, Kenya experienced the higher increase in terms of policy numbers. From 23,523 policies in 2015, 150,370 people were subscribed in 2014. Compare this with Nigeria, which in 2014 entered the agricultural insurance market with more than 540,000 policies.

[3]See http://www.cgap.org/blog/landscaping-microinsurance-africa-and-latin-america.

Table 16.2 Availability of indemnity and Index based insurance by region as a percentage of agricultural insurance products (Mahul and Stutley 2010 based on World Bank survey 2008)

Region	Countries N.	Traditional indemnity based					Index-based		
		Named peril	MPCI	Crop revenue	Greenhouse crops	Forestry	Area yield	Weather	NDVI
Africa	8	50	50	0	13	50	25	38	0
Asia	12	58	58	8	25	25	17	25	17
Europe	21	95	48	0	62	29	5	0	5
Latin America Caribbean	20	50	90	0	25	50	15	30	5
North America	2	100	100	50	50	50	100	100	100
Oceania	2	100	0	0	100	100	0	0	0
All countries	65	69	63	3	38	40	15	22	9

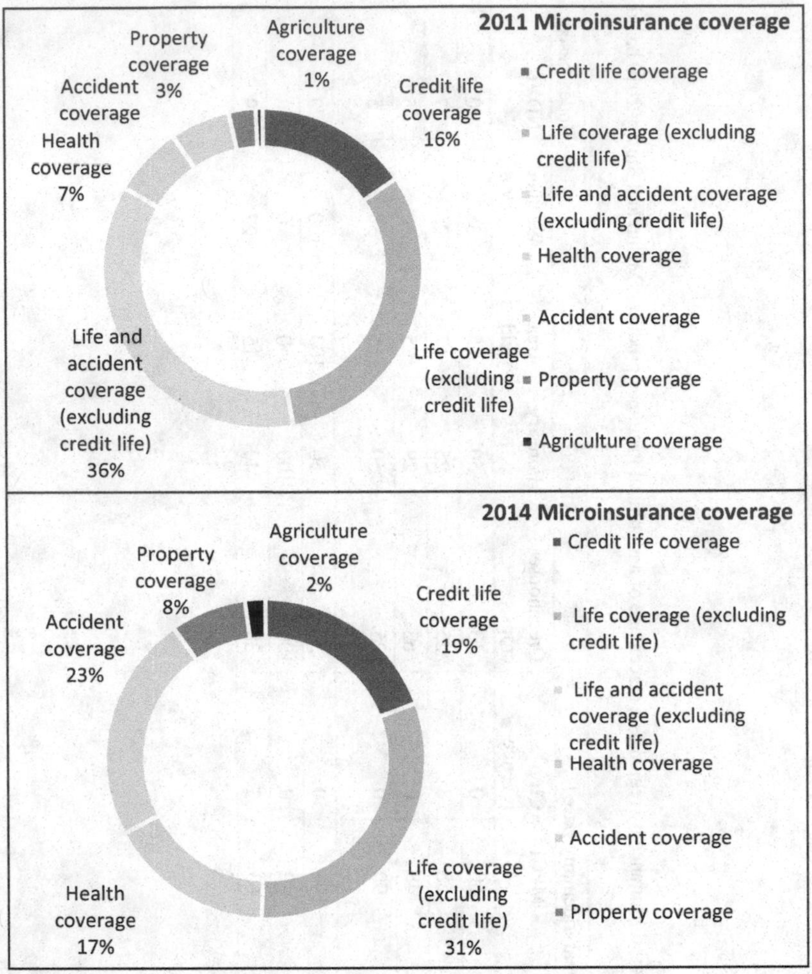

Total Coverage (2011): 44,012,010; Total Coverage excluding South Africa (2011): 16,779,368; Total Coverage (2014): 60,695,180; Total Coverage excluding South Africa (2014): 26,138,446

Fig. 16.2 Microinsurance penetration in SSA excluding South Africa (author's elaboration based on www.microinsurancecentre.org/landscape-studies.html)

16.5 Index-Based Insurance Products

Whilst indemnity-based agricultural insurance continues to be the reference in the agricultural sector, over the past ten years, there has been a growing interest among researchers, international multilateral and non-governmental organisations, and national governments in exploring the possibility of using a particular form of microinsurance—insurance tailored to the needs of the poor—to cover the potential losses of smallholder farmers associated with weather shocks (Patt et al. 2008). This

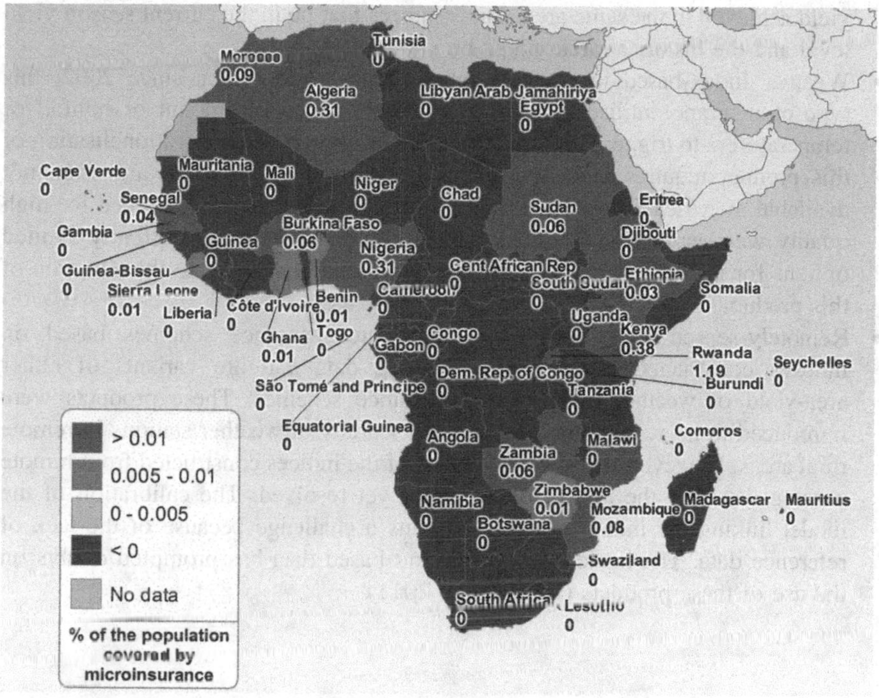

Fig. 16.3 Geographic distribution of total population covered by Agricultural microinsurance (figure for 2014) (www.microinsurancecentre.org/landscape-studies.html)

alternative form of insurance, known as index-based insurance, has been offered to stimulate rural development by allowing smallholder farmers to better adapt to climate change (Dercon et al. 2008), and remove some of the well-known structural problems associated with conventional agricultural insurance, including moral hazard, adverse selection, and systemic risk.

In contract to traditional crop insurance, index-based insurance product does not require a formal claim from the insured nor an individual check of the loss to process indemnification. Within this product, payouts are triggered by an independently monitored weather index that is based upon an objective event that causes loss (i.e. insufficient rainfall) and that is strongly correlated with the variable of interest (for example, crop yield). Based on the underlying data and information on which an index is based, we can distinguish three main types of products:

- Area-yield index insurance: which was first developed in Sweden in the early 1950s and which has been implemented on a national scale in India since 1979 and in the United States since 1993. The average yield over a large area, e.g. a district, serves as index. Indemnities for farmers are determined as a function of the difference between the current season area yield and the longer-term average yield achieved in the same area. This requires that both, the current season yield level and the historical area yields be known.

- Weather Index-based insurance: commercially underwritten since 2002, this type of insurance utilises a proxy (or index)—such as amount of rainfall or temperature—to trigger indemnity payouts to farmers. The operationalisation of this product requires intensive technical inputs and skills that are often not available in Africa. The concentration on rainfall indices and the need for high quality weather data and infrastructure, combined with the currently limited options for insurance products, present additional challenges to the adoption of this product.
- Remotely-sensed index-based insurance: are insurance schemes based on indexes constructed using remote sensing data and are variants of either area-yield or weather index-based insurance schemes. These products were introduced to try resolve the problems of scarcity of weather stations in remote rural areas. However, the low correlation of the indices constructed from remote sensing data and the actual losses is not yet resolved. The calibration of the model linking the index to losses remains a challenge because of the lack of reference data. The inadequacy of ground-based data has prompted doubts on the use of these products (Rojas et al. 2011).

16.5.1 Pilot Projects in Africa

Index-based insurance has been sold as the most promising approach to minimising ex-post verification costs (IFAD 2011). Despite its multiple advantages (i.e. removal of asymmetric information, low administrative costs once the product structures have been standardised, timeliness in payment), the penetration of this product has not produced the expected results and, after more than a decade, is still largely in the pilot stage, with several projects operating around Africa. Different alternatives in terms of product design, delivery mechanisms, pricing, and target population have been tried, but no long-term solution has yet been reached.

Difficulties in achieving positive results (World Bank 2005) have not discouraged many from exploiting the market by promoting several pilot tests. While these initiatives have helped explore the possibility of creating a market for this product, they have not yet clarified the real set of benefits for consumers. Table 16.3 summarises some characteristics of a selected number of weather index-insurance projects reviewed in other publications (Bruke et al. 2010; Carter et al. 2014; Asseldonk 2013; Hess and Hazel 2009; Skees et al. 2007; World Bank 2005).

Demand for index-based insurance is generally low. Supply and demand constraints have not yet been completely removed and results continue to be below expectations. Uptake among different products has been shown to be in the range of 20–30% (Giné 2009; Jensen et al. 2014a), with adopters usually hedging only a very small proportion of their agricultural income (McIntosh et al. 2013).

Table 16.3 Selected individual-level index insurance schemes (Bruke et al. 2010; Carter et al. 2014; Asseldonk 2013; Hess and Hazel 2009; Skees et al. 2007; World Bank 2005)

Country (ISO)	Year	Policy holder	Project name main actor[a]	Index type	Contracts N.	Notes
BFA MLI	2011	Farmers, MFI maize, cotton	Allianz	RE[b]	1471/13,843	–
ETH	2007	Smallholder farmers: teff, beans, maize, wheat, barley, and sorghum	HARITA	Rainfall index. Rain gauges	20,365	Ongoing
ETH	2010	Farmers in cooperatives: teff, sorghum, maize, wheat and barley	EPIICA	Rainfall index	2831 2464	Two products: Standalone-target 8272, Interlinked-target 5008. Closed due to limited sales
ETH	2012	Pastoralist/Livestock	IBLI	Remote sensing vegetation index	405	Ongoing
GHA	2011	Farmers Maize	GAIP	Rainfall index (rain gauge) 5 rain gauges	655	–
KEN	2009	Farmers	ACRE (Kilimo Salama)	Rainfal index (rain gauge)	67,607	Ongoing
KEN	2009	Pastorals	IBLI	Remote sensing vegetation index	256	Ongoing
MWI	2005–06	Maize, groundnut, then tobacco farmers	COIN-RE	RE	3000	Stopped
MOZ	2012	Cotton, maize farmers	Guy Carpenter	Satellite-based weather index	–	–
RWA	2009	Smallholders Maize, rice, potatos	SONARWA, SORAS (Kilimo Salama)	Rainfall index (rain gauge)	15,000	Ongoing weather station-based product that provides coverage against dry spells and excess rain and weather station-based product that provides

(continued)

Table 16.3 (continued)

Country (ISO)	Year	Policy holder	Project name main actor[a]	Index type	Contracts N.	Notes
						coverage against dry spells and excess rain
RWA	2010	Maize and rice farmers	MicroEnsure	Rainfall index (rain gauge and satellite)	24,000	Ongoing
TZA	2011	Cotton farmers	MicroEnsure	RE	24,000	Stopped
SEN		Farmers, MFI maize, peanut	CNAAS	Rainfall index (rain gauge)	10,000	Ongoing
ZMB	2013	Cotton farmers	MicroEnsure	Satellite	6610	Ongoing

[a]The projects reported in the current table are not exhaustive. They represent a schematic summary of pilot projects available to the author at the time of writing
[b]Relative evapotranspiration (RE)

Correspondingly, spontaneous uptake among the non-targeted population has never exceeded 10% (Oxfam 2013).

Though some experiences outside Africa seem satisfactory (Carter et al. 2014), several physical, economic and institutional constraints make it difficult to replicate these positive results in Africa. Bruke et al. (2010) identify several supply and demand constraints common to almost all pilots. On the supply side, the most common constraints are: lack of good quality data, start-up costs and related economic support by the government and difficulty in transferring covariate risk to the international reinsurance market. Other frequent supply constraints are related to inappropriate and/or costly delivery mechanisms (Sina 2012), lack of an enabling environment[4] (Cole et al. 2009) and unfamiliarity with the insurance market Mahul and Stutley 2010).

Premium affordability (Carter 2012; Burke et al. 2010), farmers' trust in insurance providers (Cole et al. 2009), financial illiteracy (Giné and Yang 2009), cognitive failure (Skees et al. 2008), and low willingness to pay (Chantarat et al. 2009) are usually pointed out as the major demand constraints that prevent product scalability. Similarly, empirical studies conducted in Malawi and Kenya strongly supported the hypothesis that ambiguity-averse[5] agents do not value any actuarially fair insurance contracts and have a lower willingness to pay for any specific contract (Bryan 2010).

Data constraints remain the central problem for good index design. To work properly, an index must be highly correlated with losses. Studying this correlation is of particular interest because it allows insurance providers to understand the magnitude of error associated with poor information. Indeed, the higher the correlation, the lower the error of an index in predicting losses. This error (known as basis risk) is recognised as the main drawback of index insurance products (Carter et al. 2014). It basically consists in the mismatch between the payout triggered by the index and the real loss faced by the policy holders. Basis risk not only affects the insured but also the insurance company, which might be compelled to pay an indemnity even when no loss was incurred. To detect this correlation, long historical weather information and yield data are needed. It is generally assumed that a time series of at least 20 years' data is enough to study this correlation. Additionally, for rainfall-based indices, it is also conventionally accepted that a 20-km radius data point can depict the rainfall pattern of all those living within this spatial area. This rule has, however, proved to be based on an overly optimistic

[4]Public-sector interventions are important to ensure that conditions exist for private insurers to go beyond conducting pilot projects, and to start scaling up the business to reach larger numbers of smallholders (IFAD 2010).

[5]Ambiguity aversion is best understood by considering the Ellsberg paradox. Ellsberg (1961) argued that, faced with two gambling options, one with known odds and one with unknown odds, many people prefer to choose the option with known odds, even if they can choose which side of the gamble to take. Basically, he showed that individuals react much more cautiously when choosing among ambiguous lotteries (with unknown probabilities) than when they choose among lotteries with known probabilities.

assumption (Di Marcantonio et al. 2016), particularly in regions characterised by high levels of microclimatic variation (Gommes and Göbel 2013). This aspect, in combination with low density and uneven distribution of weather stations (Washington et al. 2006) and declining number of gauges (Maidment et al. 2014), lead us to rethink the suitability of current rainfall information as a good source for index insurance construction in many African countries. In addition to historical weather patterns, consistent and long-term weather time series are also essential for detecting correlations and to retrospectively estimate the frequency of extreme events, which influences the pricing of the product.[6]

Past experiences showed that, while in some cases lack of such information discouraged suppliers from implementing further projects, in other cases it led to innovative alternatives. For instance, in the case of Malawi, one of the pioneer African countries experimenting with index insurance, the low density of automated rainfall stations prevent an additional 200,000 farmers from being included in the programme. On the contrary, insufficient weather data, low quality of historical weather data, and lack of dense weather stations did not prevent Syngenta Foundation (now operating through Agriculture and Climate Risk Enterprise Ltd. (ACRE)) from further expanding the project. The problem of low quality and scarce weather information was overcome by installing new automated weather stations.[7] Whilst this allows the insurance company to keep basis risk under check, it will not solve the problem of incomplete historical weather time series, at least in the short run. In addition, the installation and maintenance of additional weather stations is often not affordable for Meteorological Services in Africa, and can be very problematic in remote or conflict-affected areas.

For this reason, many pilots use alternative information, such as satellite-based measurements Besides the numerous advantages of applying remote sensing to the insurance market (de Leeuw et al. 2014), refining the set of missing information in an efficient and timing manner might be costly and even unfeasible (Vrieling et al. 2014).[8] However, the use of this information has brought new and as yet unresolved challenges. For instance in the case of IBLI, the performance of the first index was found to perform poorly in estimating drought-related mortality (Jensen et al. 2014b). The low quality of livestock mortality data led to study a new algorithm for

---

[6]The accuracy of these information, used to estimate the parameters of the probability distribution for the underlying weather risk, clearly determinates the pure risk of the insurance contract (Makaudze 2012).

[7]See: http://www.syngentafoundation.com/db/1/1155.pdf and http://www.syngentafoundation.org/_temp/Kilimo_Salama_Fact_sheet_FINAL.pdf

[8]Regarding the NDVI, for instance, retrieving long-term consistent time series from various sources based on different sensor characteristics and algorithms could be time and resource consuming (Miura et al. 2006). In addition to this and other technical constraints (i.e. corrections for effects such as sensor degradation, orbital drift, and atmospheric variability), Turvey and McLaurin (2012) state that three relationships need to exist for an NDVI to work as an insurance index: (i) weather parameters of interest (e.g. extreme heat and/or drought) are strongly related to the NDVI; (ii) the NDVI can explain the variability of crop yields in detail; and (iii) the NDVI measures must be correlated with losses.

the index (Woodard et al. 2016). The current index no longer explicitly predicts livestock mortality rates and product now "makes indemnity payments according to an index developed using only NDVI values" (Mills et al. 2015).

All these aspects highlight the reasons why index design is so complicated to implement and no pilots is currently scaled up in Africa to a commercially viable products at a fair price attractive to poor consumers. Additionally, while the level of uptake remains important for understanding the potential of scaling up the product, more emphasis should be given to other related aspects such as: (a) the market discovery effect (new purchaser compared to renewed insurance), (b) proportion of full cash compared non-cash purchaser (either those who pay with work or those who mainly use coupons or other form of subsidy), (c) quantity insured versus quantity owned.

16.6 Conclusion

The challenge of risk management in agriculture is to find the proper balance between taking on risk and preparing for it with ex-ante actions, and management of the consequences only after the event has occurred. Loss reduction and protection of livelihoods are the main goals of risk management actions. As there is no unique recipe to deal with shocks in agriculture, risk management strategies should pursue this goal by combining the capacity to prepare for risk with the ability to cope with the effects. The development of a comprehensive framework would help in this sense.

In the context of agricultural risk management, interest has moved recently towards risk transfer mechanisms in the form of crop and livestock insurance. A particular form of insurance, known as index-based insurance, has received growing attention, attracting substantial resources which have resulted in a large number of pilot programmes to test the effectiveness of this product to manage covariate risk in agriculture.

Although index-based insurance has been developed as an instrument to avoid consumption smoothing or depletion of valuable assets among other social welfare benefits, ambiguous results feed the debate on how much this product represents an opportunity for development, especially in a dynamic and changing environment.

The effectiveness of this instrument is still uncertain, but many lessons can be learned from past experience. Particularly, within the wide range of pilot experiences we revised, some key messages are clear and self-explanatory. In recent years the level of uptake has increased but is still insufficient to make the product commercially viable. On average, the uptake ranges from 30 to 40%, but in the majority of the cases this is mainly driven by the availability of subsidies to reduce the insurance premiums. Reasons behind the low uptake rates come from both the demand and supply sides. From the demand side, price affordability, cognitive failure and the economic behaviour of farmers have been found to be common factors that dampen the demand for the product. In many cases, these challenging aspects have called for government intervention, a solution that in the long term would not be sustainable. On the supply side, one of the major problems is the basis risk. This problem, which is intrinsic to the nature of the product, mainly steam from the inadequacy of the data and it represents a critical limitation to the upscaling of this product.

Whilst few analyses of the impact of basis risk on a product demand exist, statistical assessment of its magnitude is lacking. Understanding the "tolerable error" attached to this product would also clarify the effectiveness of the tool and confirm or reject the current trends. Similarly, considering that insurance is just one among a number of complementary instruments, it is important to understand to which extent and at what cost it is possible to design an affordable insurance scheme that responds to the real needs of the vulnerable farmers and pastoralists.

References

AGRA-Alliance for a Green Revolution in Africa. 2014. Africa agriculture status report 2014: Climate change and smallholder agriculture in Sub-Saharan Africa. Nairobi, Kenya. Issue No. 2.

Asseldonk, M. Van. 2013. Market outlook for satellite-based RE index insurance in agribusiness. LEI memorandum 13-085.

Awel, Y.M., and T.T. Azomahou, T.T. 2014. Productivity and welfare effects of weather index insurance: Quasi-experimental evidence. Working Paper, UNU-MERIT and Maastricht University.

Bryan, G. 2010. Ambiguity and insurance. Yale working paper.

Burke, M., de Janvry, A., J. Quintero. 2010. Providing index based agricultural insurance to smallholders: Recentprogress and future promise.Available at: http://siteresources.worldbank.org/EXTABCDE/Resources/7455676-1292528456380/7626791-1303141641402/7878676-1306270833789/Parallel-Session-5-Alain_de_Janvry.pdf.

Carter, M.R., A. de Janvry, E. Sadoulet, A. Sarris. 2014. Index-based weather insurance for developing countries: A review of evidence and a set of propositions for up-scaling. Working paper.

Carter, M.R., P.D. Little, T. Mogues, and W. Negatu. 2007. Poverty traps and natural disasters in Ethiopia and Honduras. *World Development* 35 (5): 835–856. doi:10.1016/j.worlddev.2006.09.010.

Carter, M.R. 2012. Designed for development: Next generation approaches to index insurance for smallholder farmers. *ILO/MunichRe Microinsurance Compendium* 2.

Carter, M.R., M. Ikegami, S.A. Janzen. 2011. Dynamic demand for index-based asset insurance in the presence of poverty traps. Working paper.

Castellani, D. 2010. Microfinance and risk sharing arrangements: Complements or substitutes? Theory and evidence from Ethiopia. Ph.D. thesis Catholic University of Milan.

Chantarat, S., A. Mude, C. Barrett. 2009. Willingness to pay for index based livestock insurance: Results from a field experiment in northern Kenya. Working paper.

Cole, SA, X. Gine, J. Tobacma, P. Topalova, R. Townsend, I.J. Vickery. 2009. Barriers to household risk management: Evidence from India. Working paper No. 09–116. Harvard Business School Finance; FRB of New York Staff Report No. 373.

de Leeuw, J.D., A. Vrieling, A. Shee, C. Atzberger, K.M. Hadgu, C. Biradar, H. Keah, and C. Turvey. 2014. The potential and uptake of remote sensing in support of insurance: A review. *Remote Sensing* 6 (11): 10888–10912. doi:10.3390/rs61110888.

Dercon, S., R.V. Hill, D. Clarke, I. Outes-Leon, and A.S. Taffesse. 2014. Offering rainfall insurance to informal insurance groups: Evidence from a field experiment in Ethiopia. *Journal of Development Economics* 106: 132–143. doi:10.1016/j.jdeveco.2013.09.006.

Dercon, S., M. Kirchberger, J. Gunning, J. Platteau. 2008. Literature review on microinsurance. Geneva: International Labour Organization. http://www.ilo.org/public/english/employment/mifacility/download/litreview.pdf. Accessed 28 Jan 2017.

Dilley, M., R.S. Chen, and U. Deichmann. 2005. *Natural disaster hotspots: A global risk analysis*. World Bank Publications.

Di Marcantonio, F., A. Leblois, W. Göbel, and H. Kerdiles 2016. Effectiveness of weather index insurance for smallholders in Ethiopia. Submitted to *Applied Economic Perspectives and Policy*.

Dixit, A., A. Pokhrel, and M. Moench. 2008. *From risk to resilience: Costs and benefits of flood mitigation in the lower Bagmati basin: Case of Nepal Terai and North Bihar*. Kathmandu, Nepal: ISET-Nepal and ProVention.

Elabed, G., and M. Carter. 2014. Ex-ante impacts of agricultural insurance: Evidence from a field experiment in Mali. Working paper University of California Davis Agricultural and resource economics. https://are.ucdavis.edu/people/faculty/michael-carter/working-papers.

Ellsberg, D. 1961. Risk, ambiguity and the savage axiorms. *Quarterly Journal of Economics* 75 (4): 643–669.

Gautam, M. 2006. Managing drought in Sub-Saharan Africa: Policy perspectives. Invited paper for an invited panel session on drought: Economic consequences and policies for mitigation, at the IAAE conference, Gold Coast, Queensland, Australia, August 12–18, 2006.

Giné, X. 2009. Experience with weather insurance in India and Malawi. In *Innovations in insuring the poor*. Washington DC: International Food Policy Research Institute.

Giné, X., and D. Yang. 2009. Insurance, credit, and technology adoption: Field experimental evidence from Malawi. *Journal of Development Economics* 89 (1): 1–11. doi:10.1016/j.jdeveco.2008.09.007.

Gommes, R, and W. Göbel. 2013. Beyond simple, one-station rainfall indices. In *The challenges of index-based insurance for food security in developing countries*. Publications Office of the European Union.

Hazell, P.B.R. 1992. The appropriate role of agricultural insurance in developing countries. *Journal of International Development* 4 (6): 567–581. doi:10.1002/jid.3380040602.

Hess, U., and P. Hazel. 2009. *Innovations in insuring the poor. Focus 2020, Brief 14*. Washington D.C.: International Food and Policy Research Institute (IFPRI).

Hill, R.V., J. Hoddinott, and N. Kumar. 2011. Adoption of weather index insurance: Learning from willingness to pay among a panel of households in rural Ethiopia. Discussion paper, 01088, International Food Policy Research Institute.

IFAD-International Fund for Agricultural Development. 2010. *The potential for scale and sustainability in weather index insurance*. Rome: International Fund for Agricultural Development.

IFAD-International Fund for Agricultural Development. 2011. *Weather index-based insurance in agricultural development: A technical guide*. Rome: World food program and IFAD.

IPCC. 2012. Managing the risks of extreme events and disasters to advance climate change adaptation. A special report of working groups I and II of the intergovernmental panel on climate change. In ed. Field, C.B., V. Barros, T.F. Stocker, D. Qin, D.J. Dokken, K.L. Ebi, M. D. Mastrandrea, K.J. Mach, G.-K. Plattner, S.K. Allen, M. Tignor, and P.M. Midgley. Cambridge, UK and New York, USA: Cambridge University Press.

Jaffee, S., P. Siegel, C. Andrews. 2010. Rapid agricultural supply chain risk assessment, a conceptual framework. Agriculture and Rural Development Discussion Paper 47, The World Bank.

Janzen, S.A., and M.R. Carter. 2013. The impact of microinsurance on consumption smoothing and asset protection: Evidence from a drought in Kenya. University of California at Davis.

Jensen, N., A. Mude, and C.B. Barrett. 2014a. How basis risk and spatiotemporal adverse selection influence demand for index insurance: Evidence from Northern Kenya. *Working paper*, Cornell University.

Jensen, N.D., C.B. Barret, and A.G. Mude. 2014b. Basis risk and the welfare gains from index insurance: Evidence from Northern Kenya, MPRA Paper No. 59153.

Lunde, T. 2009. *Escaping poverty: Perceptions from twelve indigenous communities in Southern Mexico*. Thesis (Ph.D.) Johns Hopkins University.

Maidment, R.I., D. Grimes, R.P. Allan, E. Tarnavsky, M. Stringer, T. Hewison, R. Roebeling, and E. Black. 2014. The 30 year TAMSAT African rainfall climatology and time series (TARCAT) data set. *Journal of Geophysical Research: Atmospheres* 119 (18): 10619–10644. doi:10.1002/2014JD021927.

Mahul, O., and C. Stutley. 2010. *Government support to agricultural insurance: Challenges and options for developing countries*. Washington DC: World Bank.

Makaudze, E. 2012. *Weather index insurance for smallholder farmers in Africa. Leassons learnt and goals for the future*. Stellenbosch: Sun Media.

Matul, M., M.J. McCord, C. Phily, and J. Harms. 2010. *The landscape of microinsurance in Africa*. ILO.

McCord, M.J., R. Steinmann, C. Tatin-Jaleran, M. Ingram, and M. Mateo. 2013. The landscape of microinsurnace in Africa 2012. GIZ-Program promoting financial sector dialogue in Africa and Microinsurance Centre Munich Re Foundation.

McIntosh, C., A. Sarris, and F. Papadopoulos. 2013. Productivity, credit, risk, and the demand for weather index insurance in smallholder agriculture in Ethiopia. *Agricultural Economics* 44: 399–417.

McIntyre, B.D., H.R. Herren, J. Wakhungu, and R.T. Watson. 2009. *Agriculture at a crossroads. International assessment of agricultural knowledge, science and technology for development. Synthesis report.* Whashington, DC: Island Press.

Mills, C., N. Jensen, C. Barrett, and A. Mude. 2015. Characterization for index based livestock insurance. *Working Paper*, Cornell University.

Miranda, M.J., and K. Farrin. 2012. Index insurance for developing countries. *Applied Economic Perspectives and Policy* 34 (3): 391–427. doi:10.1093/aepp/pps031.

Miura, T., A. Huete, and H. Yoshioka. 2006. An empirical investigation of cross-sensor relationships of NDVI and red/near-infrared reflectance using EO-1 Hyperion data. *Remote Sensing of Environment* 100: 223–236. doi:10.1016/j.res.2005.10.010.

Oxfam America. 2013. R4 rural resilience initiative quarterly report April-June 2013.

Patt, A., N. Peterson, M. Carter, M. Velez, U. Hess, A. Pfaff, and P. Suarez. 2008. Making index insurance attractive to farmers. Paper presented at a workshop on 'Technical issues in index insurance', held October 7–8, 2008 at IRI, Columbia University, New York.

Rojas, O., F. Rembold, J. Delincé, and O. Léo. 2011. Using the NDVI as auxiliary data for rapid quality assessment of rainfall estimates in Africa. *International Journal of Remote Sensing* 32 (12): 3249–3265. doi:10.1080/01431161003698260.

Sandmark, T., J-C. Debar, and C. Tatin-Jeleran. 2013. *The emergence and development of agricultural microinsurance*. Microinsurance Network Discussion paper. Luxembourg

Sina, J. 2012. Index-based weather insurance—international & Kenyan experiences. *Adaptation to climate change and insurance* (ACCI), Nairobi, Kenya.

Skees, J.R., A.G. Murphy, B. Collier, M.J. McCord, and J. Roth. 2007. Scaling up index insurance: What is needed for the next big leap forward? Weather insurance review prepared by the Microinsurance centre and global Agrisk for Kreditanstalt für Wiederaufbau (KfW).

Skees J.R., and B. Collier. 2008. *The potential of weather index insurance for spurring a green revolution in Africa*. Lexington, KY: The Watkins House, www.globalagrisk.com.

Turvey, C.G., M.K. Mclaurin. 2012. Applicability of the normalized difference vegetation index (NDVI) in index-based crop insurance design. *Weather, Climate, and Society* 4: 271–284. doi:10.1175/WCAS-D-11-00059.1.

Venton, C.C., and P. Venton. 2004. Disaster preparedness programmes in India: A cost benefit analysis. HPN Network Paper No. 49. London, UK: Overseas Development Institute.

Vrieling, A., M. Meroni, A. Shee, A.G. Mude, J. Woodard, C.A.J.M. Kees de Bie, and F. Rembold. 2014. Historical extension of operational NDVI products for livestock insurance in Kenya. *International Journal of Applied Earth Observation and Geoinformation* 28: 238–251. doi:10.1016/j.jag.2013.12.010.

Washington, R., G. Kay, M. Harrison, D. Conway, E. Black, A. Challinor, D. Grimes, R. Jones, A. Morse, and M. Todd. 2006. African climate change: Taking the shorter route. *Bulletin of the American Meteorological Society* 87: 1355–1366. doi:10.1175/BAMS-87-10-1355.

Woodard, J.D., A. Shee, and A. Mude. 2016. A spatial econometric approach to designing and rating scalable index insurance in the presence of missing data. *Geneva Papers on Risk and Insurance Issues and Practice* 41 (2): 259–279. doi:10.1057/gpp.2015.31.

World Bank. 2005. *Managing agricultural production risk innovations in developing countries.* Washington DC: World Bank.

World Bank. 2008. *World development report 2008: Agriculture for development.* Washington DC: World Bank.

Chapter 17
Addressing Climate Change Impacts in the Sahel Using Vulnerability Reduction Credits

Karl Schultz and Linus Adler

Abstract Adaptation projects may be difficult to prioritize and finance, as the results of projects are difficult to quantifiably measure and compare across project types, and no singular "unit" for adaptation outcomes exists. The Higher Ground Foundation is developing the Vulnerability Reduction Credit (VRC™), which incorporates cost/benefit analysis and per capita vulnerability equalization tools to measure the outputs of climate adaptation projects. The VRC quantifies in a singular unit measures to reduce vulnerability to climate change. This chapter summarizes the structure and utility of VRCs and shows through a case study from Talle, Niger, how VRCs are created and integrated into Sahelian community adaptations to heterogeneous climate risks such as flooding and droughts. VRC analysis and crediting may serve as a monitoring and evaluation tool and as an instrument to help secure project finance while supporting sustained adaptation. The chapter further considers the potential benefits to governments, donors and economies. VRC financing has advantages over standard development assistance models, particularly for project risk management, project preparation, enhanced transparency of adaptation spend, and scaling of successful pilot projects throughout an economy.

Keywords Vulnerability reduction credit · Adaptation finance · Sahel · Monitoring and evaluation climate adaptation · Climate resilience

---

K. Schultz (✉) · L. Adler
The Higher Ground Foundation, 18 Northchurch Terrace, London, UK
e-mail: karl@climateadaptationworks.com

L. Adler
e-mail: linus@climateadaptationworks.com

M. Tiepolo et al. (eds.), *Renewing Local Planning to Face Climate Change in the Tropics*, Green Energy and Technology,
DOI 10.1007/978-3-319-59096-7_17

17.1 Introduction

Climate change is happening and it is impacting communities around the world. While all nations will be impacted by climate change, primarily for the worse, the poorest countries face the most human vulnerability.

Sub-Saharan Africa is both particularly vulnerable to climate change and lacking sufficient adaptive capacity to address many of the impacts on agriculture, the built environment, health, and other sectors. The economic impact of climate change will be considerable; by the end of the century climate change is estimated to cost 10% of Africa's GDP, and the costs of effective adaptation could be between $US 10 billion and £UK 30 billion per year by 2030 (Pan African Climate Justice Alliance 2009). Africa will bear the highest costs per capita in terms of GDP (Watkiss et al. 2016).

In the context of planning processes, research has shown few examples of climate information being integrated into the planning of long-term development. Reasons given for this include short-term development challenges focusing decision-makers' attention on shorter timescales, a lack of both serviceable medium- to long-term climate information and integrated assessments of climate impacts, vulnerability, or adaptation, and a communication mismatch between the producers and users of climate information (Jones et al. 2015).

Although expenditures are generally viewed as insufficient, a considerable amount of adaptation investment is already taking place in developing countries. Overseas development assistance is considerable, but the traditional approach (as typified by the Paris Climate Agreement) for financing climate adaptation in developing countries is to set a global monetary target, rather than focus on vulnerability reduction as the measure of results. Unfortunately, many development assistance projects that are labelled "climate adaptation" have little to do with climate adaptation (Junghans and Harmeling 2012). Meanwhile, governments in Africa are diverting much of their own resources to address climate change; for instance, the Overseas Development Institute found that 5 and 14%, respectively, of Tanzania's and Ethiopia's annual national budgets were spent on addressing climate change (Bird 2014). And both governmental and development assistance spend also suffers from a lack of transparency and misallocation of funds (Alabi 2012).

The role of the private sector is often taken as critical in securing sufficient finance to meet the global adaptation investment requirements. However, a major challenge for getting the private sector involved is finding sufficient justification to undertake adaptation. As rational, profit-seeking bodies, private companies may recognize the threats climate change may bring to their business, through disrupted supply chains and harm to assets and even markets. But, for these same organizations to play a role in developing or financing projects that have limited or no direct impact on corporate returns, (as is the case with many community-level adaptation projects in Africa) a revenue stream, and related price signal must be established.

While governments might subsidize some private involvement via tenders, and while there are a number of cases where adaptation may bring clear financial returns to investors, the nature of much climate vulnerability is that, while economic returns from adaptation measures (that may be represented variously through avoided damage to buildings, etc.) are usually possible, financial returns (represented in direct project-level revenue streams) are much more difficult to achieve. This is particularly the case for adaptation projects that address the vulnerabilities of poorer communities. And so the challenge can be seen as identifying and deploying mechanisms that can convert economic need into financial investment.

When resources are allocated to adaptation, there is the risk that they may be ineffective or inefficiently invested, owing to a lack of understanding of adaptation's benefits. In part, this risk is created because of the lack of a recognized, general approach to evaluate and compare projects undertaken to reduce vulnerabilities to climate change.

17.2 The Need for New Instruments to Address Local Vulnerabilities

Without a mechanism that can allow projects to be compared, and potentially incentivized, that is free from local or other political considerations, many of the great adaptation challenges facing Africa will persist. With such a mechanism, in particular one that is fungible, has a single metric, and can be certified as a quantity of recognized "vulnerability reduction", it is possible to:

- Better prioritize projects (thus bringing in efficiencies that increase the potential for effectively using limited resources),
- Serve as a means for leveraging finance from the revenue streams created, by setting a price on a quantified level of vulnerability reduction,
- Allow for more transparent, "bottom-up" decision-making in adaptation investment, as communities, private and public adaptation technology and service providers, and project developers have a fair chance at gaining credits,
- Serve as a positive feedback mechanism as the "market" for adaptation technologies and effective project investment and operations improves through the incentive to optimize project vulnerability reduction.
- Create incentives for sustainable projects, as credits are issued only if projects can prove that vulnerability reduction has been ensured for a (past) period of time.

The challenges to creating such an instrument that may result in these benefits, include ensuring that it transparently, efficiently, and flexibly provides quantifiable and verifiable incentives, resulting in real and additional climate vulnerability reduction for poor communities. This entails ensuring that the instrument:

- is robust: that baselines are clear, that projects establish that they are going beyond this baseline and that the monitoring and verification of the outputs is unambiguous,
- is quantifiable, in order to permit price setting around the units of vulnerability reduction, and to compare projects, and
- is sustained over time, which suggests a system design wherein projects are only awarded credits after the results are verified.

17.3 Vulnerability Reduction Credits (VRC™)

The proposed instrument, designed to meet the above requirements, is the climate Vulnerability Reduction Credit (VRC™). A VRC represents avoided impact cost, normalized with an income equalization factor. Downscaled climate projections are used to estimate increased impact costs with climate change, and projects are then assessed for how they decrease these costs. In addition, projects wishing to be awarded VRCs must meet certain social and other criteria. The VRC is a credit for work done to avoid damages or losses owing to climate change (Schultz 2012). There are three fundamental assumptions in VRC analysis that, if accepted, validate their value in measuring vulnerability reduction:

1. Economic conditions are a valid measure of human wellbeing and can proxy for adaptive capacity,
2. Economic impacts can often be quantified, and,
3. Impacts can be equalized for poorer communities.

The first assumption is that while there are many very important non-economic values, and that while economic wellbeing is not equivalent to human wellbeing, everyone needs to eat and be sheltered and economic conditions are a universal, and often easily measurable, index related to human well-being, certainly as it relates to most people living in developing countries.

The second assumption is that we can most easily assess vulnerability reduction using economic cost-benefit analysis tools, which represent an established approach (Atkinson and Mourato 2008). There are a variety of methods for monetizing non-financial values (e.g., hedonic pricing), and many of the climate vulnerabilities communities face may be easily monetized (e.g. loss in agricultural production).

The third assumption, that we can equalize impacts by factoring in per capita income, is important if equity is a consideration, or more fundamentally, if human vulnerability is the main concern rather than protection of economic assets or incomes. By considering a vulnerable community's per capita income, the VRC can take into account the reduced value, and thus economic impact, of poorer communities, focusing rather on the amount of human vulnerability per capita.

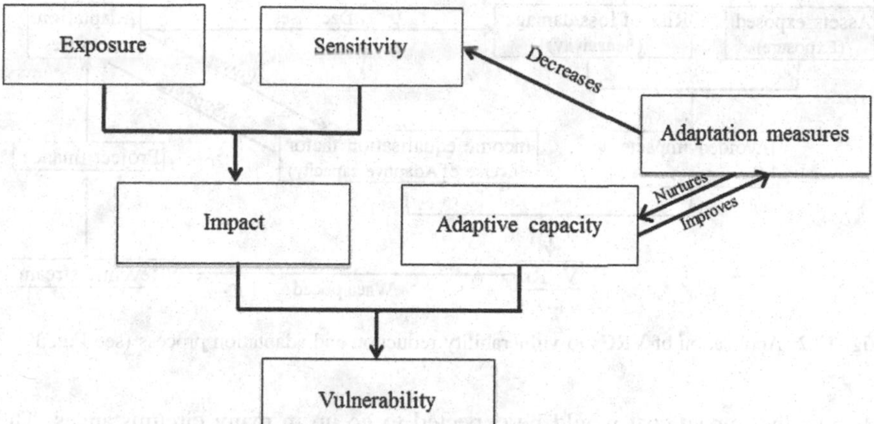

Fig. 17.1 Relationship between vulnerability, exposure, sensitivity, and adaptive capacity

The theory behind the analytical construction of VRCs is adapted from the IPCC's Fourth Assessment Report (IPCC 2007), which states that vulnerability is a function of exposure, sensitivity, and adaptive capacity. Adaptation measures decrease sensitivity and nurture adaptive capacity, and as a result vulnerability may be reduced (Fig. 17.1).

17.3.1 How Do VRCs Work?

Essentially, a Vulnerability Reduction Credit represents €50 of avoided impact cost. It consists of:

- The Avoided Impact Cost (AIC), and;
- An Income Equalization Factor (IEF).

Dividing the product of the AIC and the IEF by €50 gives the number of credits a project earns (Fig. 17.2):

$$\# VRCs = (AIC \times IEF) \div €50 \quad (1)$$

As discussed previously, this formulation is analogous to the IPCC's definition of climate change vulnerability as a function of Exposure (of a system to climate change), Sensitivity (of the exposed system to climate change) and the Adaptive capacity of the system.

It might be useful to further unpack the terminology in the VRC formula. The first term—the AIC—is analogous to exposure and sensitivity, which can be looked at through the lens of expected damage or loss of income. Under anticipated climate

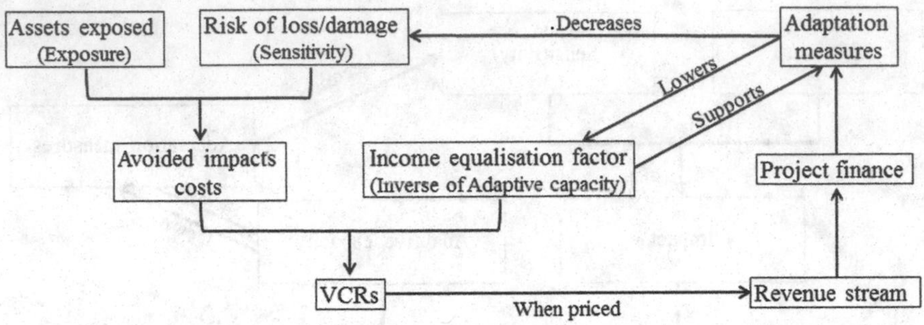

Fig. 17.2 Application of VRCs to vulnerability reduction and adaptation process (see Fig. 17.1)

change, the impact cost would be expected to go up in many circumstances. The avoided impact cost is therefore simply the expected value restored to a vulnerable system through either lessening the chance or extent of adverse climate consequences. The second term in the formulation—the IEF—is used to normalize potential project value in communities with different income levels. A project in, say, a poor village in Bhutan would be valued quantitatively differently than a similar project in, say, Colombia or the UK where avoided impact costs may be higher given the higher economic asset levels and economic productivities; this would have the distorting effect of favoring projects in wealthier communities and countries. The IEF works as follows—the local per capita income is divided into the World Bank's Gross National Lower to Upper Middle Income per capita income figure (World Bank 2016) producing a dimensionless IEF multiplier (if the per capita income is higher than the World Bank threshold figure, then the IEF is set as one). This reflects research indicating that levels of human well-being decouple from incomes as they reach this level (Wilkinson and Pickett 2009). Hence, vulnerability reduction is created and quantified based on avoided impact costs, adjusted for poorer communities by a factor inversely related to the vulnerable community's per capita income. VRC generating projects interpret and apply the relationship of different factors in reducing vulnerability, with assets exposed being the proxy for exposure, risk of loss/damage of sensitivity, and the income equalization factor the inverse of adaptive capacity (Fig. 17.2).

It's important to consider how over time VRC generation will result in enhanced adaptive capacity, as reduced vulnerability may contribute to increased per capita income and correlated enhanced capacities, including adaptation know-how, and physical resources (such as dikes, better crop varieties, etc.).

17.4 How Do VRC Projects Work?

In generating VRCs, a project employs a cascading chain of results projection (Fig. 17.3).

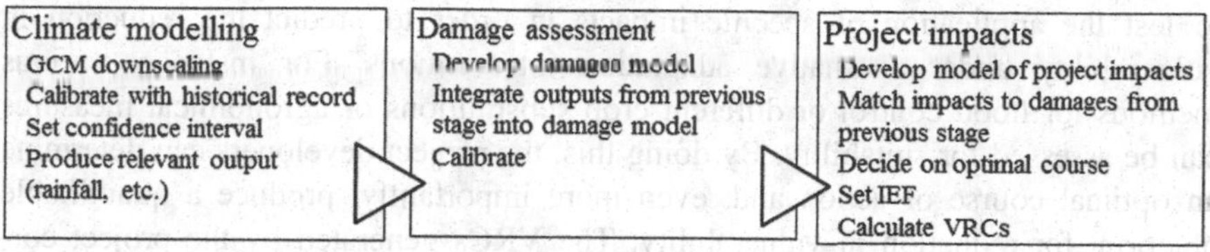

Fig. 17.3 VRC project flow

17.4.1 Climate Modelling

The climate modelling stage forms the basis for understanding the effects of climate change. So-called downscaled climate modelling outputs are often critical to quantifying climate factors such as rainfall and temperature in terms of their effects at a local level. Climate downscaling involves applying the outputs of one or more global models (more technically called "general circulation models", or GCMs) at their lowest scale into a downscaled regional to local model (USAID 2014; McSweeney 2010). There are a variety of dynamic and statistical modelling packages available (UK MET Office Hadley Climate Centre 2016), and while the cost of undertaking an independent model run can be considerable, the use of existing, validated model runs is permitted. As time passes, off-the-shelf modelling results suitable to a particular project's scope will be more likely to be available, although it is always necessary to verify their applicability to specific project purposes.

17.4.2 Impacts Estimation

Once local downscaled climate outputs have been developed or obtained, they then are used to estimate impacts. There are a wide variety of approaches to estimating impacts; to consider just two impacts from one climate output vector, rainfall projections can be used as an input in crop models or for estimating drainage, while hydrological models and civil engineering methods can be used to predict and avoid potential flooding.

17.4.3 Modelling the Intervention

Prior to the impacts modelling process, the project designers may have already identified or planned adaptation interventions, or this may be undertaken once the climate impacts are better understood. It is likely that a combination of the two approaches will apply; regardless, once an impact model exists, it will be possible to test the application of specific impacts in order to predict the reduction in vulnerability under alternative adaptation interventions. For instance, various methods for flood control or different crop substitutions or agronomical measures can be assessed for suitability. By doing this, the project developers can determine an optimal course of action and, even more importantly, produce a quantifiable blueprint for reduction in vulnerability. The VRCs generated by the project correspond therefore to the difference in net impact value from the status quo (i.e., no project, or a project that ignores expected climate change in its design).

17.4.4 Income Equalisation

The Income Equalisation Factor (IEF) (Sect. 17.3) is integral in the VRC issuance calculation and is an indicator of the adaptive capacity of communities. Income equalization ensures that VRCs are not exaggerated in high income communities (owing to potential damages being valued higher per capita) and, likewise, not understated in poorer communities. As such, projects must establish with confidence the current or recent past per capita income of the people living within the project boundary. To determine an appropriate project IEF, project developers can draw upon a number of sources, including approved government or third party per capita income data, or use sampling of the population within the project boundary to estimate incomes following standard approaches to remove bias.

17.5 Applying VRCs to Local Climate Vulnerabilities

VRCs may be used to address a number of challenges that Sahelian communities may face as they attempt to adapt to climate changes. These can be broadly divided into aiding the processes of:

- Adaptation planning and methodology development,
- Project finance, and
- Monitoring and evaluation.

These challenges cross different scales, from the very local to national and even international, and also consider the potential for involving the broadest array of potential solution providers—cities, community groups, non-governmental organizations, local private project developers, adaptation technology providers, to name some of the most important. Climate adaptation in Africa faces a number of difficulties, including that of prioritizing investments in a transparent, and ideally, effective way (Table 17.1). VRCs can be a way to plan and set targets for governments, in a diverse set of circumstances, from local allocations of public funds to submissions of nationally determined contributions that may trigger funding

Table 17.1 Climate vulnerability challenges in Sahelian Africa

Challenges in Sahelian context	VRC opportunities
Lack of fungible, quantifiable, cross-sectoral means of prioritising and setting targets for governments (e.g., Nationally Determined Contributions)	VRCs allow governments means of seeing impacts across sectors, and determining what may be most effective
Lack of approaches to mobilize international climate finance (e.g., Green Climate Fund)	Could create VRC buying pools and encourage donors to buy VRCs
Challenge going from pilot project to scaling for entire economies, identifying good projects	Successful pilots could be then turned into national programs with VRCs the basis for budgeting and evaluating
Lack of robust, transparent means of identifying and evaluating projects	Transparent methodologies and project review/validation/monitoring and verification required, linking projects with climate change
Lack of means to engage with private sector and mobilize private finance	If price put on VRCs, stimulates innovation by private sector seeking to find (and find value) in most effective vulnerability reduction measures, gives revenue stream to back debt/equity

through mechanisms such as the United Nations' Framework Convention on Climate Change's (UNFCCC's) Green Climate Fund. Funders could develop buying pools, and projects could be selected based on their potential for reducing climate vulnerabilities as reflected in the number of VRCs generated.

Such a scheme offers an opportunity for much needed private finance to be leveraged (Sect. 17.8). Pilot projects, which often are great examples of what is possible, may be difficult to replicate throughout an economy owing to insufficient access to project finance. Scaling up project types can result if a price is put on each VRC a project could be awarded.

Local community groups, NGOs, and private companies are often best suited to participate in climate adaptation projects at the local level, but may lack the funds, or the incentive to be involved in community adaptation. Through a program that prices VRCs, these groups can gain the resources required for them to act. This includes entities that have specialist skills, technologies, or resources that would otherwise have to be funded through what may be less flexible government tendering, and the private financial sector could now be directly involved in project finance as it could see a revenue stream in the anticipated issuance of VRCs.

Monitoring, evaluating (and comparing) how effective different projects, and project types, have been at reducing vulnerability, is another way VRCs may be used. The requirements of projects to develop publically available project documents and monitoring plans results in a high degree of transparency.

Project owners and beneficiaries have a direct incentive to adhere to the VRC project monitoring plan, as this is a requirement for VRC issuances. As such, VRC generating projects, especially those that secure revenue from the credits, are likely to be sustained over long periods in order to secure the maximum number of VRCs.

17.6 Climate Change in the Sahel and Impacts on Communities

The Sahel (from the Arabic "sahil," meaning "coast" or "shore") is a zone of geographic, cultural, and climatic transition stretching longitudinally across the continent of Africa between the Sahara desert and the Sudanian Savanna. The region touches or covers parts of the countries of Senegal, Mauritania, Mali, Burkina Faso, Algeria, Niger, Nigeria, Chad, Sudan, South Sudan, and Eritrea over approximately 3,800 km^2 (IRIN 2008) (Fig. 17.4).

The Sahel is a semi-arid region, receiving about 100–200 mm of rain annually, The region has a very rapidly growing population, which is expected to reach 100 million in 2020 and to double from there by 2050 (IRIN 2008).

The climate stresses on the Sahel region are generally defined by water (typically, the lack thereof, although, as discussed below, high precipitation can bring its own problems). The hydrology of the western part of the Sahel is dominated by the Niger River, the second-largest in terms of flow on the African continent. The region has also undergone significant fluctuation in rainfall over time, most recently in the period since 1970, which has been characterized as "the most dramatic example of inter-decadal climatic variability ever measured quantitatively on the planet since instrumental records have been kept" (Rasmussen and Arkin 1993; Grijsen et al. 2013; Hulme 2001; Redelsperger et al. 2006). Combined with the general difficulty in accurately modelling the effects of climate change on regional hydrological systems, this background natural variability makes future climate risks and stresses difficult to assess.

Modelling based on a common "baseline" global emissions scenario (closest to Representative Concentration Pathway 4.5 in IPCC's Technical Assessment Report 5, 2014) has been found to be consistent with a projected rise in average temperatures in the Niger Basin region of between +1.0 and +3.0 °C by 2050 (Grijsen et al. 2013).

Rainfall, however, is highly model-specific and is predicted to either increase or decrease depending on the model, with no clear directionality of this measure seen

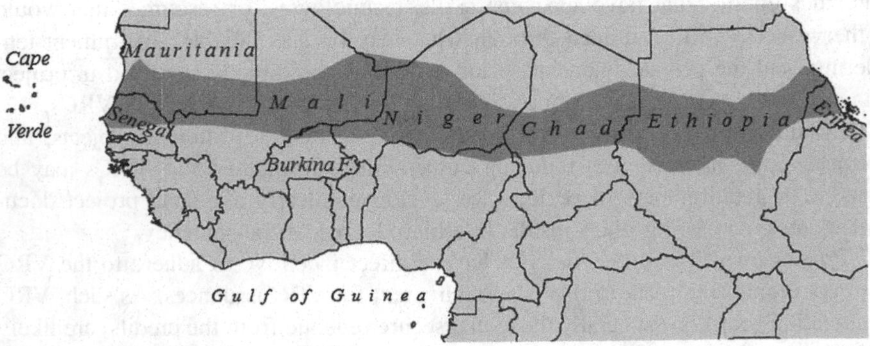

Fig. 17.4 Map showing Sahelian region in orange

until after 2050. This suggests that, in the future, communities will have to be prepared to withstand the region's characteristically complex and fluctuating climate modalities, perhaps with more variability in terms of intensity and timing, and with a likely added temperature signal contributing to heat stress and increased evapotranspiration. Even if annual rainfall decreases or remains stable, changes in rainfall pattern and timing can lead to increased incidence of flooding events, as is already being reported in parts of the region (see next section).

Despite the potential for increased rainfall in the Sahel, many climate observers caution against an assumption that climate change is, overall, "helping" West African agriculture, noting that, aside from the added heat stress dangers owing to the anticipated temperature rise, an associated rise in evapotranspiration could negate any net hydrological gain from increased precipitation (Carbon Brief 2015). Despite their demonstrated resilience, rural populations in the region are still vulnerable to increases in variability and in recent years have begun to observe a growth in seasonable irregularity that is making it increasingly difficult to plan and carry out plantings (Thomas 2013). Other impacts likely to follow from changes in rainfall frequency, timing, and intensity include shrinking of wetlands, decrease in species variety, water quality degradation, salinization, water table reduction, and erosion (German Federal Ministry for Economic Development and Cooperation and KFW Entwicklunbbank 2010).

17.7 A Case Study: Tillaberi

Recently, the ANADIA project (Tarchiani and Tiepolo 2016), a collaboration between the Italian Ministry of Foreign Affairs IBIMET-CNR, DIST-Politecnico and University of Turin, and the National Meteorological Service of Niger, was active in Tillaberi Region, Niger in developing a climate risk assessment for two subsistence farming villages located on the river Sirba at a distance of about 60 km from the capital, Niamey (Table 17.2).

With a combined population of about 7,200, these neighboring villages have faced varying stresses in the form of increased and increasingly erratic pluvial flooding, early onset of the growing season, and drought. Although as discussed above it is particularly difficult in the Sahel region to attribute seasonal or even multi-seasonal variance in weather patterns to an overarching anthropogenic forcing signal, the projected increase in temperatures in West Africa is likely to be associated with increased chaos in regional and local weather systems.

The communities, Talle and Garbery Kourou, grow a number crops both for subsistence and market sales; the crops include millet, sorghum, rice, peanuts, and tomatoes. Per capita income is approximately $200 per year. The villagers also maintain some livestock for subsistence and market purposes. Extreme flooding events, as have occurred in 2007, 2008, 2010, 2012, 2013, 2014, and 2015, have had severe impacts on livelihoods in both villages. Flooding in 2014 affected 60,000 people in Niger, with the worst effects occurring in Tillaberi. In that year, 38

Table 17.2 Climate assessment (ANADIA)

ANADIA project		
	Case study 1: Talle	Case study 2: Garbery Kourou
Background		
Geographic location	001° 37′ 697E, latitude 13° 45′ 647N	001° 36′ 589E, latitude 13° 44′ 431N
Population	2603 (2012), 2235 (2001)	4634 (2012), 3990 (2001)
Families	372	520
People per family	7	9
Distance from the capital	About 57 km	About 59 km
Main economic activities	Agriculture and livestock	Agriculture and livestock
Population perception of climate change	Seasons arriving early	Seasons arriving early
Geographic location	001° 37′ 697E, latitude 13° 45′ 647N	001° 36′ 589E, latitude 13° 44′ 431N
Climatic data		
Rain per year	530 mm (2012); 441 (2014)	530 mm (2012); 441 (2014)
Extreme events that have affected the area	Flooding 1966, 1987, 2008, 2012, 2013	Flooding 1959, 1987, 2007, 2010, 2012, 2013
Climatic variations in recent years	Seasons arriving early	Seasons arriving early
Agricultural data		
Type of agriculture (subsistence, sale, etc.)	Subsistence, local market	Subsistence, local market
Main products cultivated	Millet, sorghum, rice, peanuts, tomatoes	Millet, sorghum, rice, peanuts, tomatoes
Soil type (sand, clay, etc.)	Dune, clay and rocky	Dune, clay and rocky
Agricultural equipment available	Animal traction	Animal traction
Method of irrigation	Californian, watering can	Californian, watering can
Methodology for collecting water	Adaptation measure	Adaptation measure
Main agricultural challenges	Runoff	Runoff

people died as a direct result of the flooding, which also helped contribute to a cholera outbreak following the summer rainy season. The following year, more than 20,000 people were affected by flooding (Floodlist 2015). One major impact of flooding events is damage/destruction to houses and related built infrastructure. As in many West African villages, the residential and commercial housing stock of Talle and Garbery Kourou are primarily of an inexpensive mud type characterized (albeit in a more elaborate manner) by the distinctive "digging banco" mud architecture seen in Timbuktu. ANADIA divided the communities' housing into two typologies: the simpler "maisons en banco" ("houses of mud"); and "maisons semi-dur" ("semi-durable" or hardened structures). Maison banco structures are particularly susceptible to complete damage in flooding events and rebuilding has become a major cost factor in the wake of recent floods. ANADIA has characterised potential damages in a typical flood reaching up to approximately one-third of the real estate value in an affected community: in the case of Talle and Garbery Kourou, for instance, this would amount to a 20-year expected loss of value of more than 30 million CFA (51,000 USD). An increased incidence of flooding events with climate change might eventually multiply this expected annual loss several times.

17.7.1 Approaches to Mitigating Climate Damage

To protect village structures and crops from flooding events, ANADIA proposed several earthwork measures (Table 17.3).

- Cordons pierreux—("stone barriers") are a simple, flexible, and efficient method for water and soil control. Essentially, the cordon pierreux is a double line of fitted stones following a curved arc. This low-tech infrastructure can be quickly put into place and can help in reducing erosion, retaining water, and maintaining soil organic content.
- Gabions—These are embankments composed of rocks bound by or piled within cage structures that slow but do not stop water flow; they serve the dual purpose of controlling flood action while conserving water by permitting it infiltrate into the ground.
- Demi-lunes—Demi-lunes, or "half-moon" catchments, are semi-circular areas of dug-out terrain used to concentrate rainfall. This technique is used primarily for increasing pasture production, rehabilitation of degraded land, and crop production (although the latter use is not common in Niger).
- Banquettes—Banquettes are raised "benches" or dikes of terrain that are also useful for concentrating rainfall in order to assist planting and concentrate soil nutrients.

Table 17.3 Potential ANADIA adaptation measures and costs (ANADIA)

Proposed and current adaptation measures	Costs (US$)/unit (if available)
Single cost structural measures	
1. Gabions (gabion)	250/unit
2. Cordons pierreux (stone barriers)	65/200 m
3. Demi-lunes (half-moon catchments)	230/ha
4. Plantations	1100/ha
5. Banquettes (dikes)	500/ha
6. Building upgrades (mud houses to semi-durable)	4000/structure

Only the simplest of these measures (i.e., cordons pierreux) have been developed by local people; others are implemented by international organizations

In addition, most village dwellings are currently of the less sturdy maisons en banco type. To reduce potential catastrophic damage and cleaning costs during flooding events, structures can be rebuilt in the more permanent semi-dur style at an approximate cost of $4,000 for a typical family dwelling.

While some of these measures have already been adopted in the villages, there exist capital and institutional barriers that have prevented them from autonomously adopting the optimal range of adaptation measures. Although demi-lunes, for example, are a demonstrated technology for regenerating soil and containing rainfall, they have so far not been taken up autonomously at scale in Africa (UNEP 2016). This might be a result of cost, as the UNEP reports that demi-lune installation costs the equivalent of $150/ha to construct (ibid.); given how high this figure is relative to the typical per capita income of the region (ca. $100–200), this represents a significant barrier to autonomous implementation.

Changes and augmentation to the current crop mixture may also be necessary in order to maintain productivity as the growing season changes both in terms of increasing daily maximum and monthly average temperatures and increasingly erratic hydrological conditions.

Other potential adaptations include institutional measures such as the development of early-warning weather systems, awareness training, and the delimitation of flood zones. While such interventions are much less physically capital intensive, they require investments in intangible, time-intensive factors such as human expertise, research, policy/planning formulation, and institutional development that have to be maintained and augmented over the project lifetime and thus entail more significant recurring costs.

17.7.2 Converting Vulnerability into Adaptation: Producing VRCs in Tillaberi

The alternatives outlined above to mitigating climate-related damages in communities such as Talle and Garbery Kourou are generally not difficult to implement, but they require capital investments that could prove prohibitive. In the case of the Tillaberi ANADIA project, for instance, initial estimates based on the cost elements amount to an initial capital investment of approximately €800,000 plus recurring annual operation and maintenance costs of more than €2,000 per year over a 20+ year project lifetime. Given the low income level of the communities and the current lack of access to capital markets, such cost levels are a barrier to implementation. The ANADIA project study, for example, concluded that measures for reducing risk to the villages of Talle and Garbery Kourou by up to 23% would cost from €23 to 34 per capita (out of a per capita income of c. €160), representing an "unsustainable" expenditure for rural families or the municipalities. (Tiepolo and Braccio 2017).

17.7.3 VRC Registration, Certification, Issuance, and Monitoring

Using a VRC™ calculator, potential VRCs were estimated (Tables 17.3 and 17.4) for a project in the villages of Talle and Garbery Kourou, Niger. In this case, the VRC analysis centred specifically on measures that protect against river flooding, with interventions consisting of items 1–5 (Table 17.3) for crop and infrastructure protection via water flow control and enhanced irrigation and item 6, representing an investment in more flood-resistant village structures. Note that this represents an indicative case and that, while it relies on some figures reported by the ANADIA project, estimates of many of the driving climate factors have been made for illustrative purposes. Essentially, the calculations can be broken into three scenarios for comparison. The first scenario, labelled "Baseline" (Table 17.4), represents a continuation of the conditions in the first year of the project with a small expected growth in village size and output added to represent change in conditions assumed if the overall climate conditions did not change from that of the first project year. These calculations are useful for comparison with the indicative effects of climate change, as shown in the "Climate Change Scenario" section (Table 17.4), which represents the true project baseline with no actions taken specifically to adapt to anticipated enhanced climate damage (in this case, increased incidence of flooding of the Sirba river).

Table 17.4 ANADIA project sheet showing climate scenarios

	2018	2019	2020	…	2037
Baseline					
Total crop income CFA*000	272,090	275,894	279,774		364,512
Capital loss costs					
Building repair cost CFA*000	*340*	*340*	*340*	*0*	*368*
Grand total CFA*000	271,750	275,554	279,434		364,144
Exchange rate CFA*000/€	666.7	666.7	666.7		666.7
Grand total (€)	408	413	419		546
Climate change scenario					
Total crop income CFA*000	218,270	222,635	227,088		324,338
Capital loss costs					
Building repair cost CFA*000	*1020*	*1020*	*1020*	*0*	*1104*
Total net income CFA*000	217,250	221,615	226,068		323,233
Exchange rate w/€	666.7	666.7	666.7		666.7
Grand total (€*000)	326	332	339		485
Project implementation					
Total crop income CFA*000	290,030	295,831	301,747		430,969
Capital loss costs					
Building repair cost	*257*	*257*	*257*	*0*	*279*
Grand total	289,773	295,573	301,490		430,691

(continued)

Table 17.4 (continued)

	2018	2019	2020	…	2037
Exchange rate w/€	666.7	666.7	666.7		666.7
Grand total (€*000)	435	443	452		646
Post project aggregate productivity (€*000)	435	443	452		646
Net productivity change (€*000)	82	81	80		61
Income equalization factor					
Average per capita income (CFA*000)	731	742	752		980
Exchange rate $/CFA	0.0020	0.0020	0.0020	0	0.0020
Average per capita income ($)	183	185	188	0	245
Lower middle income threshold (GNI) ($)	4085	4085	4085		4085
Income equalization factor	22.34	22.34	22.34	#	16.68
Monetized vulnerability reduction (€*000)	82	81	80		61
Total income equalized vulnerability reduction (€*000)	1826	1807	1788	…	1023

Italics = negative cash flows
*000 = units of 100,000

The third scenario, "Project Implementation" (Table 17.4) shows the effects (reduced damages) based on the implementation of the project, as outlined above. The data and results are compared in a parallel manner in a year-by-year manner (from 2018 to 2037 in this case). Based on the apparent crop income per capita, the IEF-Income Equalization Factor is calculated, with the income adjusted vulnerability reduction calculated (Table 17.4) as the product of the IEF and the adaptation-related reduction in impact costs. Costs and revenues gives a break-down of capital and operating expenses for the project. The section under "Management Costs and VRC" (Table 17.5) lists project administrative and VRC registration costs, while "Physical project incremental. "Net Cash Flow" row demonstrates the funding barrier to project implementation, with a net negative balance in the initial years owing to investment and setup costs (as well as VRC registry costs). However, as the project begins earning VRCs starting in the third year of operation, the balance sheet becomes positive. Indeed, a net present value analysis over a 20-year lifetime at a reasonable social discount rate of 3% produces an added discounted project value of approximately €2.6 million, which, given an IEF of 22, equates to the generation of approximately 1,200,000 VRCs.

As for the financial results of the project (Table 17.6), the calculated 20-year climate damage is given by comparing the second with the first rows, and the effects of implementing the project are given by comparing the third with the second row —the avoided impact cost of about 2.6 million Euros compares favorably with the project costs. Note that the revenue stream here assumes a price per VRC sold of €5; given the break-even cost of €1.82 per VRC based on the project costs; this indicates a plausible investment.

Table 17.5 ANADIA Project sheet project costs and cash flows

	2018	2019	2020	…	2037
VRCs generated, VRC management costs and VRC pricing					
Nominal VRC value (€)	50				
VRCs/year*000	0	32	32		32
VRC monitoring costs €*000	0	5	5		5
PDD + validation €*000	50				
Verification €*000	0	5	5		5
Registration fee €	10	0	0		0
Issuance fee (@0.35 €/VRC), €*000	0	11	11		11
Total investor VRC generation costs €	60	21	21		21
VRC price (€)	5				
VRC income €*000	0	159	159		159
Physical project incremental costs and revenues					
Project incremental physical O and M costs €*000	2	2	2		2
Project CAPEX structure protection €*000	53	53			
Project CAPEX crop measures	374	374			
Project CAPEX €*000	426	426			
Revenue productivity savings €*000	82	81	80		61
Project costs €*000	62	23	23		23
Net cash flow €*000	*488*	*209*	216		197

Italics = negative cash flows
*000 = units of 100,000

Table 17.6 ANADIA VRC generation figures totals (see Tables 17.3 and 17.4)

Scenario	Total discounted net income/worth over 20 years (discount rate = 3%)
No climate change, no project €·000	9403
Climate change, no project (V0) €·000	7951
Climate change, with project (V1) €·000	10,600
Costs and benefit elements	
Project costs	2155
AIC €·000	(V1) − (V0) = 2648
Initial IEF	22 [based on initial p.c. income of $183]
No. of VRC's	(AIC · IEF)/€50 = 1,183,246
Project cost per VRC €·000	2155/1183 = €1.82/VRC
Breakeven: VRC price of €1.82+ required	

Such analysis helps assure that the generation of VRCs by a project equates to tangible creation of climate adaptation value. The VRCs produced by the implementation of verifiable, measurable adaptation measures in Tillaberi provide potential investors with an inducement to invest in these measures. The priced VRC streams produced by the project help enable it by, for example, serving as collateral to help secure a loan to cover capital costs.

17.8 Direct Benefits to Vulnerable Communities

The benefits of undergoing the above process can be considerable, and possibly transformational for the vulnerable community. By itself, of course, going through the rigor of establishing a clear and quantified vulnerability baseline, understanding how a project may reduce these vulnerabilities, and setting in place a clear monitoring framework will take time and possibly considerable expense. However, both the knowledge gained and made transparent, and the potential for VRC generation may provide a vulnerable community with:

- Enhanced Adaptive Capacity: A much better understanding of the community's climate adaptation needs, and improved community decisions on the technologies, practices and timing for investment, and
- A way for the community to secure project finance that is aligned with its interests in sustained vulnerability reduction.

The vulnerable community (along with possible development partners) are able to understand and then articulate their needs through developing a VRC baseline and being able to assess the expected results (in terms of VRCs, and by extension of vulnerability reduction) of different project alternatives. In our case study, this occurs through the clear quantification of the expected economic impacts of flooding and agricultural loss through both flooding and increased droughts. The community can then identify where the most vulnerable assets are found, and focus on identifying measures (in the ANADIA project case, for example, this involves the implementation of water control infrastructure, such as demi-lunes, that would not be constructed without outside funding, and the adoption of crop switching and planning technologies) that will reduce these vulnerabilities. While the community is free to prioritize however it wishes, it receives VRCs in proportion to the amount of avoided impacts resulting from adaptation.

As such, the community can use this clear and quantifiable understanding of its vulnerabilities as both a decision-making tool and to articulate their needs and justify donor funding. Once a community knows what particular levels of vulnerability it has in different assets and production areas (in this case, crop production and buildings/crop losses), it can seek adaptations that are most appropriate and will offer the most effective results, and though the process of developing a project document understand which will be most cost effective.

Donors will then be able to engage with the community and its adaptation partners, with the community able to describe and justify its proposed interventions.

What is perhaps the most innovative result that VRCs can offer, once a community and its partners have registered their project, is that they can then "sell" the potential vulnerability reduction to a government or donor. The vulnerability, and its reduction, transitions into being an asset for the community in its engagement with funders, and, most critically, it can offer funders a clear and relatively de-risked means of offering contingency-based finance.

There are a variety of possible financing structures that VRC generating projects could lend themselves to, with perhaps the simplest being a conventional project finance model. In this scenario, a community (perhaps aided by a project developer with technical, financial, and project development resources) would offer a donor (which could be any public or private body with an interest in reducing the community's vulnerability) to transfer some or all of its VRCs generated in return for a price per VRC. With a signed purchase agreement, the project could then go to a bank or other investor (public or private) that could then assess the project risks (noting that the community has a direct stake and incentive to reduce its own vulnerability), and offer finance.

Finally, the process of undertaking a VRC generating project can result in greater adaptive capacity as the community has tools and more immediate incentives to improve its "climate resilience". The community can profit from the priced VRCs, incentivizing a greater awareness of how to best reduce its vulnerability,

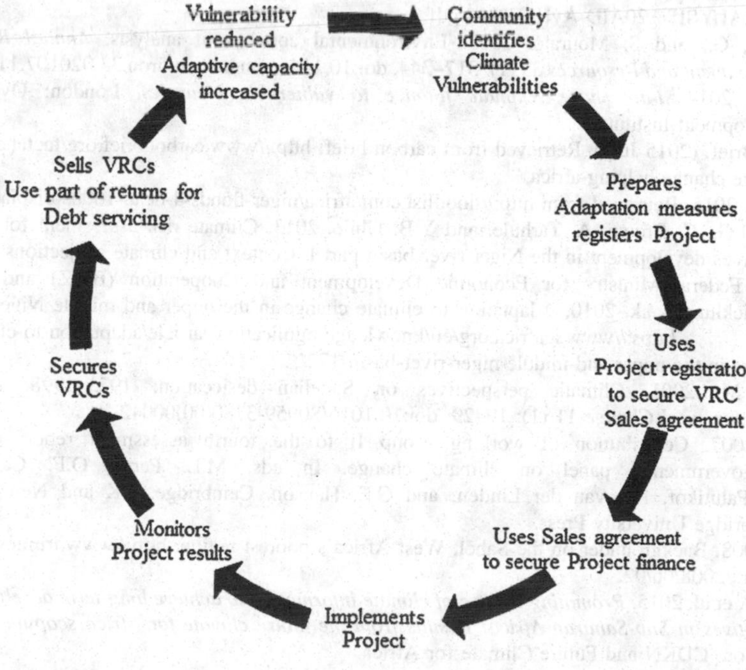

Fig. 17.5 A potential process for using VRCs to leverage project finance for climate adaptation

taking away the possibly long-off, uncertain, and thus maybe perceived to be less real threat of climate change, to be replaced with a knowledge that its adaptation actions can bring results in a growing season through VRC sales. In turn, incomes may increase, and this, as noted above, is a proxy for and can correlate with an improved adaptive capacity (Fig. 17.5).

17.9 Conclusion

Climate Vulnerability Reduction Credits (VRCs™) offer communities, donors, and governments in the Sahel and around the world with a process that can help overcome a number challenges related to effective climate adaptation. As the case study shows, VRCs offer to provide a needed source of funds and expertise to local communities, while helping donors and governments deliver scalable, transparent and capacity growing vulnerability reduction where it is needed most. There is considerable work required, however, to encourage adoption of the VRC.

References

Alabi, R.A. 2012. Sectoral analysis of impact of foreign aid in Nigeria: A dynamic specification. Downloaded on 3 June 2016 at http://www.iwim.uni-bremen.de/Siakeu/Alabi-SECTORAL%20ANALYSIS%20AID-AvH-3-2012.pdf.

Atkinson, G., and S. Mourato. 2008. Environmental cost-benefit analysis. *Annual Review Environment and Resources* 33 (1): 317–344. doi:10.1146/annurev.environ.33.020107.112927.

Bird, N. 2014. *Fair share: Climate finance to vulnerable countries*. London: Overseas Development Institute.

Carbon Brief. (2015 July). Retrieved from carbon brief: http://www.carbonbrief.org/factcheck-is-climate-change-helping-africa.

Floodlist. 2015. Retrieved from http://floodlist.com/africa/niger-floods-4-dead-1000s-displaced.

Grijsen, J.G., C. Brown, A. Tarhule, and Y.B. Ghile. 2013. Climate risk assessment for water resources development in the Niger river basin part I: Context and climate projections.

German Federal Ministry for Economic Development and Cooperation (BMZ) and KfW Entwicklungsbank. 2010. Adaptation to climate change in the upper and middle Niger river basin. https://www.icafrica.org/en/knowledge-publications/article/adaptation-to-climate-change-in-the-upper-and-middle-niger-river-basin-172/

Hulme, M. 2001. Climatic perspectives on Sahelian desiccation: 1973–1998. *Global Environmental Change* 11 (1): 19–29. doi:10.1016/S0959-3780(00)00042-X.

IPCC. 2007. Contribution of working group II to the fourth assessment report of the intergovernmental panel on climate change. In eds. M.L. Parry, O.F. Canziani, J.P. Palutikof, P.J. van der Linden, and C.E. Hanson. Cambridge, UK and New York: Cambridge University Press.

IRIN. 2008. Backgrounder on the Sahel, West Africa's poorest region. http://www.irinnews.org/feature/2008/06/02.

Jones, L., et al. 2015. *Promoting the use of climate information to achieve long-term development objectives in Sub-Saharan Africa: Results from the future climate for Africa scoping phase*. London: CDKN and Future Climate for Africa.

Junghans, L., and S. Harmeling. 2012. Different tales from different countries, a first assessment of the OECD adaptation marker. Germanwatch Briefing Paper, September.

KFW Entwicklungsbank. 2010. Adaptation to climate change in the upper and middle Niger river basin: River Basin Snapshot. Draft for discussion: http://ccsl.iccip.net/niger_river_basin.pdf.

McSweeney, R.J. 2010. Selecting members of the 'QUMP' perturbed-physics ensemble for use with precis. UK Met Office: http://www.metoffice.gov.uk/media/pdf/e/3/SelectingCGMsToDownscale.pdf.

Pan African Climate Justice Alliance. 2009. The economic cost of climate change in Africa.

Rasmussen, E.M., and P.A. Arkin. 1993. A global view of large-scale precipitation variability. *Journal of Climate* 6: 1495–1522. doi:10.1175/1520-0442(1993)006<1495:AGVOLS>2.0.CO;2.

Redelsperger, J.L., Thorncroft, C., Diedhiou, A., Lebel, T., Parker, D. and Polcher, J. (2006). African monsoon multidisciplinary analayis: An international project and field campaign. *Bulletin of the American Meteorological Society* 87 (12): 1739–1746. doi: 10.1175/BAMS-87-12-1739

Schultz, K. 2012. Financing climate adaptation with a credit mechanism: Initial considerations. *Climate Policy* 12 (2): 187–197. doi:10.1080/14693062.2011.605563.

Tarchiani, V., and M. Tiepolo. 2016. *Risque et adaptation climatique dans la région Tillabéri, Niger. Pour renforcer les capacités d'analyse et d'évaluation*. Paris: L'Harmattan.

Thomas, A. 2013. Sahel villagers fleeing climate change must not be ignored. *The Guardian*, August 2.

Tiepolo, M., and S. Braccio. 2017. Local and scientific knowledge in multirisk assessment for rural Niger. In *Renewing local planning to face climate change in the tropics*, eds. M. Tiepolo, A. Pezzoli, and V. Tarchiani. Springer.

UK Met Office Hadley Climate Centre. 2016. *An introduction to the PRECIS system*.

UNEP-United Nations Environment Program. 2016. Sourcebook of alternative technologies for freshwater augmentation in Africa: 1.1.2 Demi-lunes or semi-circular hoops. Newsletter and Technical Publications: http://www.unep.or.jp/ietc/Publications/TechPublications/TechPub-8a/hoops.asp.

USAID. 2014. *A review of downscaling methods for climate change projections*. Washington, DC: USAID.

Watkiss, P. et al. 2016. The costs of adaptation. In The adaptation finance gap report 2016. Nairobi: United Nations Environment Programme. http://web.unep.org/adaptationgapreport/sites/unep.org.adaptationgapreport/files/documents/agr2016.pdf.

Wilkinson, R., and K. Pickett. 2009. *The Spirit Level: Why equality is better for everyone*. London: Penguin Press.

World Bank. 2016. World Bank country classifications. http://data.worldbank.org/about/countryclassifications.

Chapter 18
Renewing Climate Planning Locally in the Tropics: Conclusions

Maurizio Tiepolo, Alessandro Pezzoli and Vieri Tarchiani

Abstract In the Tropics, a significant increase in the number of cities provided with climate plans by 2020, as announced in the 11th Sustainable development goal of the United Nations, requires an unprecedented effort. To achieve it, we have to simplify the planning process and improve the quality of the plans. The aim of this book was to collect methods and experiences to inspire the simplification of the planning process and increase the quality of climate planning. We focused attention on the three critical phases of the planning process: analysis, decision making in planning, climate measures. Sixteen case studies from Ethiopia, Haiti, Malawi, Mexico, Niger, Senegal, Tanzania and Thailand cover automatic weather stations in remote areas, rainfall estimation gridded datasets, open data for vulnerability index to climate change, early warning systems, quality of climate plans index, multi-risk local assessment, flooding risk evaluation method, backcasting, spatial dimension in disaster risk reduction and resilience, gasification stoves, index-based insurance and vulnerability risk credit. After indicating the possible analyses, 19 recommendations were supplied to the United Nations SDGs monitoring system, the national weather services and those responsible for natural risks, to the Development banks, Official development aid and the research institutions.

Keywords Climate trends · Climate measures · Climate planning · Decision making · SDG · Vulnerability tracking · Tropics

M. Tiepolo is author of 18.1, 18.2.2 and 18.3 sections. A. Pezzoli is author of 18.2.1 and 18.2.3 sections. V. Tarchiani is author of 18.2 and 18.4 sections.

M. Tiepolo (✉) · A. Pezzoli
DIST, Politecnico and University of Turin, Viale Mattioli 39, 10125 Turin, Italy
e-mail: maurizio.tiepolo@polito.it

A. Pezzoli
e-mail: alessandro.pezzoli@polito.it

V. Tarchiani
National Research Council—Institute of BioMeteorology (IBIMET),
Via Giovanni Caproni 8, 50145 Florence, Italy
e-mail: v.tarchiani@ibimet.cnr.it

M. Tiepolo et al. (eds.), *Renewing Local Planning to Face Climate Change in the Tropics*, Green Energy and Technology,
DOI 10.1007/978-3-319-59096-7_18

18.1 From the State of Climate Planning to the Book Approach

This book contributes to "substantially increase the number of human settlements adopting and implementing integrated policies and plans towards resource efficiency, mitigation and adaptation to climate change, resilience to disasters, and disaster risk reduction", in short, to what we have called "climate planning". That announced is one of the targets of the 11th Sustainable development goal of the United Nations, "Make cities inclusive, safe, resilient and sustainable" which the signees undertake to achieve by 2020. We proceeded by 4 steps.

First of all, we decrypted the concept of climate planning expressed in the New urban agenda (2016) and in the Sendai framework for disaster risk reduction (2015) of the United Nations.

Secondly, we identified the category of cities which are expected to prevail in the Tropics in 2020 and 2030 (Chap. 1).

Third, we ascertained the state of climate planning in large and medium-sized cities in the Tropics, identified local access to the key information necessary to climate planning and the most common climate measures (Chaps. 1 and 10).

Four, we compared vision (UN SDGs) and reality (urbanization trends, state of climate planning, key-information access, most frequent climate measures). The comparison highlights three critical points of the process which the 11th SDG aims to put in place.

18.1.1 New Tools and Measures

The climate plan proposed by the United Nations is that usually applied in the large cities of OECD and BRICS countries, where stand-alone plans or the acclimatisation of long-term general, comprehensive and master plans prevail. But in 2030, the urban population in the Tropics will be concentrated mainly in cities under 0.3 million pop. (46%), even if large cities will see the biggest increase (from today's 38–44%). Medium-size tropical cities mainly use medium-term municipal development plans because they do not have sufficient resources to prepare and run additional tools or because no other plans are required by the national laws. Consequently, the plans that should be produced from now on and the measures that should be implemented to limit climate change and its impacts on human settlements should be different from those that local administrations are currently implementing in application of the single national legislations. This requires a regulatory adjustment that cannot be achieved in the short term. The alternative is to continue producing stand-alone plans that are detached from the local legal system and which have little chance of being appropriated and implemented (becoming an expenditure item in the municipal budget) once external financing ends.

18.1.2 Increasing the Production of Plans

The pace sustained by climatic planning in the last seven years has more than doubled the number of climate plans in the Tropics compared to the previous seven years. Today, just 104 cities in the most advanced tropical countries (OECD and BRICS) still don't have one. But no fewer than 950 cities in tropical DC and LDCs still have no tools to reduce emissions or cope with the impacts of climate change, and half of these are in countries with no climate planning experience. If the production of plans could keep up this pace until 2020, not even half of the large and medium-sized tropical cities would have a climate plan. And this can certainly not be considered a substantial increase in climate planning.

18.1.3 Increasing the Quality of the Plans

Our survey of 338 tropical cities revealed an even lower quality of the plans, which is inevitably translated into a poor impact of the measures envisaged. Two percent of climate plans estimate the potential impact of the planned measures, only 6% of the plans characterizes the local climate change and only one out of five climate plans specifying the quantity of measures to be taken. This uncertainty creates further inaccuracies: source of funding (indicated by one plan out of four) and the cost of measures (indicated by 42% of the plans only). The analysis and decision making tools are therefore two weaknesses of the examined plans. Knowledge of climate change and its impacts is often lacking. And when knowledge exists it is acquired in a way that would radically renewed. For example, the preliminary analyses of planning (hazard, hazard-prone zones, vulnerability) are occasional, rarely become vulnerability tracking and are almost never locally accessible. This aspect is not considered by the SDGs monitoring system. We see this as a limit: what is not monitored cannot be accomplished.

We have reached the conclusion that, in order to achieve the above-mentioned target a radical renewal of local climate planning is necessary.

Producing new plans, more than in the past and of better quality, requires the simplification and acceleration of certain steps of the current planning process. It is with this intention that the book uses sixteen case studies to tackle three key-steps of the planning process:

(1) Analysis for local climate planning
(2) Decision making tools for local planning
(3) Innovation in climate measures

18.2 Lessons Learned from the Book

18.2.1 Analysis for Local Planning

Analysis for local climate planning includes characterisation/probability of hazard in the contexts exposed to climate change, in the medium-long term, which is the horizon of climate plans (from 5 to about 20 years). In these cases, the analysis is called up to characterise the climate, taking into account the needs for agricultural and livestock production, protection of key infrastructures and, more generally, of human settlements. In greater detail, we have seen (Chap. 2) that it is necessary to ensure a technical capacity in maintaining and operating AWS-automatic weather stations safety equipment. AWS are particularly important for flood monitoring, which is often carried out in remote watersheds compared to the built-up areas on the plain, which are frequently large and medium-sized cities.

We then have to consider the new threats due to climate change: intense rainfall and dry spells also in drought treated countries (Chap. 3). Importance of tailored climate analysis (at local scale) with daily resolution (not seasonal trends like IPCC) superimposing natural hazard likelihood with other indicators.

For farming activities (e.g.: urban and peri-urban) it is useful to have high resolution daily precipitation gridded datasets: probability of unfavourable conditions for crops (according to critical periods in growing stages) and favourable trends (late end of the rainy season) which can, however, trigger conflicts with herders (Chap. 4).

The analysis also implicates the ascertainment of vulnerability to climate change, which can be necessary on different scales: regional, using the municipality as a minimum unit of analysis, or restricted to parts of the smallest administrative jurisdiction. Depending on the contexts, these are occasional assessments, as in the case of the Ethiopian districts (Chap. 5), or tracking, where open data exist, as in Haiti (Chap. 6). The open access of the single indicators (e.g.: that on flood exposure) guaranteed by a national service, would allow every local authority to directly access information on exposure and vulnerability relating to its jurisdiction on a scale sufficient to its local planning needs, accelerating the local analysis process and cutting spending, which could be addressed towards measures for adaptation. The diagnosis for planning includes the Early Warning System. This device also belongs to the measures but is considered here to ascertain that the hazard prone areas are effectively covered by EWS. Also the device for extreme precipitation alerts in Malawi uses open source information (Chap. 7).

When monitoring cropping season, it is necessary to plan things with caution, paying attention to translating information products into services that should be simple but effective for farmers, establishing a communication and collaboration between scientific institutions, national technical services and rural communities for an effective dissemination and adoption of the advice (Chap. 8). Climate services for farmers should be integrated into a concerted and participatory strategy (including content and format of advice, communication channel, partnership with other local stakeholders and training), which enables final users to adopt scientific innovation in ways that are socially acceptable and environmentally sound. The approach needs to pay special attention to gender: in many cases, women are the backbone of the farming community, however they are disadvantaged in accessing climate and weather information as well as training programs.

18.2.2 Decision Making Tools for Climate Planning

The huge effort that has gone into climate planning in the last seven years has been carried out largely by the large and medium sized cities of the member countries of the OECD and BRICS, but much less than by those of the DC and LDCs (Chap. 10). The quality of planning measured with the QCP Index composed of 10 indicators remains low.

Climate planning is expected to spread, but it should also rise in quality, something which can be appreciated with the QCPI-Quality of Climate Planning Index (Chap. 10). This requires an improvement in methods, integrating the concept of risk for example. The MLA-Multirisk Local Assessment (Chap. 11) presents various advantages compared to the Community Risk Assessment. For example, it identifies the potential impacts of climate change and those of the risk reduction methods. The cost of potential damages compared with risk reduction costs allows the identification of the benefits of the single measures, also in the frequent cases of communities exposed to multiple risks. Lastly, the method combines technical knowledge and local knowledge. The FREM-Flooding Risk Reduction Method integrates in participated methods for flood risk management, a robust analysis of local condition in a context where local capacities on climate and prediction are absent (Chap. 12).

The use of forecasting in the case of measures to increase access to water in Dar es Salaam can be integrated with the identification of shared community objectives (backcasting). The first method is effective in establishing the technical limits of an intervention and can be helped by the definition of possible scenarios with the second method (Chap. 13).

These methods however require integration into local procedures. An example is supplied by disaster risk reduction and resilience in the case of La Paz, Mexico. The method proposed allows an assessment of urban resilience with a spatial dimension (Chap. 14).

18.2.3 Innovation in Climate Measures

As far as the measures are concerned, it is a matter of innovating the range available, focusing on those more capable of mitigating emissions and risk, and then financing them. Two measures are almost completely absent from the 346 plans for tropical cities considered by our survey.

The first regards the use of alternative fuels to wood and coke for cooking. The "three stone" is still the most frequently used stove in many large tropical cities. The effects of this device on emissions of CO_2 and on deforestation (reduction of carbon sinks) around cities are clear. It alters the plant coverage, increasing run-off and soil erosion, which aggravate floods. The use of crop residues to produce pellets and special gasification stoves is an answer to deforestation and to environmental alteration, which could impact on climate change mitigation (Chap. 15). Its use in an urban context would allow the limitation of the incessant process of deforestation which has exceeded a radius of 100 km around Niamey, the capital city of Niger, affecting most of the watersheds responsible for the disastrous floods in 2010, 2012 and 2016.

The second measure is the index-based insurance to share risk in the case of adverse weather events to which farming is exposed (Chap. 16), rarely provided for in local climate plans. This, however, is a measure that has experienced an impasse but which has potential if accompanied by measures for risk reduction.

Lastly, the question of financing the measures. The VRCs-Vulnerability Reduction Credits system (Chap. 17) presented and tested in the case of Niger can allow the channelling of funds towards plans which are likely to have the biggest impact.

18.3 Areas for Future Research

Climate plans are a way of mainstreaming climate change mitigation and risk reduction measures. However, they run the risk of remaining unimplemented and of not changing the conducts of people and key players. This is why it is necessary to act in other ways at the same time.

Key infrastructures (production and distribution of energy, roads, railways, ports, dams or important urban services, from education to health) could have a considerable impact on the reduction of emissions and the reduction of risk. The possible impact of infrastructures on climate change and, vice versa, of climate change on the construction and exercise of such infrastructures, if considered right from the concept of the project stage, can be substantially reduced. The critical entrance point are the Development banks. The request for specific climatic requirements in the proposals of financing submitted to the Development banks has an impact of the entire growth machine: designers, surveyors, construction firms and maintenance firms, suppliers of machinery, construction material, local administration and sectorial administration of the Government.

A second sector is that of technical standards for the construction of acclimatized infrastructures.

Attention should be extended to these two sectors in the future.

18.4 Recommendations

Nineteen recommendations to different stakeholders are triggered by the book.

United Nations SDGs monitoring system

1. Consider the percentage of large and medium-sized cities with climate plans as an indicator for the monitoring of the 11th SDG.
2. Consider the quality of the climate plans as an indicator for the monitoring of the 11th SDG (QCPI).
3. Use a database on cities and climate planning, similar to that created for this book, and update it.

National Hydro-meteo Services

4. Integrate the different sources of climate data. In LDCs and specially in remote areas, the observation networks are few and far between. Climate analysis on a local scale (municipal) can be practiced successful, integrating data observed through the network of automatic weather stations and rainfall estimation gridded datasets like TRMM, CHIRPS, etc.
5. Publish the data on floods so that local planning can be based on the analysis of past events also in semi-arid countries, where floods are becoming a high risk, both in urban and rural areas. Developing flood reporting systems based on geo-referenced data and on the longest possible timescale.

National Services Responsible for Natural Risks

6. Switch from vulnerability assessment to vulnerability tracking, possibly through open data with a reduced number of indicators and make access to information open.
7. Make all the information useful to local climate planning freely available, from climate data to the areas exposed to flood, landslide, sea level rise and drought, to farmland, the prices of farm products on the markets, forestry coverage, victims and damage after disaster, through to school enrolment by gender.
8. Consider the capacity to react to vulnerability.

Federal/State Authorities Responsible for Local Development Planning Guidelines

9. Increase the number of climate measures in the municipal development plans.
10. Prepare lists of recommended measures, depending on the potential impact.
11. Make climate plans more detailed, both in analysis and in the measures, which should be structural and non-structural (e.g.: index-based insurance, climate services, etc.).
12. Introduce into municipal development plans the assessment of the residual risk as a tool to make decisions regarding the measures that can have the biggest impacts on the reduction of risk.

Research Institutions

13. Switch from monitoring the climate plans to tracking plan implementation.
14. Ascertain the state of climate planning in towns (under 0.1 million pop.).
15. Extend the survey on the other entry points of climate action: infrastructure projects and construction standards.
16. Ascertain the impact of the most important measures to reduce risk, in detail.

Official Development Aid

17. Assess adaptation measures in projects in terms of VRC-credits.

Insurance sector

18. Relaunch index-based insurance.

Development Banks

19. Finance projects for transition to the gasification stove fuelled by pellets, in urban and rural areas.

Printed in the United States
By Bookmasters

Printed in the United States
By Bookmasters